Hex, Inside and Out

The Full Story

Hex, Inside and Out

The Full Story

Ryan B. Hayward
Bjarne Toft

CRC Press is an imprint of the
Taylor & Francis Group, an **informa** business

CRC Press
Taylor & Francis Group
6000 Broken Sound Parkway NW, Suite 300
Boca Raton, FL 33487-2742

Printed on acid-free paper
Version Date: 20181214

International Standard Book Number-13: 978-0-367-14425-8 (Hardback)
International Standard Book Number-13: 978-0-367-14422-7 (Paperback)

Library of Congress Cataloging-in-Publication Data

Names: Hayward, Ryan B., 1958- author. | Toft, Bjarne, author.
Title: Hex, inside and out : the full story / Ryan B. Hayward with Bjarne Toft.
Description: Boca Raton, Florida : CRC Press, [2019] | Includes bibliographical references and index.
Identifiers: LCCN 2018047156| ISBN 9780367144258 (hardback : alk. paper) | ISBN 9780367144227 (pbk. : alk. paper) | ISBN 9780429031960 (e-book : alk. paper)
Subjects: LCSH: Game theory. | Hex (Game) | Combinatorial analysis.
Classification: LCC QA269 .H39 2019 | DDC 519.3--dc23
LC record available at https://lccn.loc.gov/2018047156

Visit the Taylor & Francis Web site at
http://www.taylorandfrancis.com

and the CRC Press Web site at
http://www.crcpress.com

In memory of Claude Berge,
who loved to play Hex.

Contents

Prologue

In the summer of 1983 my Ph.D. supervisor Vašek Chvátal told me that he was going to Paris to edit a book on perfect graphs with Claude Berge. I replied that I liked France, and he kindly allowed me and fellow students Bruce Reed and Chính Hoàng to join him. And so I met Claude and learned Hex.

From January 1984, for the next several months, Vašek, Bruce, Chính, my girlfriend Julie Bates and I met regularly at Claude's office in *Maison des Sciences de l'Homme* (Home for Humanities and Social Science) at 54 Boulevard Raspail in Paris. On a typical day we would arrive before noon, head downstairs with Claude and other visitors[1] for two hours of lunch and coffee, then head back up to Claude's office to play games.

Claude was a generous host, and he loved to play Hex. He would often say aaaaaaaaaaaaaaha with a twinkle in his eye as he placed a stone on the board, suggesting that he had found a miracle move that would upend the game. We played Hex often, and he taught me about virtual connections and the must-play region. One day he gave me a copy of his manuscript *L'Art Subtil du Hex* (The Subtle Art of Hex). I was disappointed that this elegant introduction to Hex theory was never published. From that time, I have been thinking about writing a book on Hex.

Thirty-five years later, here it is.

Ryan B. Hayward, Edmonton, Canada

My interest in Hex started during Christmas 1969. As a gift I got a Hex board from my parents. It was a beautiful teak board, produced by the designer Skjøde Knudsen in Skjern, Denmark. In my family we often played games: Monopoly, Scrabble and various card games, in particular during Christmas when we were all there. My mother had seen the Hex board advertised as a special offer to readers of the women's magazine *Flittige Hænder* (Diligent Hands) and ordered one. I still have it and love to use it in play. The small wooden balls are nice to hold in your hand and put on the board.

[1]Visitors included Frédéric Maffray, Henri Meyniel, Michel Las Vergnas, Yahya Hamidoune, Neil Grabois and Jean-Marie Pla. Pla and Berge coined the term *hypergraph*; Berge wrote a book on hypergraphs in 1970.

During the following years I was busy with research and teaching, and I did not play much. I always liked teaching and started also to give lectures for general audiences. Mathematics is about problem solving, and for my general talks I tried to find simply formulated unsolved problems. Hex was a good source.

In 2004 I met Ryan Hayward at the conference in memory of Claude Berge in Paris. Claude Berge was a hero of mine, and at the meeting I lectured about his achievements in game theory, graph theory, literature and art. Back home I suggested to Thomas Maarup, a bright student at the University of Southern Denmark, that he might take up Hex as a research topic. He got hooked and wrote a beautiful master's thesis, still available online at `http://maarup.net/thomas/hex`. Thomas studied the original sources and found among other things Piet Hein's manuscript from December 1942 outlining the newspaper article that introduced the game.

Another hero was Martin Gardner. In 2013 I had the privilege to be at the H-STAR Institute at Stanford University in Palo Alto, California, for four months. The university library hosts the Martin Gardner Archive of his *Scientific American* columns, where I studied Gardner's Hex correspondence, including with Piet Hein and John Nash. This supplemented the findings Ryan and I had made at the Piet Hein archive of Anni and Hugo Hein in Denmark that we have included in this book, and made it possible for me to contribute to a coffee-table book about Piet Hein that Gyldendal published in Denmark in 2015.

Ryan and I eventually discovered that another bright student, Jens Lindhard, at the University of Copenhagen, had been involved with Piet Hein in promoting Hex in 1943. Lindhard later became a famous physicist, a professor at Aarhus University in Denmark and president of the Danish Royal Society. I doubted that there were any documents left from his involvement, but asked a science history friend, Henry Nielsen, at Aarhus University, about it. The answer was that Lindhard had left a huge collection of papers, deposited in a basement room of the university. It turned out to be a rich source, showing Lindhard's life and many interests, including his Hex notes and letters from 1943.

Since the conference in memory of Claude Berge in Paris 2004 I kept in contact with Ryan, sharing our fascination of Hex and encouraging him to finish his book. It is finally here.

Bjarne Toft, Odense, Denmark

FIGURE 1: Claude Berge (left), Jean-Marie Pla and Neil Grabois playing Hex in 1974 on a teak Skjøde Skjern board bought by Claude at *Maison du Danemark* on the Champs-Elysées in Paris. Photo © Michel Las Vergnas.

Preface

Hex — first called Polygon — was invented by the Danish poet and designer Piet Hein. In 1942 he showed the game to some friends, including chess prodigy Jens Lindhard. They encouraged him to market it, so he wrote a series of columns — with puzzles by Lindhard — that appeared in the newspaper Politiken from December 26, 1942 until August 11, 1943. Then Polygon disappeared.

Six years later, the game mysteriously reappeared in America when John Nash described it in abstract form to David Gale. Thinking it might be fun to play, Gale built a game board that he left in the common room of the math department at Princeton University.

In 1950 Parker Brothers heard about Hein's game and published it with the name Hex. And in 1957 Martin Gardner introduced it to the world via his *Scientific American* column, *"Concerning the game of Hex, which may be played on the tiles of the bathroom floor."*

The rest, as they say, is history.

☼ ☼ ☼

Hex is fun for players of all ages: it takes seconds to learn, but years to master. The board size can vary, altering game complexity. To allow a balanced game against a strong player, a weak player can be allowed extra opening moves.

The mother of all connection games, Hex is the subject of books by Martin Gardner and Cameron Browne. Hex theory touches on graph theory, game theory and combinatorial game theory, with elegant proofs that the game has no draws and that the first player can win. From machines built by Claude Shannon to agents using Monte Carlo Tree Search, Hex is often used in the study of artificial intelligence.

Written for a wide audience, this is the full story of Hex, inside and out, with all its twists and turns: Hein's creation, Lindhard's puzzles, Nash's proofs, Gale's Bridg-it, the game of Rex, Shannon's machines, Bridg-it's fall, Hex's resilience, Hex theory, the hunt for winning strategies, and the rise of Hexbots. We also include puzzles — with solutions — and a Hex chronology.

The publication of this book coincides with the 76th anniversary of the publishing of *Vil De laere Polygon?* — Would you like to learn Polygon? — the first published article on the game now called Hex.

Ryan B. Hayward and Bjarne Toft

Acknowledgments

Research for this book was supported in part by a Natural Sciences and Engineering Research Council of Canada Discovery Grant.

☼ ☼ ☼

A large number of people have been extremely helpful, providing — and giving access to — information. To mention just a few: Thomas Maarup (Denmark); Anni and Hugo Piet Hein (Middelfart, Denmark); Anker Tiedermann (Virum, Denmark); Jotun Hein (Oxford, UK); Søren Galatius, Persi Diaconis, Jim Gardner, Timothy Edward Noakes and Stan Isaacs (Stanford University); Henry Nielsen and Hans Buhl (Aarhus University). I also wish to thank my wife Bente and my family for sharing my interest.

Bjarne Toft

I would not have finished this book without the support of many friends.

The project had a few false starts. Around 1992 I wrote an outline but found I didn't have a lot to say. In 1999 Jack van Rijswijck and I started background research — including a long phone conversation with John Nash — but then learned from Martin Gardner that Cameron Browne's book *Hex Strategy* was about to come out. Later Philip Henderson and I started again. but his time was eventually taken away by Silicon Valley.

I met Bjarne in Paris in 2004 at the graph theory conference in memory of Claude Berge, and again — with my father — at a graph theory meeting in Nyborg in 2005. A shared passion for Hex led to a family visit to Denmark in 2010. In 2018, several visits and football matches later, we finally finished.

Thank you, Vašek Chvátal for sharing your passions for math and writing. Thank you, Vašek, Bruce Reed, Chính Hoàng and Julie Bates for your camaraderie — and Hex games — at the start of this journey.

Thank you, Liz Greenaway, Peter Taylor, Robin Wilson, Cameron Browne, Colin McDiarmid, Frédéric Maffray, Myriam Preissmann, András Sebö, Darse Billings, Timo Ewalds, Sylvia Nasar, Stephen Kennedy, John McCleary and Peter Winkler for your advice and support.

Thank you, Thomas Maarup for locating and translating Danish documents, and Cameron and Jack for sharing your board-drawing scripts.

Thank you, Yngvi Björnsson, Michael Johanson, Morgan Kan, Nathan Po, Jack van Rijswijck, Philip Henderson, Broderick Arneson, Aja Huang, Jakub Pawlewicz, Brad Thiessen, Henry Brausen, Jesse Huard, Kenny Young, Noah Weninger, Chao Gao and Martin Müller for your work on the computer Hex project and Jonathan Schaeffer for your support.

Thank you, John Nash, Katharine Gale, Marian and Terry Titus, Jerry Weaver, Hans Weinberger, Eve Siegel, Micah Beck, Ron Evans, Bert Enderton, Jing Yang, Ingo Althöfer, Donald Crowe, Kate Jones, Birgit Bock and Kathie Cameron for sharing your stories and photos.

Thank you, Anni and Hugo Piet Hein for your hospitality and access to the Piet Hein papers, Jim Gardner for permission to quote from Martin Gardner's papers, and Aarhus University for access to Jens Lindhard's papers.

Thank you, Phil, Martin and Noah for your meticulous feedback and Sarfraz Khan, Fraser Callum and Robin Lloyd Starkes for shepherding this project through its final stages.

Thank you, Jack and Phil and Bjarne for sustaining this project. Thank you, Bjarne and Bente for your generous hospitality. And thank you Liz, Annie, Em, Audrey, Herb and Owen for your patience.

Ryan B. Hayward

Permissions

The authors gratefully acknowledge the following:

Mudcrack-Y board images © Craige Schensted.

Kaliko game images © Kadon Enterprises, Inc.

Permission to quote from Martin Gardner correspondence in the Stanford University Martin Gardner Papers archive and to use the photo of Martin Gardner, courtesy James Gardner, Managing Partner, Martin Gardner Literary Interests.

Permission to quote from Piet Hein correspondence, courtesy Hugo Hein, Piet Hein Trading APS.

Software for drawing game boards based on software by Cameron Browne and Jack van Rijswijck, used with permission.

Reproduction of Piet Hein columns *Vil De laere Polygon?* (Politiken December 26 1942, page 4) and *POLYGON Opgave Nr. 49* (Politiken August 8 1943, page 4) courtesy Politiken, Rådhuspladsen 37, DK-1785 København V, translations by Bjarne Toft.

Photos of Jens Lindhard and his Polygon papers © History of Science Archives, Center for Science Studies, Aarhus University, Denmark.

Photo of Niels Bohr with Piet Hein © Niels Bohr Archive, Copenhagen.

Photo of Shannon's machines © MIT Museum.

Every effort has been made to secure the appropriate permissions. If you feel your copyright has been infringed, please contact the publisher, who will attempt to address this at its earliest opportunity.

Rules of Hex

The board is an $n \times m$ array of hexagons, usually with $n = m$. Two opposing sides are black. Two opposing sides are white. One player has black sides and black stones (or a black marker). The other player has white sides and white stones (or a white marker). Players alternate turns. On a turn, a player puts a stone in (or marks) an empty cell. The winner is whoever joins her two sides.

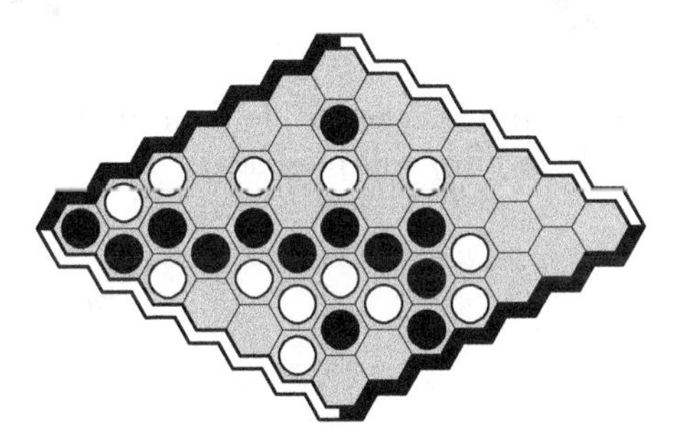

FIGURE 2: A finished game on a 7×7 board. Black wins.

Swap rule

The swap rule is an optional extra rule that offsets the advantage of playing first. The first player moves. The second player then either plays next in the usual manner — with stones and sides of the other colour — or swaps stones and sides with the first player but does not move (so the first player plays next). Players then alternate turns as usual.

FIGURE 3: A swap rule example. (from left) Petra (black) plays first. Jo swaps. Petra (now white) plays next.

FIGURE 4: Another swap rule example. Jo (white) plays first. Petra chooses to play a black stone. Jo (white) will play next.

Mirror-swap rule

Some people implement the swap rule in a different way, called mirror-swap. Players do not change colours: instead, if the second player swaps, the *stone* on the board *changes colour and moves* to its equivalent 'mirror' location, i.e. reflected in the straight line through the acute corners.

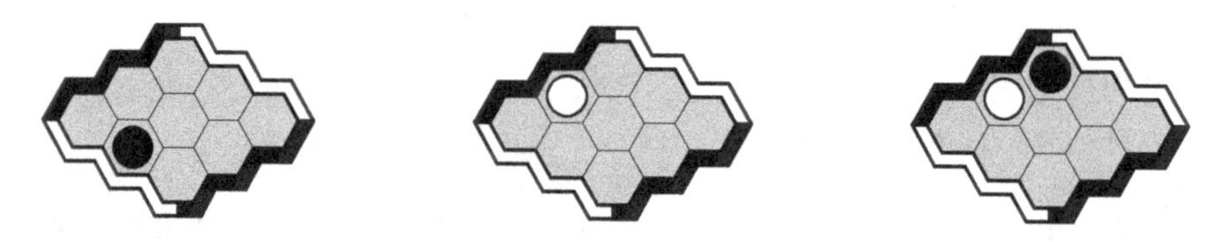

FIGURE 5: A mirror-swap example. (from left) Petra (black) plays first. Jo (white) mirror-swaps. Petra (black) plays next.

Chapter 1

Birth

Suddenly in the half-light of dawn a game awoke, demanding to be born.

Piet Hein[1]

1.1 Polygon

On December 26 1942 the Danish newspaper Politiken introduced a new game — then called Polygon, now called Hex — to its readers. Over the next two weeks, coinciding with the Christmas vacation, each day's newspaper included a fresh Polygon column, with tips, a puzzle, and an empty board on which to play. Try it yourself: copy a board, grab a pencil, and find an opponent.

Polygon's story begins with its creator, Piet Hein, born December 16 1905.[2] Hein's father and mother — engineer Hjalmar and opthamologist Estrid, nee Hansen — had an unusual marriage agreement: she would give him a son, and he would build her an eye clinic in the newly renovated top floor of their home on Gammel Torv, the oldest square in Copenhagen. The agreement was minimally satisfied: Piet was their only child.

Hein jokingly claimed to be a direct descendant of the Dutch naval hero — or pirate, depending on your point of view — Piet Pieterszoon Hein. The joke is that this would be quite an achievement, as Piet Pieterszoon was childless.

In 1922 Hjalmar died after a kidney operation. Piet and Estrid then moved permanently into their summer home — Rungsted Skovhus — in North Zealand on the 'Danish Riviera', where Estrid lived until her death in 1956.

Piet Hein was an unusual character with diverse interests. After graduating in 1924 from the Copenhagen high school *Metropolitanskolen*[3] he moved

FIGURE 1.1: (from top) The Hein family home, Gammel Torv 10-12, before renovation to make room for Estrid's eye clinic, after renovation, and today. At right is the tower of the Copenhagen Cathedral.

FIGURE 1.2: Niels Bohr (front center), Werner Heisenberg (front right) and Piet Hein (back center) circa 1930. © Niels Bohr Archive, Copenhagen.

around at university, first studying engineering and philosophy in Copenhagen, then painting in Stockholm, and finally — supervised by family friend Niels Bohr, who had won the Nobel Prize for physics in 1922 — physics at Bohr's Institute for Theoretical Physics. In 1931 Hein dropped out of university.

Hein supported himself by consulting and from royalties from opthamologic inventions. He was also fascinated by games and puzzles. In the 1930s, supposedly while struggling to stay awake during a talk by Werner Heisenberg — who received the 1932 Nobel Prize for Physics — Hein invented the Soma cube. In this geometric puzzle, seven different pieces, each formed by gluing together three or four $1\times1\times1$ cubes in all possible irregular ways — so, not a box — fit together to form the regular $3\times3\times3$ cube, a typical Piet Hein surprise. He called this puzzle the world's smallest philosophical system. Years later, for a Soma booklet, he wrote this:

> *This is a little book about a little game. One should not despise small games. Albert Einstein's bookshelves, both in his office and at home, overflowed with books — except for two shelves, which were reserved for small games that he occasionally played with and thought about.*

FIGURE 1.3: The Soma cube.

> *He said, "Small games are much like the big game one plays with nature when you try to explore its secrets. That is also a simple game, but nevertheless difficult. You can learn a lot from small games."*
>
> *One might add that it suits people to occupy themselves with small games and to master them.*

After copyrighting it in Denmark and England, Hein started marketing Soma in 1934. The cardboard package that contains the cube is another Piet Hein surprise: held together without glue or fasteners, it unfolds to reveal a sheet with instructions and puzzles.

1.2 Design of Hex: first part

Inspired by the popularity of the Soma cube, Hein sought new game ideas.[4] In a 1957 letters to Martin Gardner[5], Hein recalls when the idea behind Hex came to him:

> *...I invented [Polygon] myself in the years up [to] 1942 after having mused for several years over the possibility of making a game on that topological property of the plane and sphere and topological equivalent surfaces (but not the toric ring) which is the basis of the (unproved? R.s.v.p.) four-colour [theorem]: that you cannot connect more than four points in such a surface two and two without two connecting lines crossing each other.*

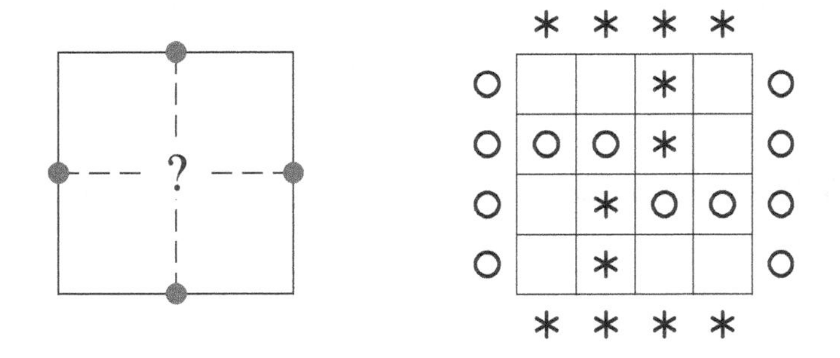

FIGURE 1.4: The first idea behind Hex: the dashed lines must cross. On a rectangular grid, armies can block each other.

So this is the first idea for Hein's new game. Imagine a four-sided territory surrounded by two armies: each army occupies two opposing borders — or fronts — of the territory. Each army wants to join its own two groups, which by the topological property will separate the enemy's two groups.

But Hein saw a problem with this idea: if the game is played on the usual rectangular grid, the armies can block each other as in Figure 1.4. And Hein did not want his game to allow stalemates.

So, for several years, Hein himself was blocked from finishing the design of this new game. Before we describe how the final design came to him, we briefly discuss the four-colour problem.

1.3 Four colours and crossing lines

In this section, we explain Hein's statement that "*the topological property* [underlying Hex ... is] *the basis of the four-colour theorem*". You can skip this section without any loss of continuity.

Consider a planar map, that is, a map drawn on a flat surface. We wish to colour the map so that region boundaries are clearly visible: any two regions whose boundaries touch (not just at a single point, but on an interval with positive length) must have different colours. How many colours do we need?

This map-colouring problem dates from 1852, when schoolboy Francis Guthrie posed the now-famous four-colour problem: when colouring regions of a planar map so that touching regions have different colours, do four colours suffice?

In his 1957 letter to Gardner, Hein refers to the "*(unproved? R.s.v.p.) four-colour theorem*". At that time the four-colour problem was still unsolved. In 1879 Alfred Kempe argued that the answer is yes, but 11 years later Percy

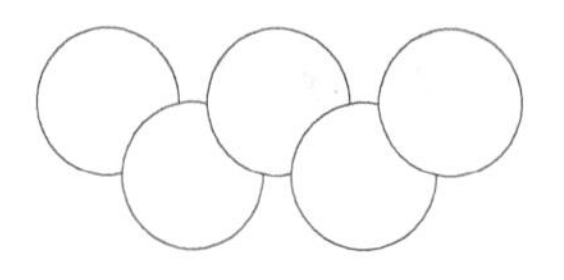

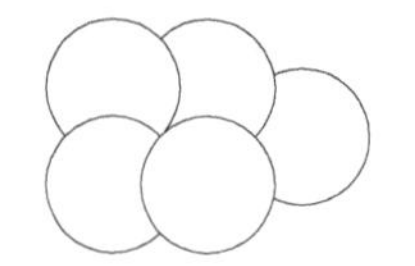

FIGURE 1.5: One map needs two colours, the other needs three.

FIGURE 1.6: One map has clique size and colour number 3, the other has clique size 3 and colour number 4.

Heawood found a flaw in his reasoning. It was not until 1976, and only with the aid of computers, that a correct proof was found: four colours do suffice.[6]

So, what does the four-colour problem have to do with the crossing lines property? Consider the two maps in Figure 1.5: one needs two colours, the other needs three. Can you see why?

The answer is that the second map has a clique — a set of regions, with each pair touching — of size three. Within a clique, each two regions need different colours, so there must be as many colours as there are regions. So the size of a map's largest clique gives a lower bound on the map's colour number, i.e. the number of colours needed to colour the map's regions so that neighbouring regions have different colours.

How big can we make this lower bound on a map's colour number? Does any planar map have clique size four? Yes, as seen in Figure 1.7.

Does any planar map have clique size five? This so-called five-princes question — can a territory be split into five regions, with each two sharing a boundary — was asked by August Möbius around 1840[7].

The answer is no: the four-colour theorem says that any planar map can be four-coloured, so each planar map has clique size at most four. But using

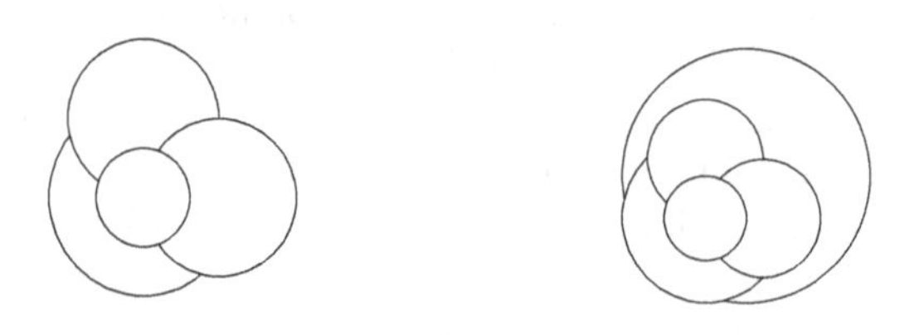

FIGURE 1.7: Regions forming a clique, and not forming a clique.

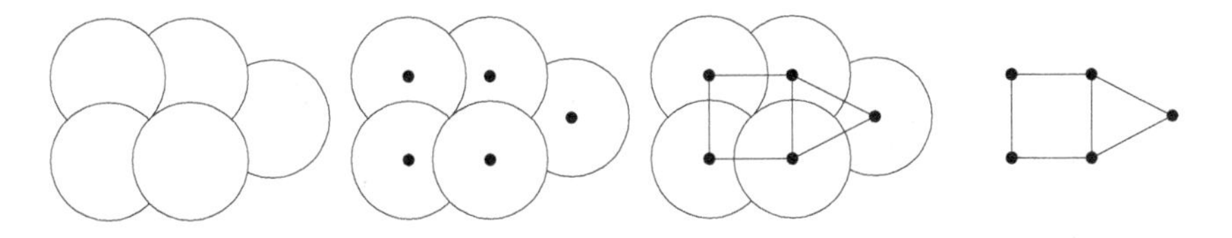

FIGURE 1.8: Constructing a map's dual diagram.

the four-colour theorem to reach this conclusion is using a sledgehammer to kill a fly. Here is a simpler argument with dots, lines, and crossings.

For any planar map, we can place a dot inside each region. And between each two dots whose regions touch, we can draw a line. By drawing lines so that they cross boundaries only where two boundaries touch, we can ensure that no two lines cross. Now erase the region boundaries, leaving only dots and lines: this diagram is the dual of our original map.

Conversely, if we have a dot-line diagram[8] on a flat surface with each pair of dots joined by a line, and no two lines cross, then — by gradually expanding each dot into a region — we can draw a planar map with this dot-line diagram as its dual. A clique in a dot-line diagram is a set of dots with each pair joined by a line.

So the five-princes problem is equivalent to this five-dots problem: can five dots be placed on a flat surface, each pair joined by a line, so that no two lines cross?

In a letter to Christian Goldbach in 1750, Leonhard Euler mentions this formula for any convex polyhedron, which also applies to planar dot-line diagrams: a diagram with (at least 3) D dots, R regions, and L lines satisfies $D + R - L = 2$. Each region is bounded by at least three lines and each line is on the boundary of two regions, so R is at most $2L/3$. Substituting for R in Euler's formula gives that L is at most $3 \times D - 6$.

So the answer to the five-princes problem is "no". If the answer were "yes", there would be a dot-line diagram with $D = 5$ and $L = 10$, since each of the 10 pairs of dots is joined by a line. But this contradicts the bound: L can be at most $3 \times 5 - 6 = 9$.

In fact, the "no" answer to the five-dots problem is mathematically equivalent to the statement that, in a game of Hex, at most one player can win.

The topological property on which Hein based Hex is that *two lines within a quadrilateral, one joining one pair of opposite sides, the other joining the other pair, must cross.*[9] This property follows from the "no" answer to the five-dots problem. Argue by contradiction: assume that the two lines inside the quadrilateral do not cross. Now join each of the four ends of these two lines to a common dot outside the quadrilateral, as shown in Figure 1.9; this gives a planar dot-line diagram with five dots, each pair joined by a line, and no lines crossings, which is impossible.

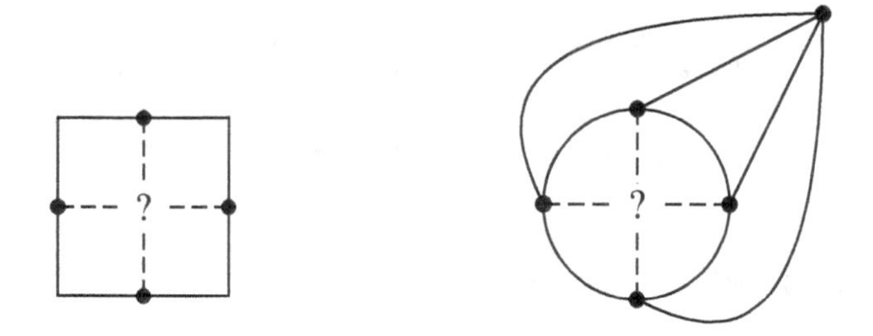

FIGURE 1.9: The dashed lines must cross, otherwise no two lines in the five-dot diagram cross, contradicting Euler's formula.

In calling this topological property the basis of the four-colour problem, Hein presumably refers to the property we have discussed: a planar map has clique size at most four. Noticing this, one can then ask — as Guthrie did — whether four colours suffice to colour any such map.

Hein's comment *but not the toric ring* refers to colouring a map drawn on a torus, or doughnut: there, seven colours are sometimes needed (and are always enough).

In 1942, the Swiss mathematician Hugo Hadwiger posed a conjecture that includes the four-colour theorem as a special case. [10] This famous conjecture remained virtually unknown until it was rediscovered in the 1950s. It is still unsolved.

1.4 War and poems

On April 9 1940 Nazi Germany invaded Denmark and Norway. From that time, Hein gradually went into hiding. In the process, he lost access to his consulting fees and royalties[11].

Seeking income, Hein wrote poems he called *gruks* in Danish, from *grin og suk*, meaning 'laugh and sigh'. Grooks first appeared in *At Taenke Sig* ['to think for oneself' or 'I am thinking about it' or 'can this be true?'], a daily satirical back-page Politiken feature that has run from 1932 to the present. Grooks described the occupiers in words that Danish readers — but not German censors — understood. The second grook below hints that those who support the occupiers might come to regret it.[12]

Kumbel Kumbell is a pseudonym. *Hein* means *sharpening stone* in old Danish. *Piet* is a form of Peter, from *stone* in Greek. And *kumbel* means *tombstone* in old Norse. So *Kumbel Kumbell* is *Piet Hein.*

The April 17 grook had this preface:

> *After the first grook by Kumbel Kumbell a few days ago, our telephones have been ringing from morning till evening. Especially our*

Vaarlig Vemod	Spring Melancholy
Dagene	The days are
laenges,	getting longer
laenges,	longer,
laenges.	longer.
Hvad laenges	What are they
de mod?	longing for?
Dagene	The days are
laenges,	longing,
laenges,	longing,
laenges	longing
imod St. Hans	for midsummer.

Trøste-gruk	Consolation grook
Den,	Losing
som taber	one glove
sin ene Handske	is certainly painful,
er heldig	but nothing
i Forhold til den,	compared to the pain,
som taber	of losing
den ene,	the one
kasserer	throwing away
den anden ...	the other ...
og finder	and finding
den forste	the first one
igen.	again.

FIGURE 1.10: First grooks. Kumbel Kumbell, April 14 and 17, 1940.

> *female readers demand that Kumbel Kumbell become a permanent writer in* At Taenke Sig. *He is so thoughtful. He has the coolest rhythm, the finest voice, which entices some light from these dreary grey days.*[13]

In December 1940 Hein published *77 Gruks*, a first grook collection.[14] By 1957, over 175 000 grook books had sold.[15]

In December 1941 Hein published a 1942 almanac with a calendar as well as sections on pastimes and games.[16] See Figures 1.14 and 1.15. Thinking about these games perhaps inspired Hein's design: he later described Hex as a game of duelling labyrinths.

FIGURE 1.11: Hein's two choices for board shape.

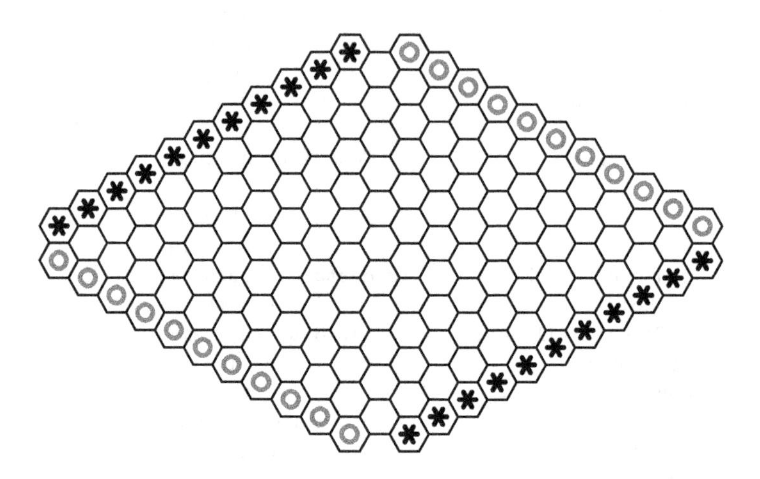

FIGURE 1.12: Hein's finished design: the Polygon game board.

1.5 Design of Hex: final part

Sometime before September 1942 Hein had a eureka moment: his game design would work if he used a hexagonal grid instead of a rectangular grid. As he later wrote Martin Gardner,

> *It took me some time (I am sorry to say) to find what is the key to the realization of the principle: that only three fields [ed.: cells] should meet in any point, because if four or more meet, the two opponents can block each other mutually. That's why an orthogonal pattern is forbidden, and a hexagonal pattern is the simplest solution. The next step was to find out that the whole board should be [diamond, i.e. rhombus] shaped. (It can also be [a rectangle]). The number of fields was a matter of experience.*[17]

For board shape, Hein chose diamond over rectangle. For grid size, he chose 11×11. Hein's design of Polygon was complete.

FIGURE 1.13: Cover of the first grook collection. © Piet Hein.

FIGURE 1.14: Cover of Hein's almanac. © Piet Hein.

2, 3 eller 4 Spillere skiftes til med en blød Blyant at trække en lodret eller vandret Streg forbindende to Punkter. Formaalet er at trække den sidste Streg, som lukker et Kvadrat. Hver Gang man lukker et Kvadrat, har man Ret til umiddelbart derefter at trække endnu en Streg. Hvert lukket Kvadrat gælder 1 Point, men der noteres kun for hver Spiller de Points han faar *mere* end sidste Spiller fik. Den, der først noteres, behøver altsaa kun at lukke 1 Kvadrat, men den næste skal saa lukke 2 for at faa 1 Point, den næste saa igen 3 Kvadrater for at faa 1 Point o. s. v. Noteres en Spiller for 0 i et Træk, saa skal næste Spiller igen kun have 1 Kvadrat for at faa et Point. Naar alle Kvadrater er lukket, sammentælles Points, Vinder er den, der har faaet flest. Saa visker man ud og begynder forfra.

75

FIGURE 1.15: Labyrinth game from Hein's almanac. © Piet Hein.

Notes

[1]Circa 1942, Hein draft of first Politiken column [51].

[2]A recent Piet Hein biography is *Piet Hein Verdensdanskeren* [27].

[3]This school was founded in 1209.

[4][51], Polygon demonstration manuscript, p79.

[5]1957.02.13, 1957.03.23 letters Hein-Gardner, collection of Hugo Hein [33].

[6]For more on this story, see Wilson's fascinating tale *Four Colors Suffice.*

[7][72]

[8]Dot-line collections are usually called graphs, and dot-line diagrams called graph drawings. Hex is one of many games that are played on a graph.

[9][36]

[10]For a graph, a *minor* is a graph obtained by removing any number of edges and/or nodes and then contracting any number of edges. Hadwiger conjectured that, for every positive integer t, each graph which does not have a clique of size t as a minor is $(t-1)$-colourable. E.g., each graph with no five-clique (in particular, each planar graph) is four-colourable. A class of maps is *topological* if taking a map in the class and erasing some borders always yields a map in the class. The conjecture implies that, over all maps in a topological class, the maximum colouring number equals the maximum clique-size (number of countries, each pair sharing a border). For planar maps the latter number is 4, so the conjecture generalizes the four colour theorem. For toroidal maps the number is 7. For more on Hadwiger's conjecture, see Toft's [69] surveys, which includes early history, or Seymour's overview [65].

[11]1957.02.13 letter Hein-Gardner [33].

[12]This grook also appeared in *77 Gruks*, the first Kumbel grook collection [34]. Kumbel grooks appeared in *At Taenke Sig* from 1940 through 1961.

[13]Hein suggests women liked his poems, but he caused women much grief. Around this time, Hein — who remarried three times — had left his first wife Gunver and was having an affair with Tove Ditlevsen, who wrote about this in her autobiography *Gift* — 'marriage' or 'poison'. Once after making love, Tove noticed that a rose in a vase beside the bed had lost its leaves. 'It does not believe in budding anymore', she remarked. Hein jumped out of bed and wrote this now well-known grook:

```
En blomst ved min elskedes seng har jeg sat
en rose der rødmende stod der i nat
Først faldt et blad, og så to og så fler,
nu tror den vist ikke på knopskydning mer.
```

```
   I gave my love a rose of purest red.
All night it stood blushing beside my bed.
   One petal fell; then two; then a score:
     It won't believe in budding any more.

        Piet Hein, Grooks IV, Borgen 1972
```

Tove quotes Nadja, an earlier lover of Hein's, as saying this about Hein: *He is a dangerous man, created to make many women unhappy.*

[14][45]

[15]1957.02.13 letter Hein-Gardner [33].

[16][35]

[17]1957.03.23 letter Hein-Gardner [33].

Chapter 2

Preparing to launch

A The proof of the pudding is in the eating.

Anonymous

Goals are pure fantasy unless you have a specific plan to achieve them.

Stephen Covey

2.1 Will Polygon sell?

Would Polygon be fun to play? Hein was a designer but not a strong game player. It would be costly to print Polygon pads if they did not sell. To gauge Polygon's popularity, Hein turned to his social circle.

On September 30 1942 Hein placed an order for 25 50-sheet Polygon pads that he presumably distributed to students at Copenhagen University and the nearby Polytechnical Institute (now Technical University of Denmark),[1] including Aage Bohr and Jens Lindhard, each 20 years old. Both later studied physics with Aage's father Niels. Aage would win the 1975 Nobel Prize for Physics. Lindhard would become a distinguished professor in Aarhus and was already a strong chess player: in 1941 he finished 3rd out of 14 players in the Copenhagen Chess Championship, losing only one game.[2]

Polygon passed the popularity test: players found the game was as intriguing as chess but much simpler to learn. So Hein started his marketing plan. On October 31 he ordered 500 more Polygon sheets. He applied for a patent and on November 16 signed a contract with Politiken: he would provide them with regular columns on the game, they would publish the columns and print 20 000 game pads, and together they would share the profits from pad sales.

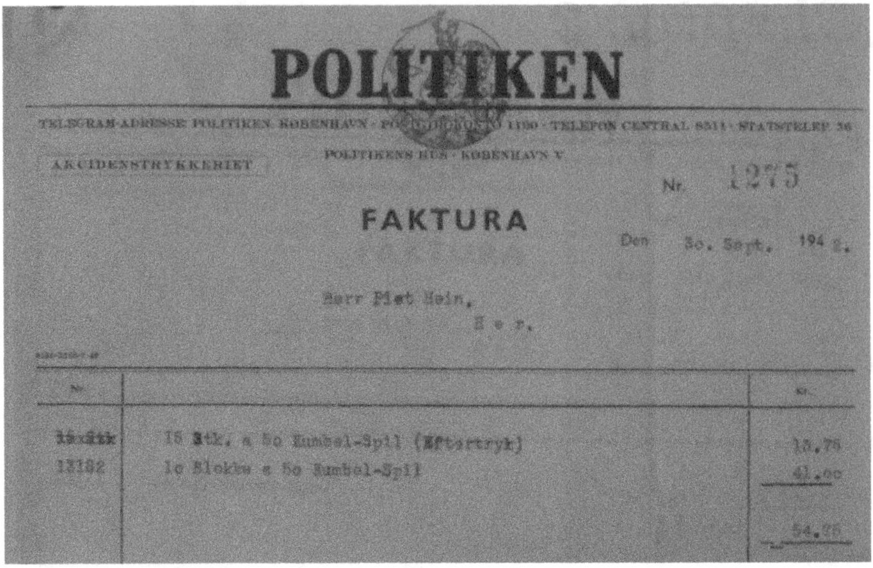

POLITIKEN

AKCIDENSTRYKKERIET

POLITIKENS HUS · KØBENHAVN V

Nr. 1275

FAKTURA

Den 30. Sept. 1942.

Herr Piet Hein,

Her.

FIGURE 2.1: September 30, 1942 invoice for printing Polygon pads.

2.2 Call for puzzles

Hein thought that *Opgaver* — tasks or problems or puzzles — would be the best way to hook the newspaper audience on Polygon, so he again reached out to his Polygon network. Starting December 4, he distributed sheets with a 7×7 board on one side and a call for puzzles on the other. Here is a translation by Bjarne Toft:

> **The Game's Only Rule** *POLYGON is played by two players: White and Black, whose symbols are respectively a circle and a star. Each player has two opposing fronts, marked with their symbols. Each player in turn marks any empty field with their symbol, The game continues until one player wins. A player wins by joining their two sides with a path, which can twist and turn.*
>
> **A Search for Puzzles** *In order to bring POLYGON to a larger audience — which we all want — we need a large supply of Polygon puzzles. So I urge anyone with an interest in the game to construct such puzzles. Some puzzles arise naturally during a game. We seek puzzles for all board sizes, from 5x5 to the standard 11x11. Puzzles for smaller boards are especially desired. Send POLYGON puzzles, with commentary, to me. Empty POLYGON gamesheets will be sent to puzzle-makers free of charge. Among the*

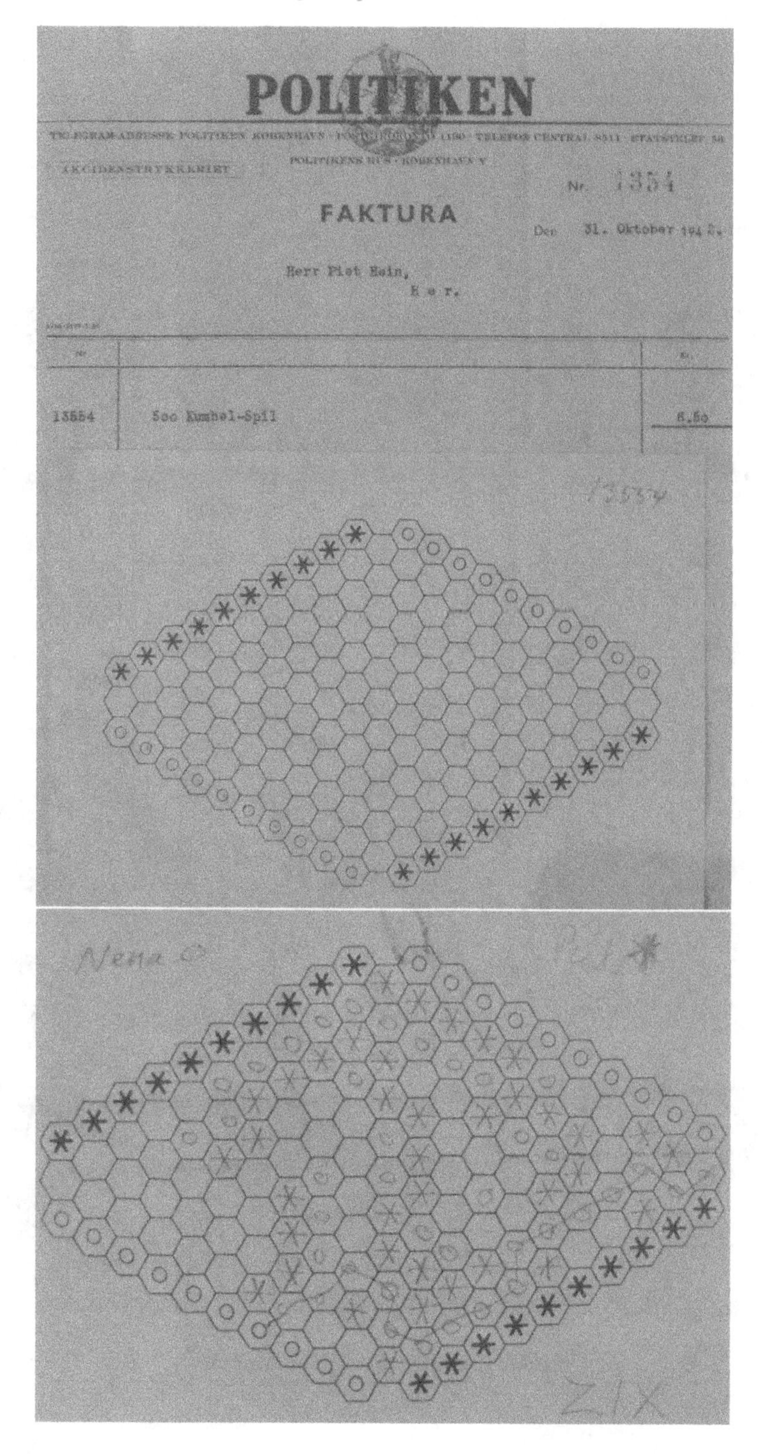

POLITIKEN

POLITIKENS HUS · KØBENHAVN V

AKCIDENSTRYKKERIET

Nr. 1354

FAKTURA

Den 31. Oktober 194[illegible].

Herr Piet Hein,

H e r.

Nr.		Kr.
13554	500 Kumbel-Spil	6.50

FIGURE 2.2: Invoice for 500 Polygon sheets, and a game between Piet Hein and his second wife Nena on a game sheet. We do not know what ZIX means.

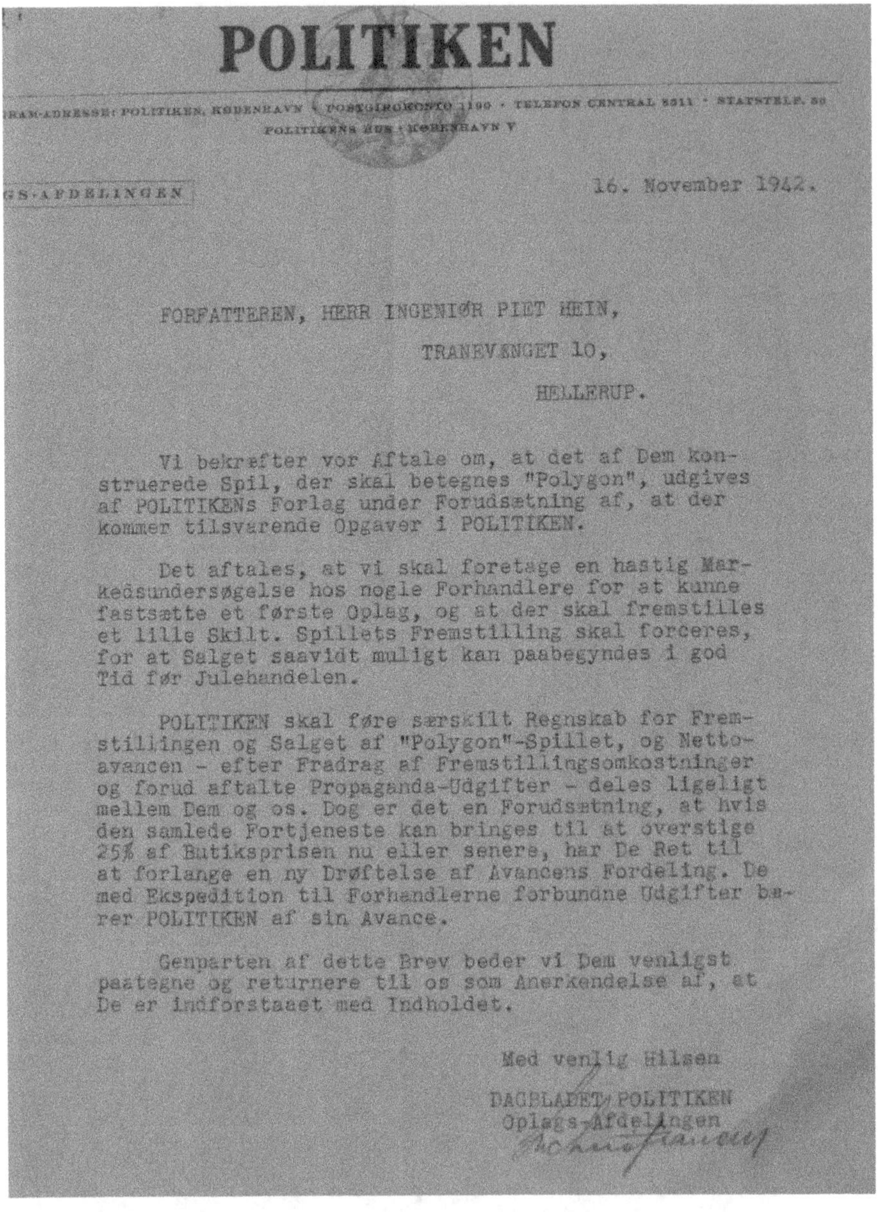

POLITIKEN

RAM-ADRESSE: POLITIKEN, KØBENHAVN · POSTGIROKONTO 1100 · TELEFON CENTRAL 8311 · STATSTELF. 80
POLITIKENS HUS · KØBENHAVN V

GS-AFDELINGEN

16. November 1942.

FORFATTEREN, HERR INGENIØR PIET HEIN,
TRANEVÆNGET 10,
HELLERUP.

Vi bekræfter vor Aftale om, at det af Dem konstruerede Spil, der skal betegnes "Polygon", udgives af POLITIKENs Forlag under Forudsætning af, at der kommer tilsvarende Opgaver i POLITIKEN.

Det aftales, at vi skal foretage en hastig Markedsundersøgelse hos nogle Forhandlere for at kunne fastsætte et første Oplag, og at der skal fremstilles et lille Skilt. Spillets Fremstilling skal forceres, for at Salget saavidt muligt kan paabegyndes i god Tid før Julehandelen.

POLITIKEN skal føre særskilt Regnskab for Fremstillingen og Salget af "Polygon"-Spillet, og Nettoavancen - efter Fradrag af Fremstillingsomkostninger og forud aftalte Propaganda-Udgifter - deles ligeligt mellem Dem og os. Dog er det en Forudsætning, at hvis den samlede Fortjeneste kan bringes til at overstige 25% af Butiksprisen nu eller senere, har De Ret til at forlange en ny Drøftelse af Avancens Fordeling. De med Ekspedition til Forhandlerne forbundne Udgifter bærer POLITIKEN af sin Avance.

Genparten af dette Brev beder vi Dem venligst paategne og returnere til os som Anerkendelse af, at De er indforstaaet med Indholdet.

Med venlig Hilsen
DAGBLADET POLITIKEN
Oplags-Afdelingen

FIGURE 2.3: Politiken contract.

SPILLETS ENESTE REGEL

POLYGON spilles af to Spillere: Hvid og Sort, som har Mærkerne henholdsvis en Cirkel og en Stjerne. Hver Spiller har to modstaaende Fronter, afmærkede i Felterne. — Man skiftes til at sætte sit Mærke i et hvilket som helst tomt Felt — og fortsætter indtil den ene har vundet. — Det gælder for hver Spiller om at besætte Felter paa saadan en Maade, at de danner en sammenhængende omend nok saa snoet Forbindelse mellem vedkommende Spillers to Fronter.

EFTERLYSNING

For at POLYGON-Spillet kan lanceres for en videre Kreds — hvilket er en Forudsætning for, at der fortsat kan fremstilles Spillediagrammer — behøves et ret stort Opgave-Materiale.

Jeg opfordrer derfor alle, som har Interesse for Spillet, til at ha Opmærksomheden henvendt paa Ideer til POLYGON-*Opgaver*. Saadan Ideer opstaar ofte under Spillet. Der er Brug for Opgaver paa Spillebrætter af alle Størrelserne fra 5×5 Felter indtil den normale Spillestørrelse 11×11 — især paa Brætter af de smaa Antal Felter. POLYGON-Opgaver med Løsninger bedes indsendt til mig. POLYGON-Spillebrætter tilsendes Opgaveindsendere gratis, saa vidt Oplaget rækker.

Blandt de bedste af de hver Uge indtil videre indkomne Opgaver udtrækkes én, som belønnes med en *Præmie* — i Ugen indtil Lørdag d. 12. December 1942: en Julegaas.

En POLYGON-*Opgave* bestaar i et POLYGON-Spillebræt, paa hvilket nogen af Felterne er besat med Hvids og Sorts Mærker. Denne Afmærkning behøver ikke at kunne være fremkommet under et normalt Spil, specielt kan der udmærket være mange flere af den ene Spillers Mærker end af den andens. Til en Opgave hører 3 Oplysninger: 1. *hvilken Spiller, der skal begynde*, og 2. *hvilken* (ikke nødvendigvis altid den samme) *Spiller, der kan vinde*, og 3. (helst) i *hvor mange Træk*, hvis begge spiller stærkest muligt. I Overenstemmelse med den Maade, hvorpaa en Løsning angives (se nedenfor), regnes Antallet at Træk efter begge Spilleres samlede Antal.

En Opgave maa være nogenlunde éntydig, dvs. naar begge spiller bedst muligt, skal der praktisk talt kun være én Fremgangsmaade, især skal det første Træk ligge fast. En Opgave bør ikke indeholde overflødige Mærker.

En Opgaves *Løsning* noteres saaledes paa Spillebrættet: de Mærker, som hører med til den stillede Opgave, sættes paa sædvanlig Maade; de Mærker, som hører til Løsningen, erstattes i kronologisk Orden med Tallene 1, 2, 3 osv., saaledes at den som trækker først altsaa har de ulige Tal, den anden de lige. For Overskueligheden kan Sorts Tal gøres særlig fede. Naar Udfyldningen af de sidste Felter i en Løsning er rent rutinemæssig, er det en Fordel, at den ikke udføres. Naar (hvad der ofte sker) to Nabo-Felter, som er blevet besat af henholdsvis Hvid og Sort, ligesaa godt kunde være blevet besat at dem omvendt, kan det angives ved, at de to Felter forbindes med dette Tegn: ⌒ .

Venlig Hilsen,

4. December 1942.

Piet Hein.

Tranevænget 10, Hellerup.

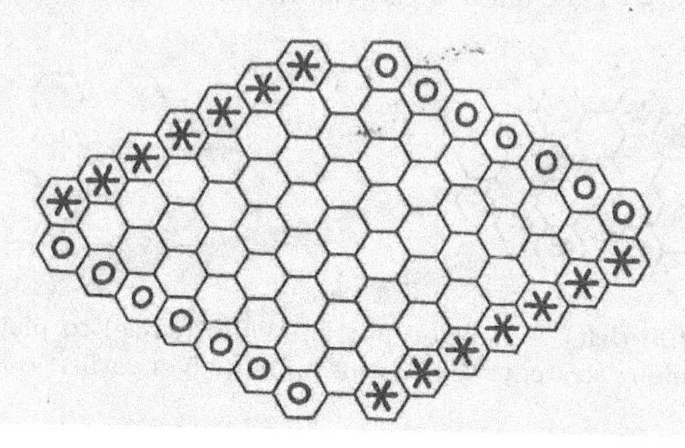

FIGURE 2.4: Hein's call for Polygon puzzles.

best of those puzzles submitted this week, one will be picked at random on December 12, 1942 to receive a special prize: a Christmas goose.

A POLYGON puzzle consists of a POLYGON board with some cells marked by White and Black symbols. Such a position need not arise during a normal game, so one player might occupy more cells than the other. For each puzzle, include this information: 1) who moves next, and 2) who wins (not necessarily the same player), and 3) (preferably) in how many total moves — count the moves for each player — assuming that each player plays perfectly.

A puzzle should have a unique solution, i.e. there should be, practically speaking, only one winning move. The winning move should be indicated. A puzzle should not include an excessive number of marked cells.

Record the puzzle solution on the game board, marked in the usual way (with stars and asterisks), and labelled 1,2,3, ... so that the player who moves next has odd labels and the opponent has even labels. Include any helpful commentary. When (as often happens) a player has an arbitrary choice of marking either of two neighbouring cells, indicate this with an S symbol linking the two cells.

Sincerely, Piet Hein

4 December 1942 *Tranevaenget 10, Hellerup*

Figure 2.5 shows a Polygon puzzle we created in response to Hein's call for puzzles. Unfortunately we are too late — by more than 70 years — to enter it in the draw for the Christmas goose! Also, there is more than one first winning move: in addition to the solution we show, White can also first attack one of the two cells marked with a black S. Since our puzzle allows more than one winning move, we would not have won the goose in any case.

As we shall see later, our best guess for the winner of the goose is Lindhard.

FIGURE 2.5: (left) A Polygon puzzle: White (circle) to play, who wins? (right) A winning strategy: White wins on move 5 or earlier.

2.3 Parenthesis talk

Aage Bohr was the chair of *Parentesen* (Parenthesis), the University of Copenhagen's science club where around December 12 — perhaps at Aage's invitation — Hein gave the talk *Mathematics as a Game, the Mathematics of Games.* At the end of the talk he introduced Polygon. In 2004 Thomas Maarup — a graduate student at the University of Southern Denmark, supervised by Bjarne Toft — discovered Hein's notes for this talk in the archive of Hugo Hein. Here are some extracts from the notes, which appear in complete form in Maarup's thesis.[3]

Scribbled in the top margin, perhaps in response to an introduction by Aage, are these words:

> *a very small contribution towards improving the miserable conditions regarding math and physics as expressed by Aage Bohr*[4]

Hein's opening remark on the interplay of math and games shows his sense of humour:

> *What I intend to provide tonight is merely a sketch of a thought to an introduction to a game. The idea is to look at mathematics as a game, and the game is a simple example of looking at games as mathematics. I am not sure how much spiritual nourishment there is in it for you, so it would put me at ease if you would continue eating and drinking. ...*
>
> *With mathematically described structures the rules can be made concrete by making a mechanical model which rearranges the symbols for you. However, in general the principle is that you yourself are obliged to manipulate the symbols and this* game *— given by the* symbols *and the rules — that* is *the model.*[5]

Hein has six game design criteria:

> *By observing a number of existing games ... I have come up with the following six requisites ... And the degree to which they satisfy them ... is an indication of the game's value ...*
>
> - **fair***:* [ed. each player has an equal chance to win],
> - **progressive***: not moving in circles* [ed. no board position reappears],
> - **finite***:* [end after a] *limited period or number of moves,*
> - **easy to comprehend***: no move is to stir up the situation, turning all advantages to disadvantages and vice versa, making it impossible to plan regularly winning tactics,*

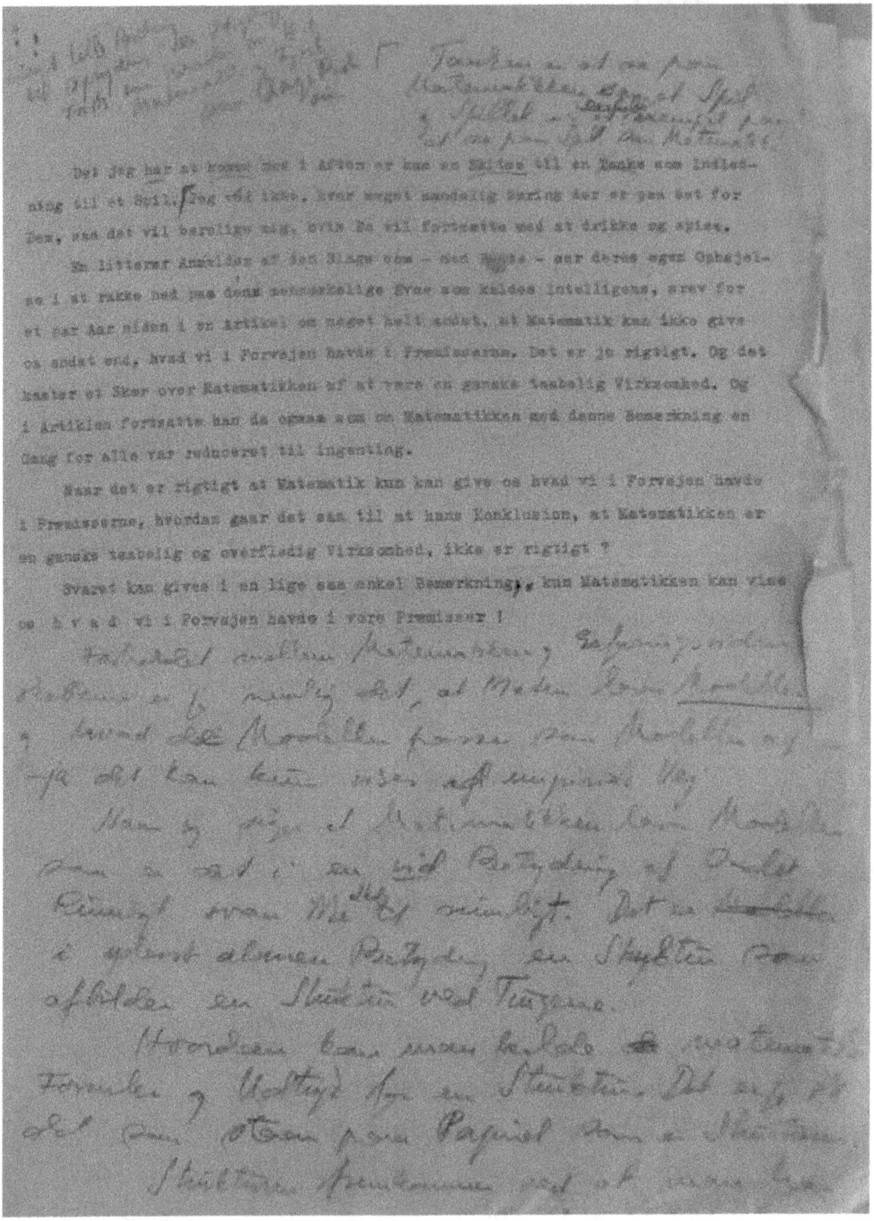

Det Jeg har at komme med i Aften er kun en Skitse til en Tanke som Indledning til et Spil. Jeg véd ikke, hvor meget maadelig Næring der er paa det for Dem, saa det vil berolige mig, hvis De vil fortsætte med at drikke og spise.

En litterær Anmelder af den Slags som – med Rette – sær deres egen Ophøjelse i at rakke ned paa den [illegible] Evne som kaldes Intelligens, skrev for et par Aar siden i en Artikel om noget helt andet, at Matematik kan ikke give os andet end, hvad vi i Forvejen havde i Præmisserne. Det er jo rigtigt. Og det hæver vi Skær over Matematikken af at være en ganske taabelig Virksomhed. Og i Artiklen fortsatte han da ogsaa som om Matematikken med denne Bemærkning en Gang for alle var reduceret til ingenting.

Naar det er rigtigt at Matematik kun kan give os hvad vi i Forvejen havde i Præmisserne, hvordan gaar det saa til at hans Konklusion, at Matematikken er en ganske taabelig og overflødig Virksomhed, ikke er rigtigt ?

Svaret kan gives i en lige saa enkel Bemærkning, kun Matematikken kan vise os h v a d vi i Forvejen havde i vore Præmisser !

FIGURE 2.6: The first page of notes for the Parenthesis talk.

- **strategic**: *versatile in its possibilities for combination,*
- **decisive**: *not end in a draw.*[6]

The next part of the notes is in point form:

- *most games: pieces force others back and forth,*
- *4-colour problem, a toric ring 7,*
- *no moves – paper and pencil, game history shown on game board,*
- *first player win provable unlike chess, I believed second player could win,*
- *strategy: offensive and counter-offensive measures.*[7]

Toric ring 7 presumably refers to colouring a map on a torus — the surface of a doughnut — where seven colours suffice and are sometimes necessary.

First player win suggests that someone in Hein's circle of Polygon players observed that there exists a winning strategy for the first player. But knowing that there exists a winning strategy is not the same as knowing a particular winning strategy. As we will see later, finding winning strategies is not easy: to date, for boards 11×11 or larger, no particular winning strategy — even a winning first move — is known.

The notes next mention *Opgaver* — so Hein presumably posed some puzzles — and close with a remark and a poem:

> *Unlike with chess,* [it is] *quite amusing that a game can be created which is so simple in concept yet impossible to master. Does not the poet say:*[8]

```
Mod Livets Problemer              Compared to life's problems
er Mennesker smaa                            people are small
isaer er Ens Aand                  especially one's mind is
af ufattelig Lidenhed                      incredibly small
Det højeste Maal                            The highest aim
man kan onske at naa                one can strive to attain
er at faa lidt Facon                     is to grasp even a bit
pa sin store Uvidenhed.             of one's own ignorance.

Piet Hein                             trans. Thomas Maarup
```

The Parenthesis talk — where Hein perhaps distributed some call-for-puzzles sheets — ends Hein's pre-publication marketing plan. Two weeks later, Polygon appeared in Politiken.

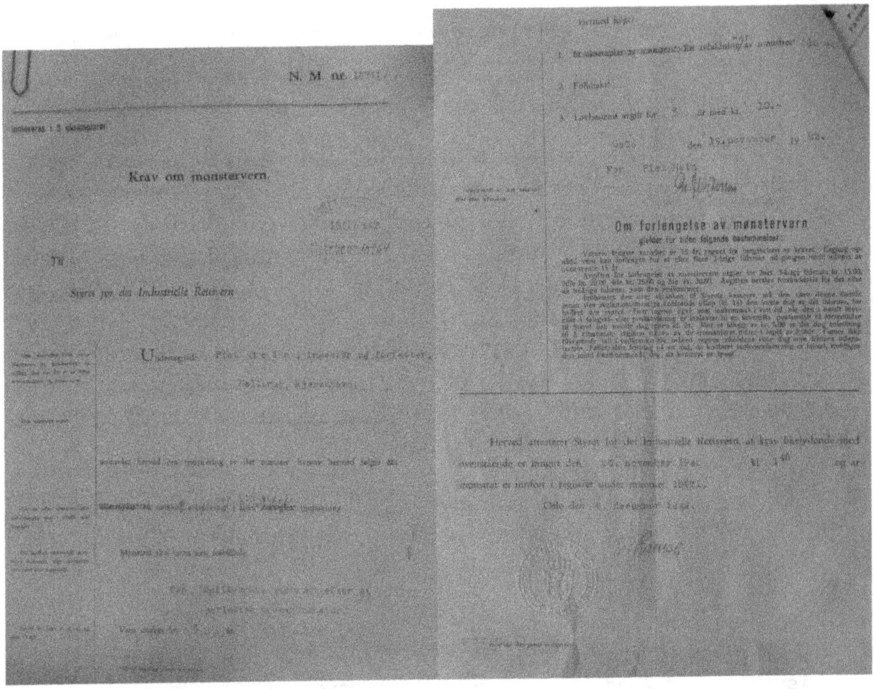
N. M. nr.

Krav om mønstervern.

Styret for det Industrielle Rettsvern

Om forlengelse av mønstervern

FIGURE 2.7: Norwegian Polygon patent application.

FIGURE 2.8: October, 1942. Niels Bohr Institute staff celebrate Bohr's birthday. Detail below: Lindhard (back left), Aage Bohr (back 2nd from right) and Niels Bohr (front right). The institute's tradition of marking Bohr's birthday with a group photo started in 1935 and continues to this day. © History of Science Archives, Center for Science Studies, Aarhus University, Denmark.

Notes

[1] 1957.04.12 letter Hein-Gardner [33].

[2] `http://danbase.skak.dk/turneringer.php?id=177` `http://phys.au.dk/om-instituttet/historie/jens-lindhard-1922-1997/nekrolog/`

[3] [51] pp 76-78.

[4] Parenthesis notes [33].

[5] Parenthesis notes [33].

[6] Hein's Parenthesis manuscript lists only the name of each property: the definitions here are from Hein's Politiken salon manuscript [33, 51].

[7] Parenthesis notes [33].

[8] Parenthesis notes [33].

Chapter 3

Polygon in Politiken

A Polygon game is two labyrinths in combat.

Piet Hein[1]

3.1 Vil De laere Polygon?

On Saturday December 26 1942, the article *Vil De laere Polygon?* [Would you like to learn Polygon?] on page 4 of Politiken introduced the game to the general public. Over the next eight months, Politiken published about 50 Polygon columns and hosted two Polygon evening salons. Hein wrote the columns: none is signed, but the Hein archives has drafts of the first columns, his Parenthesis talk, and a salon demonstration.[2] As Hein had hoped, the columns were popular, and eventually 50 000 Polygon pads sold.

Due to its historical significance, we present here the complete first Polygon column. Original column © Politiken, translation BT, figures RBH based on originals by Hein.

POLITIKEN *26 December 1942*

Would you like to learn Polygon?

Piet Hein has created a game that can be played with equal pleasure by a chess expert and one barely able to hold a pencil.

Politiken today publishes a prize puzzle that will create headaches for beginners.

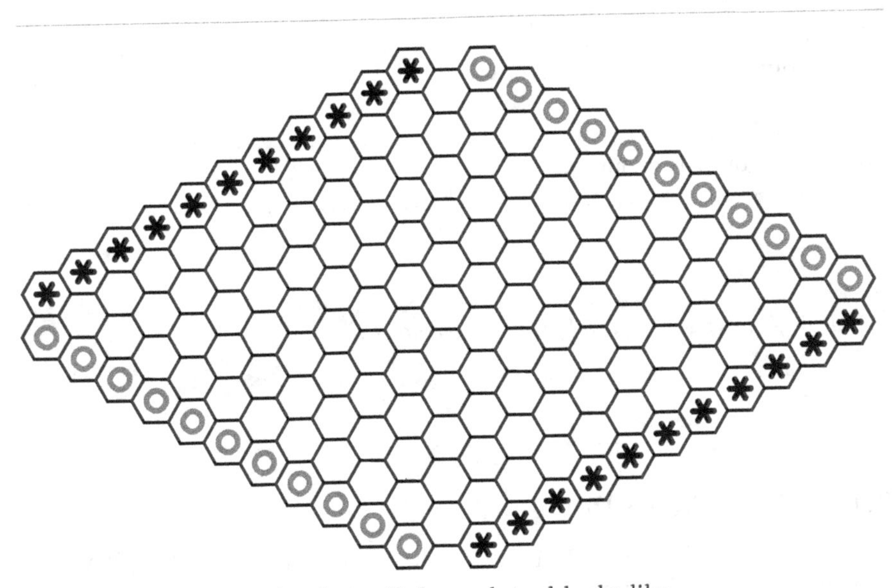

This is what a Polygon board looks like.

A new and exciting pastime has appeared. Here is its background: At a meeting two weeks ago of the Mathematical and Natural Sciences Society with the amusing name Parenthesis, with members of the Student Union Chess Club also invited, the author and engineer Piet Hein gave a lecture titled "Mathematics as a Game - the Mathematics of Games", and finished by demonstrating a board game that he has created based on a new idea. The game created immediate enthusiasm, and has since been practised by a small group of fans. They claim that we have here a simple game with the complexity of chess, yet requiring much less to get started.

Surprisingly, the new game, in spite of its mathematical origin, is simple and easy to grasp. It can be played by anyone capable of holding a pencil, yet contains an abundance of possibilities. Experienced chess players have searched in vain for a strategy by which the starting player can always win, but none has been found. Today the game, called Polygon, is introduced by Politiken to its readers.

In the following article Piet Hein gives a first lesson, so that readers can avoid all the beginner's mistakes. At the same time we present the first Polygon competition puzzle. Piet Hein poses a puzzle, and among the correct answers we will draw for a 50 crowns first prize and a 25

crowns second prize and three 10 crowns third prizes. Send solutions to Politiken's editorial office before Wednesday the 30th with the envelope clearly marked "Polygon".

Polygon's inventor, Piet Hein, presents the game.

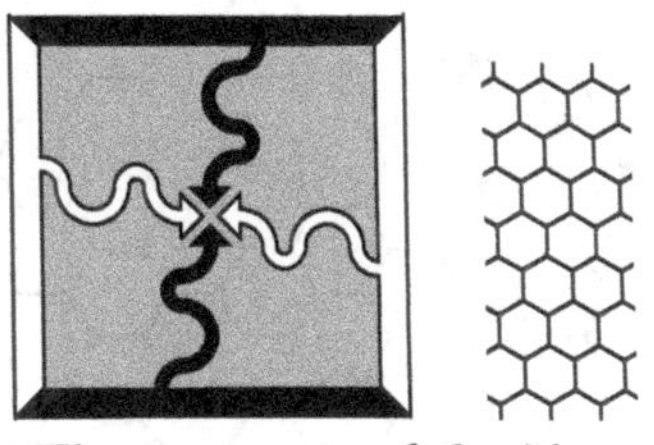

The two parts of the idea.

Now we hand the word over to Piet Hein, inventor of Polygon.

The game is based on the simple fact that two lines within a quadrilateral, each connecting a pair of opposite sides - see the figure - must cross. On this basis one should be able to construct a game where each player owns two opposite sides, with the goal that exactly one of the two players can connect their two sides.

It does not work to give the board a rectilinear grid, as on a chess board. For on a board where four or more cells meet in a point the two opponents can block each other. So neither connects their two sides. Thus one must use a board where at most three cells can meet. This is most easily realized with a hexagonal grid - see the figure again.

With a hexagonal grid the board assumes a diamond shape, the only issue left is to choose the number of cells. You can see the final result.

The only rule

The first player White has as mark a circle, the other player Black has an asterisk. Before the game, each player's two sides — or fronts — are marked. The two players take turns, one after the other, marking any empty cell with their mark. The game continues until one player wins. For each player the goal is to form a connected chain, however twisted, between their two fronts.

Now everyone can try on the board above!

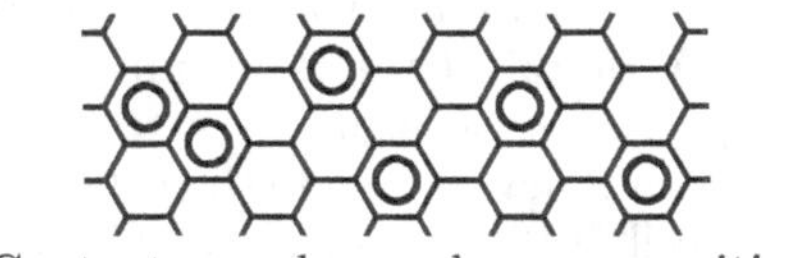

Contact-, angle-, and across-positions.

A first lesson is that you do need not play cells in direct contact with each other — see the figure — to have a safe connection. When two cells are in angle position to each other, and the two cells in between are free, then the connection is also safe: it is clearly enough to mark one of the cells in between when the opponent marks the other. The third position, the across position, is not safe since the opponent can interrupt the connection by playing the intermediate cell. Its use depends on the cells in the surrounding area: in fact, it soon becomes necessary to take a larger part of the board into consideration.

Another lesson, usually learned later, but which can help beginners if we reveal it now, is that it pays to start roughly in the middle of the board. A reasonable start to the game, but by no means forced, is this:

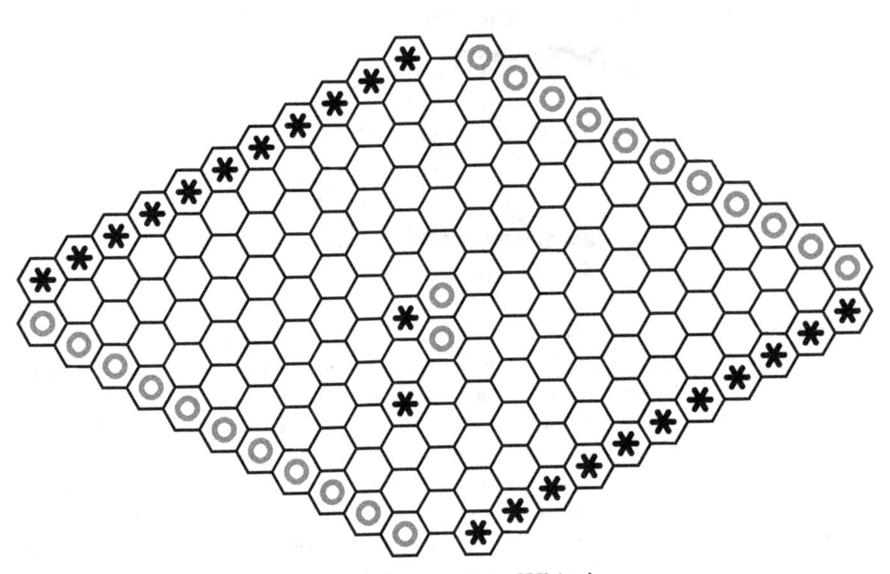

A possible opening — it is White's move.

Here White started in the center. Black then played in contact with White's cell, making two cells in angle position to White's cell now unsafe. White then played in contact with her first move. Now Black replies by taking an angle cell that would have been useful for White. Where should White go now? There are several good possibilities.

This is how this game has started. Now everyone can continue. It is White's move! You do not have to be ingenious in the beginning. There is no better way to learn the game than to play. It is useful to view a position both offensively and defensively, that is in considering both your and your opponent's threats to form a connection. A connection for one player is a barrier for the other, as we know.

A puzzle that is a game.

Finally, here is the first puzzle (see the figure). As you can see, Black has marked more cells than White. Black even has a row of cells between her two sides that is solid — except for one white cell. If Black can just extend this row around the white cell, then Black will win.

White moves next. Where should White play? (The move bringing victory soonest or defeat latest). And who wins, assuming that both play the best possible?

The best way to solve this puzzle is to give the roles of White and Black to two players - and then simply play the game.

Tomorrow and on following days we shall bring new boards, opening sequences and puzzles.

There is another lesson that you learn much later, that is tempting to whisper to new players: do not assume that you have mastered the game, even if you think that you have found a sure way to win.

The first Polygon puzzle.

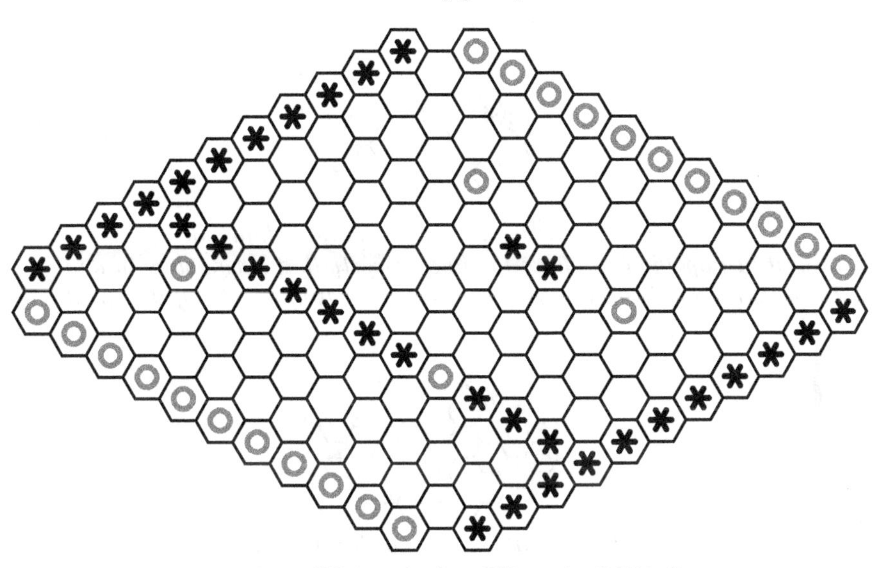

Puzzle 1. White to play. Who wins? How?

3.2 Polygon columns

Overlapping the Christmas school holidays, Politiken published a Polygon column daily up to January 11. Each column had commentary, a puzzle, and — except for December 27 — a solution to the previous day's puzzle. Polygon game pads were not yet in stores, so columns also included an empty board on which to play. On Sunday December 27, Hein discusses 'Polygon in small doses', i.e. play on small boards.[3]

> *A good way to gain an overview of the game is to start on smaller boards. The normal board has size 11×11. But nothing prevents one from playing on smaller boards.*

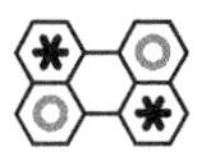

> *The smallest board has only 1 cell, so the first player wins: the other player has no move.*

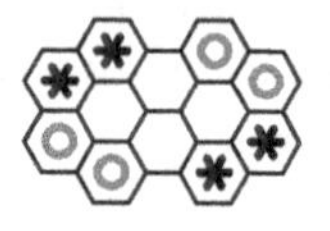

Already with the 2×2 board, the first player does not always win. But it is easy to see in which cells to make the first move to ensure a win.

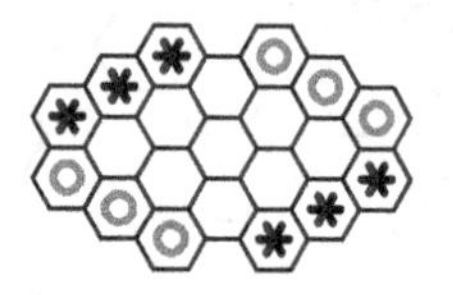

Again with the 3×3 board, it is not hard to find safe openings. But two openings . . . have the fine property that if you start in one of them, and both players play best, then all nine cells of the board will be filled.

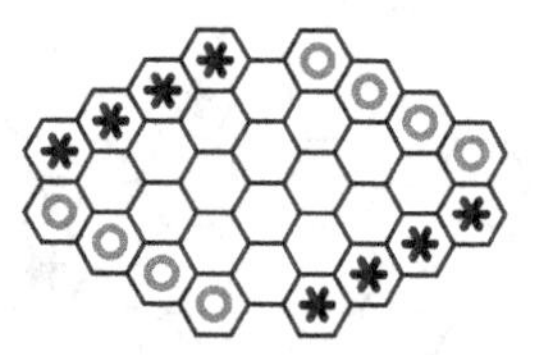

With the 4×4 board, it is easy to find two safe openings, but harder to find all of them. And with this board one can already make Polygon puzzzles: e.g. if White first claims both *acute corner cells, Black cannot win. But if Black now takes one of the two middle cells, it is not so easy for White to win. How should White play? With the 5×5 board it is already possible to have interesting games and difficult puzzles.*

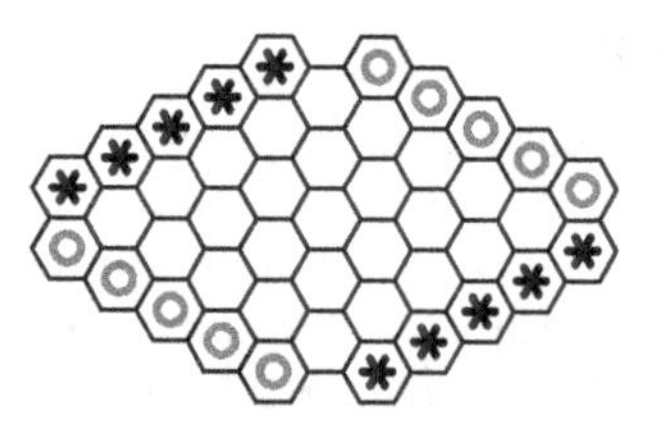

In the opening sequence from December 27, Hein refers to moves 3-8 as 'rubbing shoulders'. Today — in Hex and also in the board game Go — we call such a move sequence a *ladder*. As you can see in Figure 3.1, the ladder extends towards Black's side, so White 'breaks' this ladder with move 9.

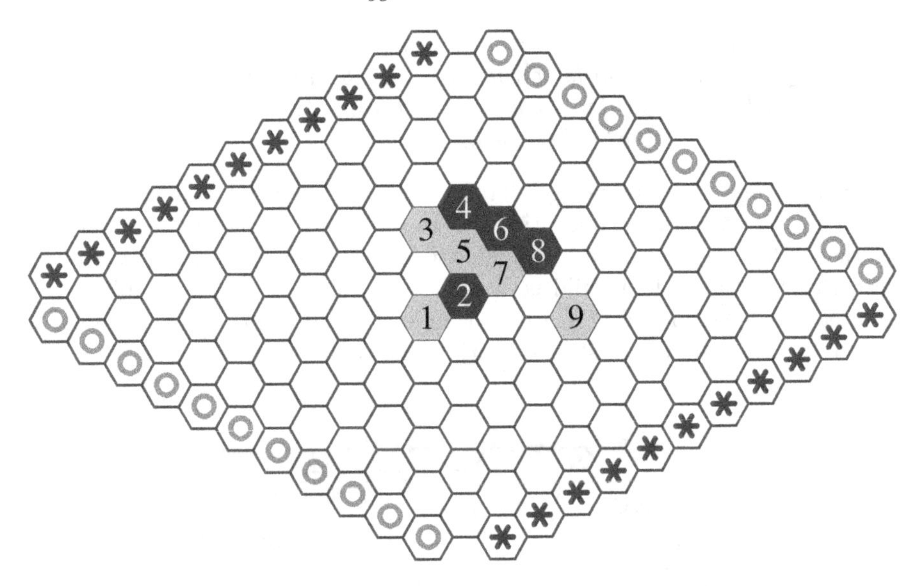

FIGURE 3.1: December 27 Polygon opening sequence.

December 27's puzzle is easier than the previous day's contest puzzle. In the solution — at the end of this section — each 'S' shows an angle connection. White's first move is the only winner, but there are different sequences depending on Black's second move.

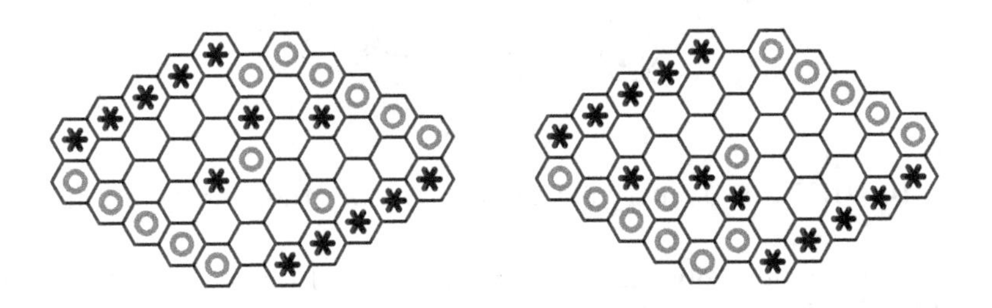

FIGURE 3.2: Polygon Puzzles 2 (December 27) and 16 (January 10) by Jens Lindhard. White to play.

New Year's Day brings the results of the contest puzzle from December 26. Most entries are correct, so the five winners — who come from across Denmark — are chosen by lot. And a second contest is announced: 100 *kroner* (Danish crowns) — today worth around 2000 *kroner* or $300 — for the best played game.

Politiken reportedly receives hundreds of contest entries, as well as other Polygon mail. Some readers claim to have found a way to play so that the first player always wins, but Hein is sceptical:

> *It is unlikely that such a method exists. Strong chess players have been playing Polygon for a couple of months and not found such a method. ... If you think you have such a method, send it to us in an envelope marked "Universal Method".*

Here is one reader's strategy: play first in the center and from there extend — directly or via angle connections — to both sides. This strategy actually works for boards up to size 5×5, but not for larger boards. On January 9 — also the day Hein's first child is born — Hein shows one way to block this strategy on the standard 11×11 board. Indeed, the opening sequence from the December 26 column also shows a way to block this strategy.

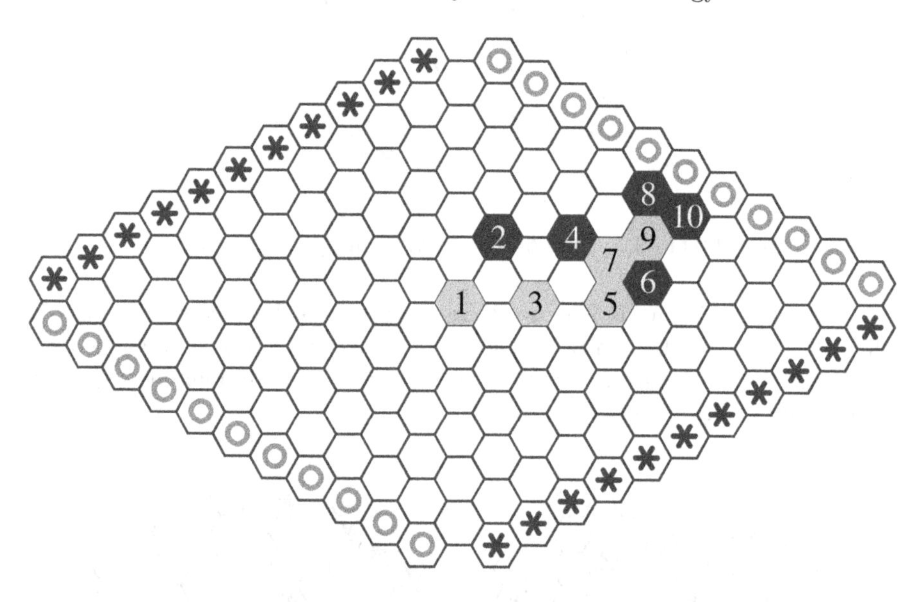

FIGURE 3.3: January 9. One way to block the center opening.

To get a feel for Polygon, try Polygon Puzzle 16 in Figure 3.2: can you find White's winning move? Black is close to joining the bottom right side: what happens if White blocks there, as in Figure 3.4? This move loses! Move 2 is Black's only winning reply. The figure shows a typical continuation: after 3, Black blocks at 4, again the only winning reply; notice how the players rub shoulders with moves 1-3, 6-8 and 10-14. We leave it to you to find White's winning move. The answer is in Figure 3.4.

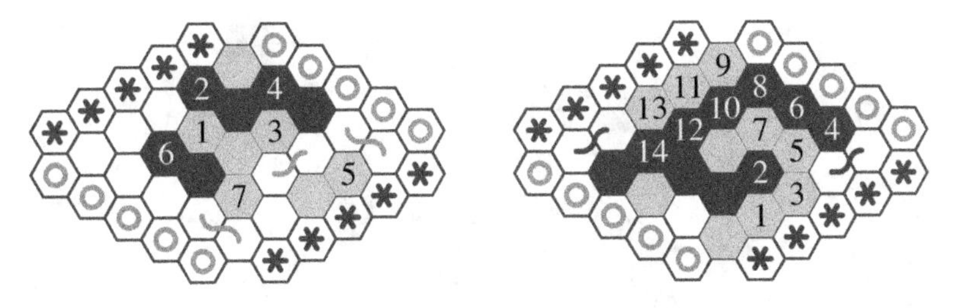

FIGURE 3.4: Puzzle 2 solution (left) and a Puzzle 16 continuation in which White 1 loses.

The Polygon puzzles that Hein's friends supply are exquisitely crafted — most have only one winning move — and Hein's Parenthesis talk and Polygon columns include many useful playing tips:

- *the game cannot end in a draw ... a barrier for your opponent is a connection for you, so view a position both offensively and defensively,*
- *it can be proved that the first player can win, but no one has found a guaranteed winning strategy,*
- *start roughly in the middle,*
- *cells that do not touch are sometimes safely connected,*
- *make moves that threaten to form two or more paths,*
- *to learn, play on small boards,*
- *to improve, play many opponents.*

3.3 Polygon pads

By January 15, the Polygon game pads — small enough for a pocket or purse — had arrived in shops and bookstores. Each pad had 50 sheets and cost 50 *øre* or .5 *kroner*, worth about 10 *kroner* or $1.50 today. So from that date, Polygon columns — now Wednesday and Saturday — no longer included an empty board.

The orange front cover was printed on paper. The back cover was printed on cardboard, with rules on the orange outside and a grook — also shown in Figure 3.28 — on the cream inside. The pads were glued along the top-left and bottom-left: in the figure, notice the traces of glue on the edges of the inside back cover.

FIGURE 3.5: Front and back cover of a 50-sheet Polygon pad.

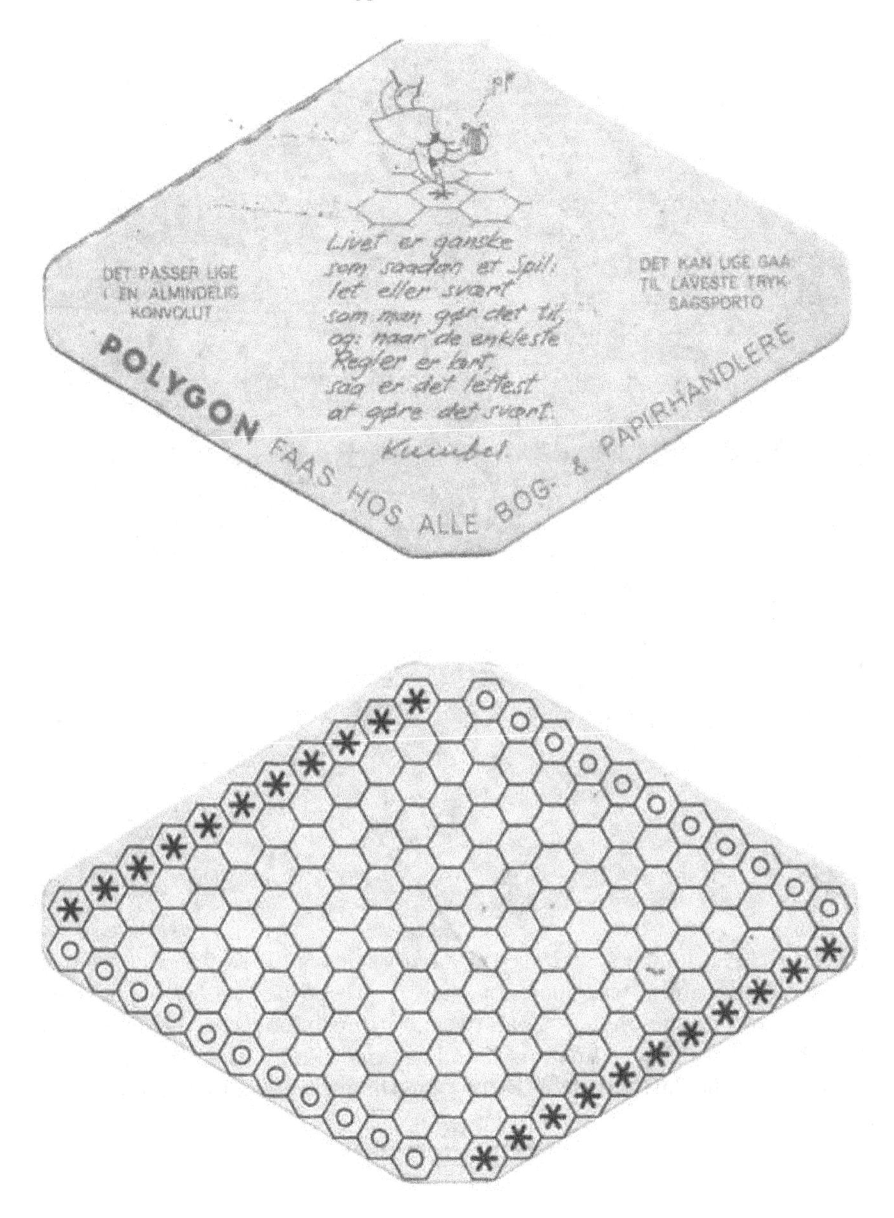

FIGURE 3.6: Inside front cover and sheet of Polygon pad.

FIGURE 3.7: January 31 Polygon advertisement. Take a break from the nerve-wracking times. Caption (not shown): Yesterday we had a bookseller from Østerbro on the phone: “I can’t stand it much longer! Polygon-fever has infected my shop. People buy a pad and ask me to play: then they play there for the rest of the day, forgetting to buy anything else.” © Piet Hein, courtesy Hugo Hein.

FIGURE 3.8: January 17 advertisement for Polygon game pads. Ladies and gentlemen: the all-clear *has* sounded! © Piet Hein, courtesy Hugo Hein.

On several dates in January, Politiken published a Polgyon game pad advertisement instead of a column. The ad for January 17 shown in Figure 3.8

— the all-clear *has* sounded! — is a reminder of the ongoing war. The ad for January 31 shows bookstore customers absorbed in a game.

The January 20 column gave the results of the best game contest. An initial weeding of entries left more than 100 games, which were pruned to 12, and finally 2 before the winner was then chosen by lot. Hein's analysis of the game follows that of a commentary by Jens Lindhard,[4] presumably one of the unnamed contest judges.

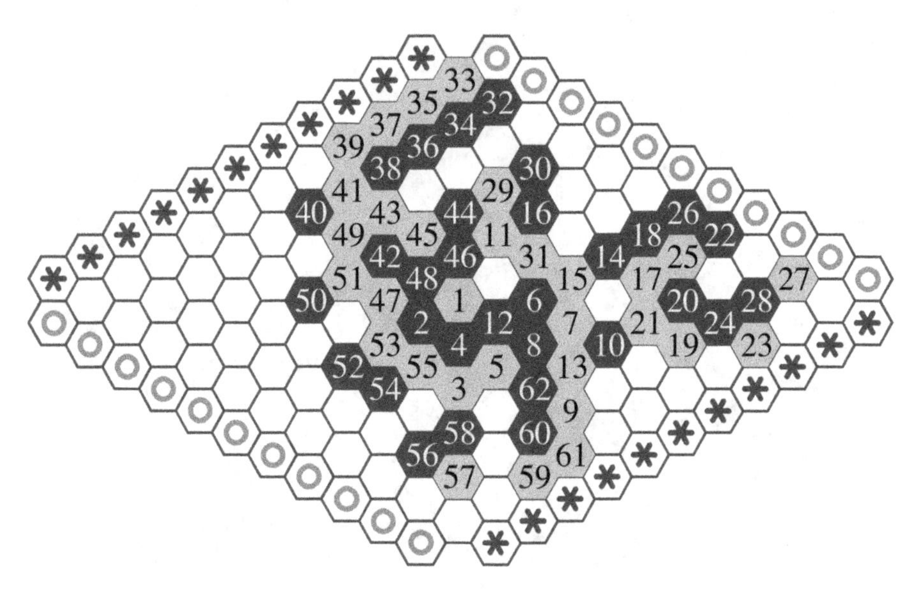

FIGURE 3.9: Best-game contest winner: J. Koefoed vs T. Sigurgeirsson.

Hein gave no criteria for a good game: presumably it is one with many twists and turns, both literally and figuratively, and with few weak moves. How can we recognize a weak move?

We will answer this question in more detail later. For now, observe that Hex has no draws, so to win it suffices not to lose, so to block all opponent threats. Often, each opponent threat is to a cell on a shortest opponent-side-to-side path. So a move that does not intersect any opponent shortest path is usually weak. By contrast, a move that intersects all, or most, opponent shortest paths is often strong.

For example, consider move 4 of the contest-winning game. After move 3, the number of empty cells on a shortest white-winning path is nine: five cells to join the upper right side and 1, one cell to join 1 and 3, and three cells to join 3 and the lower left side. So on move 4 Black plays between 1 and 3, intersecting every shortest white-winning path. Similarly, on move 5 White intersects every shortest black-winning path. Move 6 is not on any shortest white-winning path, but is it a weak move? No, because it threatens either to

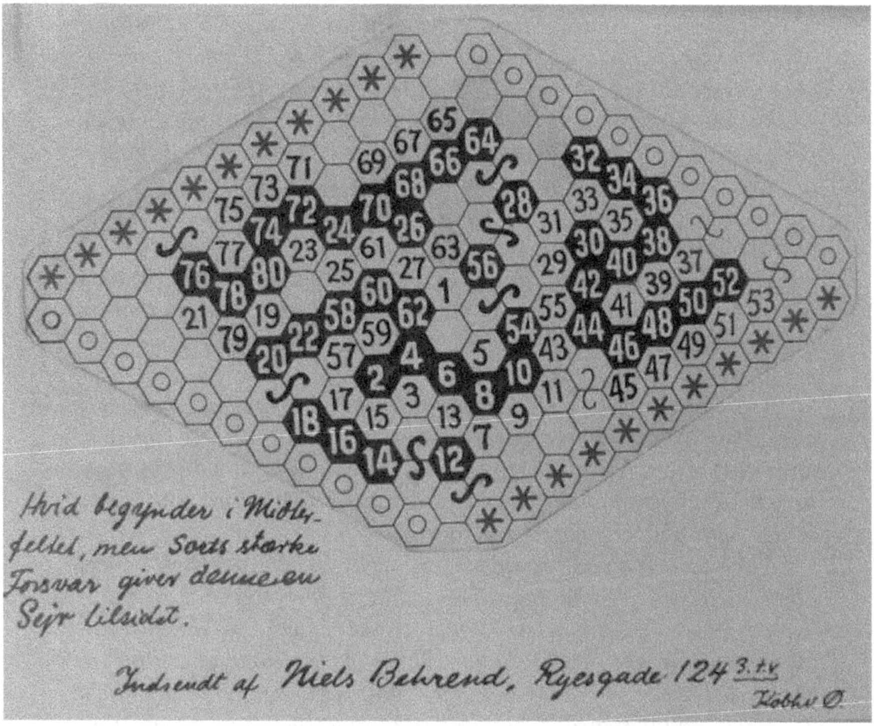

FIGURE 3.10: Another best-game contest entry. Courtesy Hugo Hein.

join the black group at 2,4 or to jump towards the upper-left side via 11, in each case neutralizing 1.

3.4 Polygon salons

To give readers a chance to improve their Polygon skills, Hein organized two Polygon salons, held in the Politiken Hall at 8pm on Monday February 1 and Wednesday February 10, 1943. Salons — public meetings where ideas are discussed — are popular in Europe. The speakers were Piet Hein and Jens Lindhard.

The Hein archive has an undated nearly-complete 17-page salon outline. Pages 0-2 have Hein's hand-written introductory remarks:

> *The purpose of this Polygon demonstration and the extravagant setting we have for us [BT: perhaps a reference to Politiken's grand hall, or to a large game board used to show moves to the audience]*

is to try to give some of <u>*that*</u> *which cannot be given a newspaper article or in printed instructions, namely direct experience with the game. Here we can discuss it, analyze it, we are* <u>*together*</u> *with it.*

Let me first tell you that I am certainly not a strong Polygon player, nor do I feel obliged to become one just because I happen to have discovered the game.

I will begin this evening by shedding some light on the game and its strategy from a theoretical viewpoint. Then, a very strong player will show you the finer points of the game, and — if time permits — play against one or more audience members, so that we might also observe characteristic strategic elements of a game in progress.

A beginner often feels that White, the first player, has a big advantage. Many have even suggested that White can always win after only a few moves. This common misconception is due to inexperienced players having failed to learn how to block an opponent's attack with a strong counter-attack.

After learning a strong counter-attack to White's first move, other players have the impression that it is better not to move first, because Black can see White's first move and so make an annoying countermove. But then White has the same advantage in her next move, and so on. It is this continuing alternating pattern of attack, counter-attack, counter-counter-attack, and so on, that brings 'play' into the play, which results with neither player having a straight path, but rather with paths that wind into each other. The possibilities for promising continuations of the game multiply with each move.

If you think that you have found a method that guarantees that White will win, let me calm you: expert players have played the game for months, and have found that in practice there is no significant advantage between playing as White or as Black, and that absolute knowledge of the game probably lies outside what is possible in practice.

But, as a little teaser, let me tell you that Polygon has this property: in contrast to most other games, it can be proved that the first player in theory always can win, that is, if she could see to the end of all possible lines of play.

The only way to become a strong Polygon player is to play with players you do not know: they will have learned Polygon independently from you, and will have learned counter-attacks that you might not have seen in your circle of play. It is impossible to appreciate the many facets of the game by always playing against the same opponents, or with yourself. The road to mastering Polygon is to play with as many different people as possible.

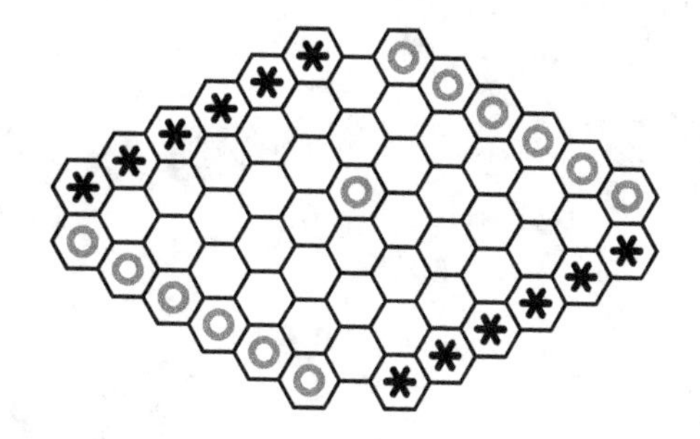

FIGURE 3.11: Salon manuscript page 5. Black plays next: who wins? And in how many moves if both players play best?

Pages 3-4 are missing: perhaps Hein here discussed play on small boards. Page 5 shows the center 6×6 opening, and asks the questions shown in the figure above. On pages 6-9, Hein reiterates game features covered in the columns, and credits Polygon players for the game's success:[5]

> *As I have said, I have no particular gift for playing Polygon. What interested me was the 'meta-game' in which I investigated new game [design] possibilities. It was at the suggestion of experienced players [e.g. Hein's chess-playing friends] that Polygon made its way into the world, and eventually the newspaper. It was because of reader demand that we published game pads. And it has greatly surprised me to see such interest in the game.*

On page 8, Hein captures the essence of Polygon:

> *The point of the game is always to block the opponent as much as possible. This is not a negative feature but part of the soul of the game, that offensive and defensive measures eventually melt together. A barricade against the opponent is always a path for you. You can consider a Polygon game as two labyrinths in combat.*

On page 9, Hein discusses the search for a winning first-player strategy which, as he mentioned, must exist:

> *Strong players of other games have taken up [11×11] Polygon and played it for the past 3-4 months it has existed, and they think that it would take years before one could consider sufficient possible lines of play to be reasonably sure to win. In any event, a winning strategy will require more than just a few moves. A game between strong players can take an hour or two. But White can win.*

Since White can win, what is a winning first move? Hein observes that beginners tend to start in the middle, but shows an opening — similar to that in the January 9 column — in which Black puts up some resistance: with moves

FIGURE 3.12: Copenhagen Police Headquarters.

4,6,8, Black prevents the White group of moves 3,5,7 from joining the upper-right side. Hein and Lindhard call this blocking sequence a *Politigaarden* or 'Police Headquarters'. The access road to the Copenhagen police headquarters approaches the building and then forks to either side; similarly, after entering the building one must turn one way or the other. As Lindhard wrote,[6]

> *this is why we call it* Politigaarden, *because there also one is forced, after entering, to turn either left or right.*

On page 10, Hein encourages players to consider the whole board and shows an unusual connection pattern. On page 11, he analyzes a game between Lindhard and Jens Peter Møller, which we discuss in the next section. He then shows Polygon Puzzle 1 and apparently discusses *side connections*, i.e. connection strategies that allow safe attachment to a side of the board: we found no manuscript page numbered 13, but an unnumbered page that fits here shows the 4.3.2 and 7.6.5.2 connections, named after the number of cells in each row of the connection shape.

On page 14 of the salon manuscript Hein answers this question: on the empty 11×11 board, which cells safely attach to a player's side? In the first row by a player's side, each cell already touches the side, so no connection strategy is needed. In the second row, the 2.1 angle connection joins each cell but one to the side. In the third row, the 4.3.2 connection joins each cell but

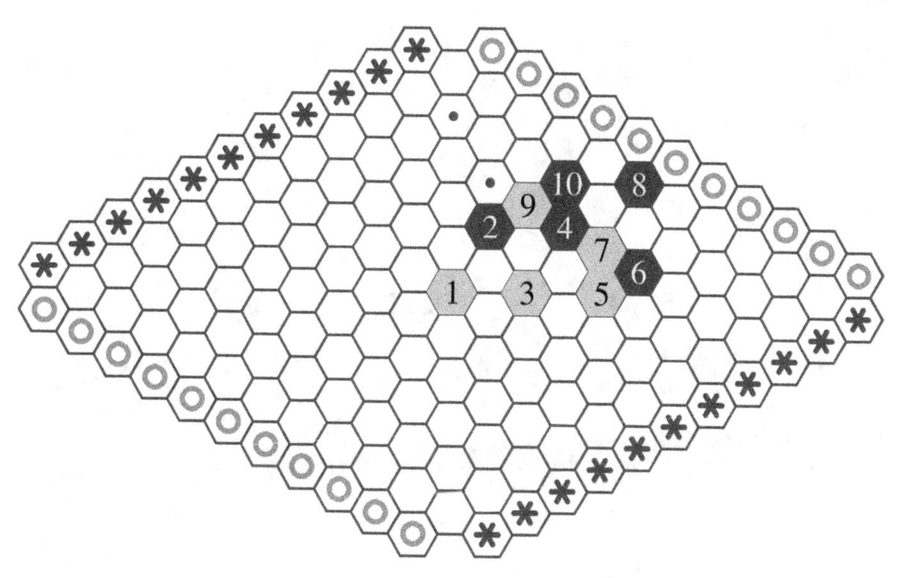

FIGURE 3.13: With 4,6,8 Black forms a 'Police Headquarters' that blocks White 3,5,7 from the upper right side. Dots are possible next moves.

two to the side. Continuing with the 7.6.5.2 connection for the fourth row and the 10.9.8.5.3 connection — which Hein perhaps mentioned — for the fifth row yields the manuscript page 14 figure: the shaded cells are exactly those that safely connect to the bottom white side.

To end the salon, Hein called for volunteers to play — simultaneously — against Lindhard. Simultaneous exhibitions are common in games such as chess and Go. An experienced chess player, Lindhard was presumably comfortable playing simultaneous games.

Below are Hein's closing remarks. Hein felt strongly about point 2, even writing a grook about it. See Figure 3.18.

> *I would like to thank everyone who has shown interest for the game this evening. I would like to reiterate the three points that I think are important to understand the game.*
>
> *1. View the game both defensively and offensively, and annoy the opponent as much as possible, because a barricade for the opponent is a path for yourself.*
>
> *2. Play with as many different people as possible, so that you will not be stuck with limited tactics.*
>
> *3. Learn to consider larger and larger parts of the board. The game occurs not in the middle, and not between two neighbouring cells, but over the whole board.*
>
> *Thank you for this evening.*

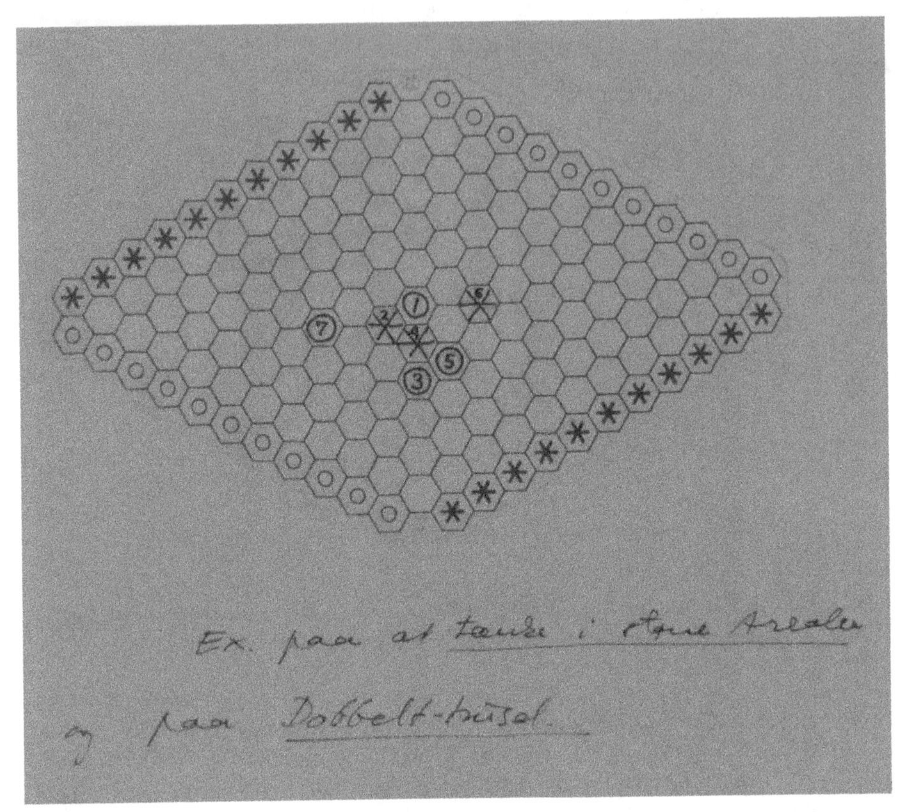

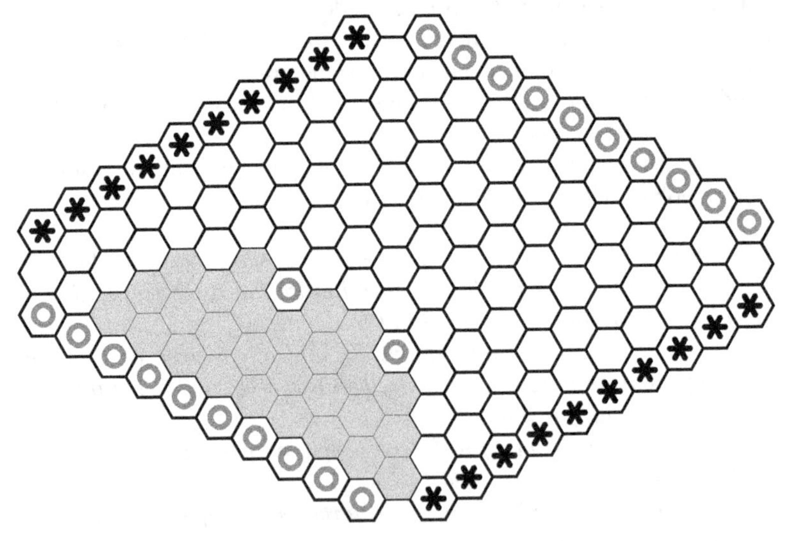

FIGURE 3.14: Salon manuscript page 10. *An example of thinking in larger areas and the double house.* (bottom) "Double house" perhaps refers to the shaded pattern: even if Black moves next, White can join both cells — not just one — to the side. Courtesy Hugo Hein.

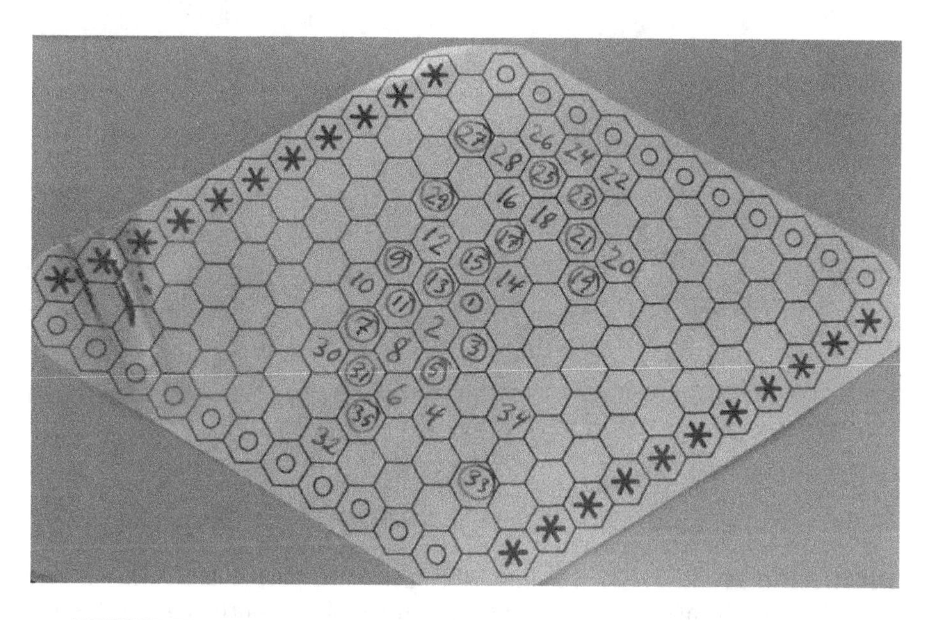

FIGURE 3.15: Lindhard-Møller game record. Courtesy Hugo Hein.

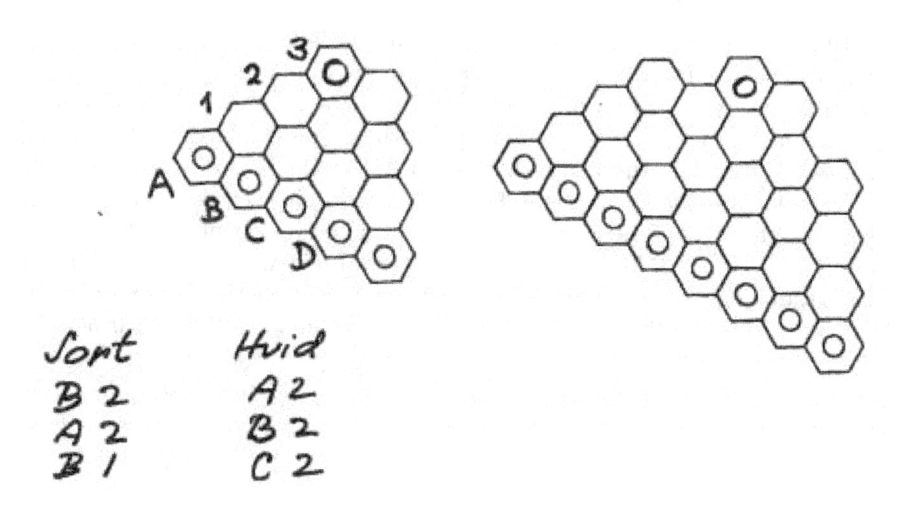

FIGURE 3.16: 4.3.2 and 7.6.5.2 side connections. Courtesy Hugo Hein.

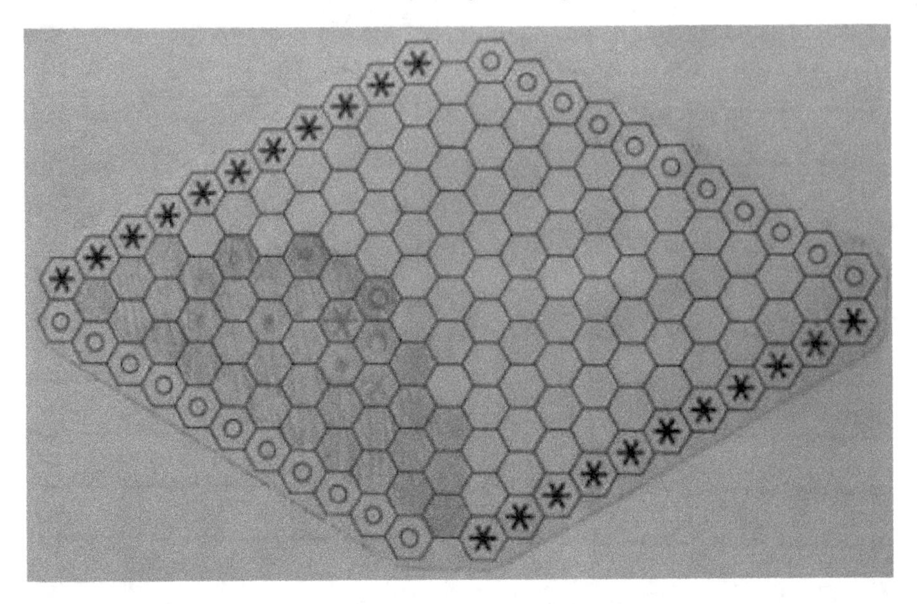

FIGURE 3.17: From Salon manuscript page 14. If played as the first move, each shaded cell safely joins White's bottom-left side. Courtesy Hugo Hein.

3.5 Lindhard-Møller game

Here is Lindhard's analysis of the Lindhard-Møller game, presumably as presented in the Politiken salons. Our source is Lindhard's two-page signed hand-written commentary from the Hein archive. We also give move suggestions by MoHex, an intermediate strength computer player with a strong endgame solver. We will discuss MoHex in a later chapter.

[BT: This might be the Jens Peter Møller born in Thoreby February 2, 1899 who graduated from high school in 1916 and from Copenhagen University (*cand. mag.*) in 1923. Møller wrote mathematical reviews that were published in 1931 and 1938, and was perhaps a math instructor at the university.]

White starts in the center field – 1.f6 – mostly by tradition, not because it is much stronger than other openings. Black replies 2.f5, probably the best response. After 3.g5 Black drops to 4.h3, deflecting White's push towards the bottom front. White turns at 5.g4 and Black rubs shoulders 6.g3. Rather than continue to rub shoulders, White jumps 7.e4. Black reasonably cuts 8.f4.

White angle-connects 9.d6. Black probes the connection 10.d5 but White restores it with 11.e5. Here we thought Black might separate the two white groups with 12.e6; instead Black plays 12.d7 and white unites with 13.e6.

[MoHex prefers Møller's 12.d7 to Lindhard's suggested 12.e6. After 12.e6 White can push towards the top with 13.d7. If Black separates with 14.e7,

Vores Ukuelige Partiskhed.	Our Unfailing Onesidedness.
Gruk om Sindets Enhed trods alt.	A gruk on closedmindedness.
Jeg sidder og hviler	I sit and rest
min livstraette Sjael	my world-weary soul
ved at spille et Slag	by playing a game
Polygon med sig selv.	of Polygon all alone.
Der er bare den Fejl	There is only one flaw
ved min ensomme Leg:	in my solitary play:
jeg kan ikke la vaer	I cannot stop siding
med at holde med m i g.	with myself all the way.
MKUBEL	(trans. BT)

FIGURE 3.18: Grook on the peril of self-play. Hein scrambled KUMBEL to indicate author as MKUBEL. © Piet Hein, courtesy Hugo Hein.

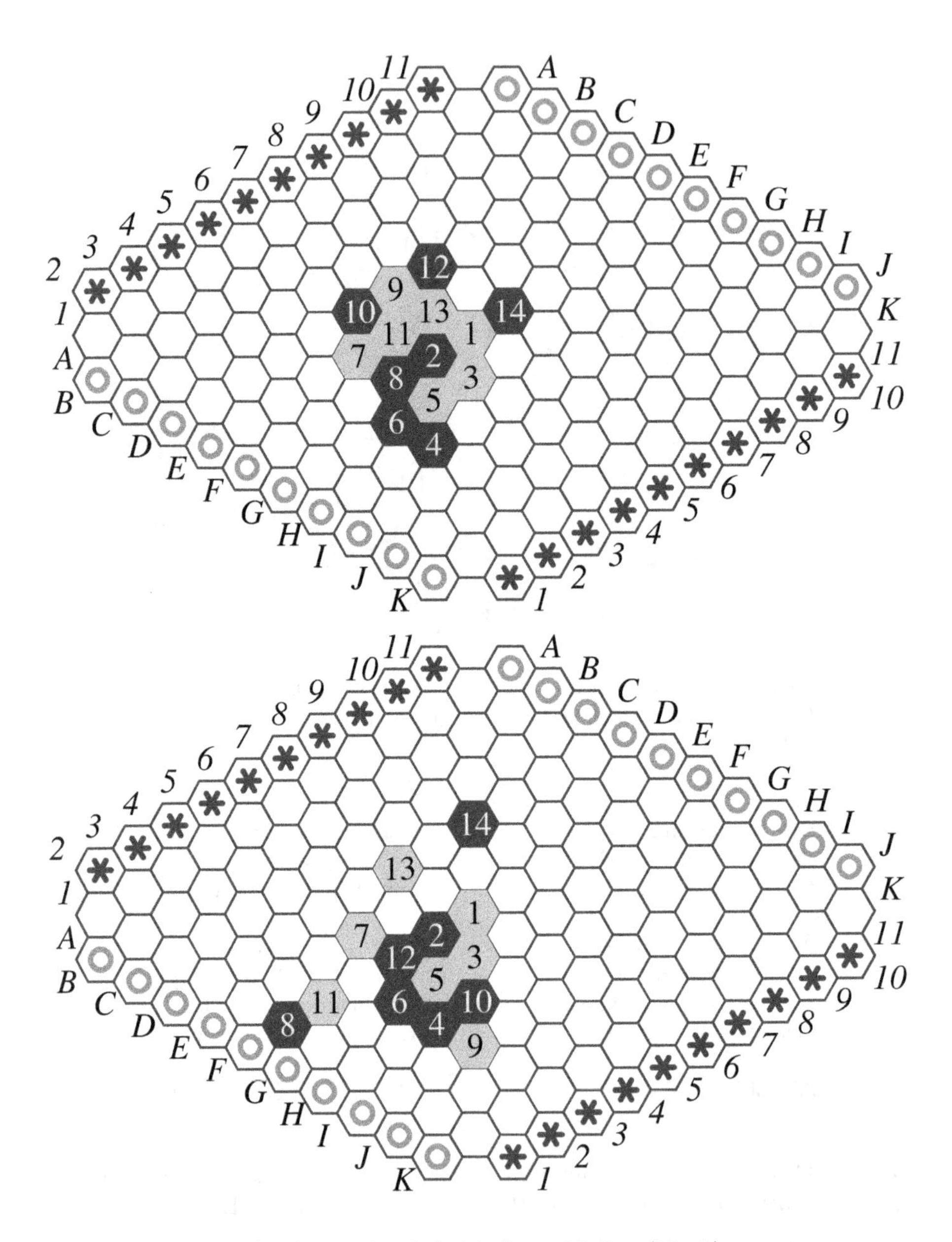

FIGURE 3.19: (top) Lindhard (white) vs Møller (black) opening moves. (bottom) MoHex prefers to block near the bottom with 8.f1 rather than 8.f4. This might be better for Black.

White angle-connects at 15.c9. If Black blocks with 14.d8, White has 15.c9. If Black blocks with 14.c9, MoHex White has 15.d9: Black can play 16.d8, but White unites with 17.e7, Black blocks at 18.e8 but replies 19.g7. But all Black moves lose: it takes our computer solver about 10 hours to confirm that 12.e6 wins. See Figure 3.20.]

Black starts a 'police headquarters' block with 14.f7 that continues 15.e7, 16.d9; it might have been better to delay this with 14.e8, 15.g7 and then block with 16.g8, 17.f8, 18.e10.

[MoHex prefers Møller's 14.f7 to Lindhard's suggested 14.e8, but then agrees with Lindhard's moves 15-18. See Figure 3.21.]

After 17.e8, 18.e9 White jumps to 19.g8, which is safely joined – either directly at f8 or via angle connections at h6 – to the white group. Black starts a police headquarters block with 20.g9.

Play continues as usual until White jumps to 27.b10. Black must cut at 28.c10 but with 29.c8 White joins the upper front to the middle group. It is [also] now impossible for Black to stop White from joining the middle group to the lower front. After 33.j2 White has a double threat: Black closes off one side so White goes the other way.

Jens Lindhard

We end our discussion of the Lindhard-Møller game with a new puzzle in Figure 3.22. This position can arise after a Black 'police headquarters' variant block 16.d10. It is White to play: can you finish Black off? Be careful!

3.6 Polygon peters out

For a while, Polygon was popular. Danish architect Anker Tiedemann, who founded the magazine *Better Living* in the 1960s and later the wine website *Vinavisen*, was 13 years old in 1942: he recalls everyone in his circle buying pads and playing Polygon.[7] Hein's Politiken columns regularly reported on Polygon mail:

- *We receive daily heaps of letters from players across the country, who have been playing the game on trains, in lunchrooms and shelters ...*
- *... we receive requests to make the puzzles more difficult — accompanied by incorrect solutions to earlier puzzles ...*
- *Polygon puzzles seem to provide glittering examples of the psychological fact that there is great similarity between solving a puzzle and not understanding it.*

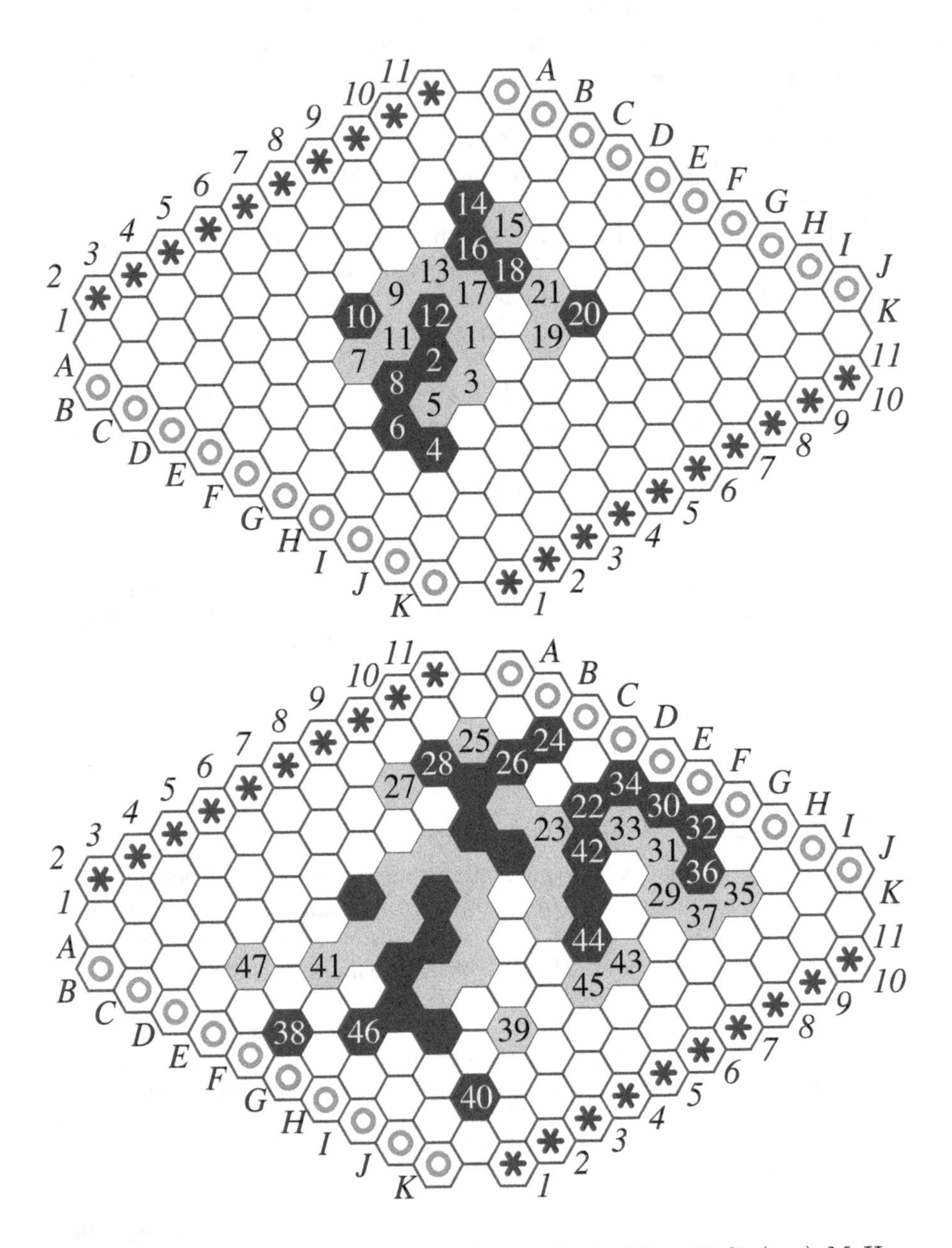

FIGURE 3.20: MoHex prefers 12.d7 to 12.e6. After 21.f8 (top) MoHex quickly finds a win (bottom).

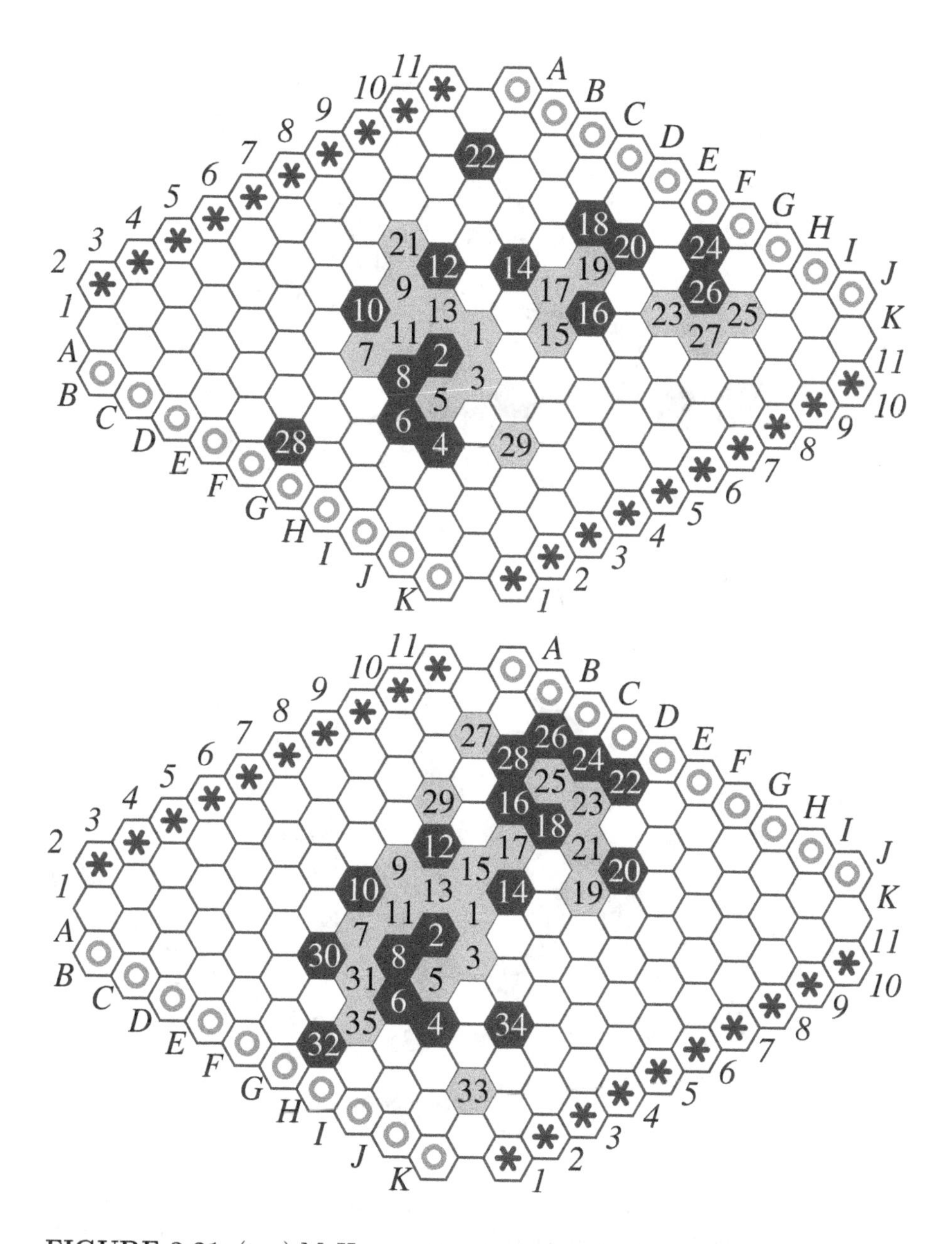

FIGURE 3.21: (top) MoHex continuation after Lindhard's suggested 14.e8, finding the win instantly on move 21. (bottom) Finished Lindhard-Møller game. White wins.

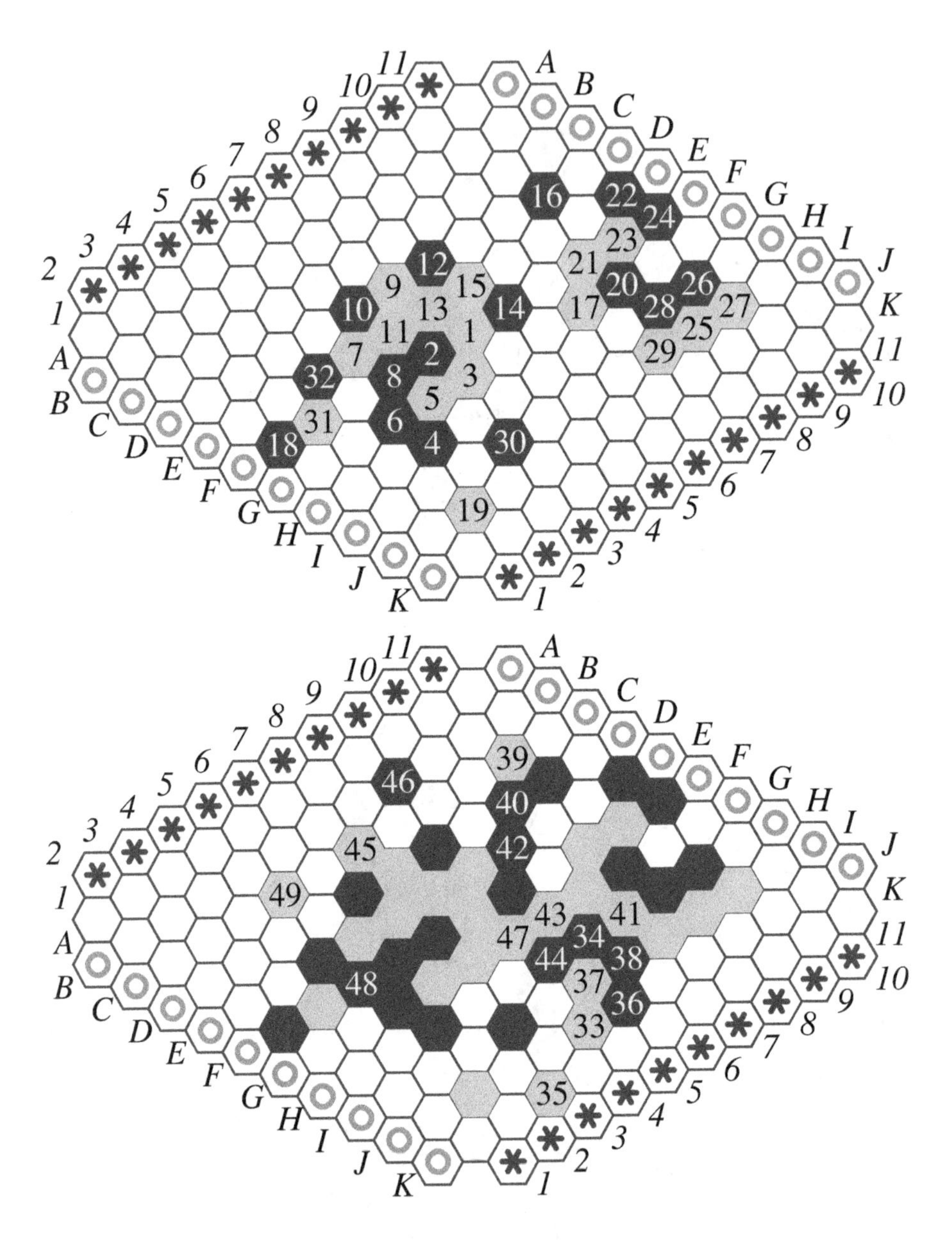

FIGURE 3.22: (top) A new puzzle from the Lindhard-Møller game. White to play. Many moves win. (bottom) Solution. 33.f3 — restore the e4-f2 angle connection that Black cut with 32.e3 — loses! E.g. 34.h8 35.j5 36.j6 37.i7 38.h6 39.h7 40.e8 41.a10 42.d8. Instead, use double threat d4-f3 to take more territory: 33.c4 indirectly restores the angle connection and wins! Moves c4,i5,i6,i7 also win. (bottom) MoHex finds winning move 33.j5, with this continuation.

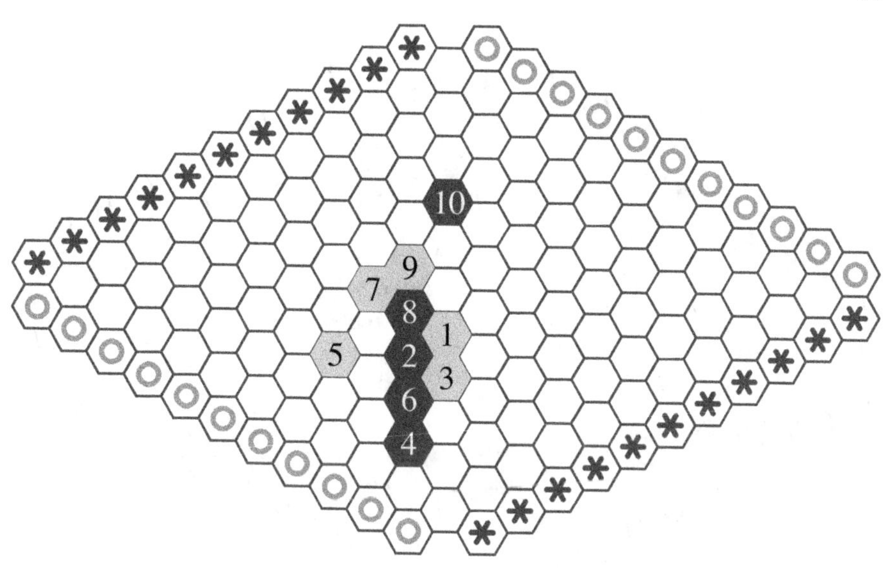

FIGURE 3.23: Opening for the best continuation contest.

- *Readers ask, is it a big advantage to play first? No!* [Earlier, Hein had mentioned that the first player has *some* winning strategy.] *No one has been able to find such a strategy. Among players of equal ability, the advantage of playing first seems to decrease as playing skill increases.* [RH: First player advantage decreases as board size increases. Playing first helps a lot on 11×11, less so on 19×19.]
- *And must the first player start in the center? No! The six neighbouring cells are also strong opening moves.*

In Denmark, Polygon was a Sudoku of its day: by March 1943, in a country with fewer than 4 000 000 people, 50 000 booklets had been printed. To celebrate, Hein announced another contest, offering a 50 crown prize for the best game continuing from the opening in Figure 3.23.

March 27 brings the results of the best continuation contest, with reportedly more than 200 entries. The winner is the game between Niels Behrend and Helge Benzin in Figure 3.25.

> *The opening was a bit to White's advantage, however by no means justifying that White wins in 80% of the entries . . . [Among the best entries] White wins about half the time. White's best move after the opening is surely 11.c8, but play in the middle might also be useful.*

Figure 3.24 shows how MoHex continues after Hein's suggested 11.c8. MoHex prefers 11.f7, but Hein's suggestion seems to bring the win more quickly.

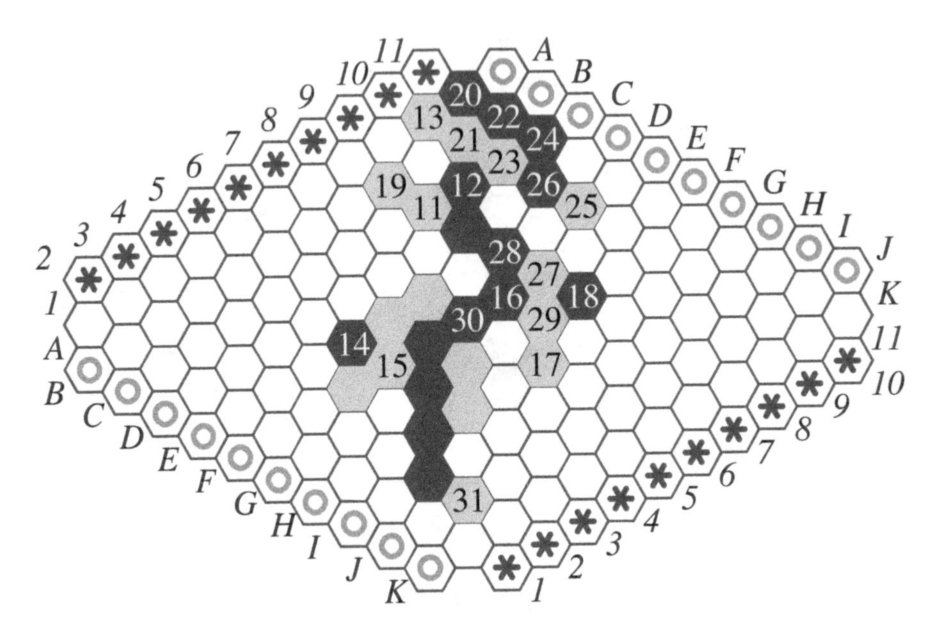

FIGURE 3.24: Best continuation after Hein's recommended 11.c8. In this MoHex (with solver) self-play, White is never in trouble and sees the win with 19.b8.

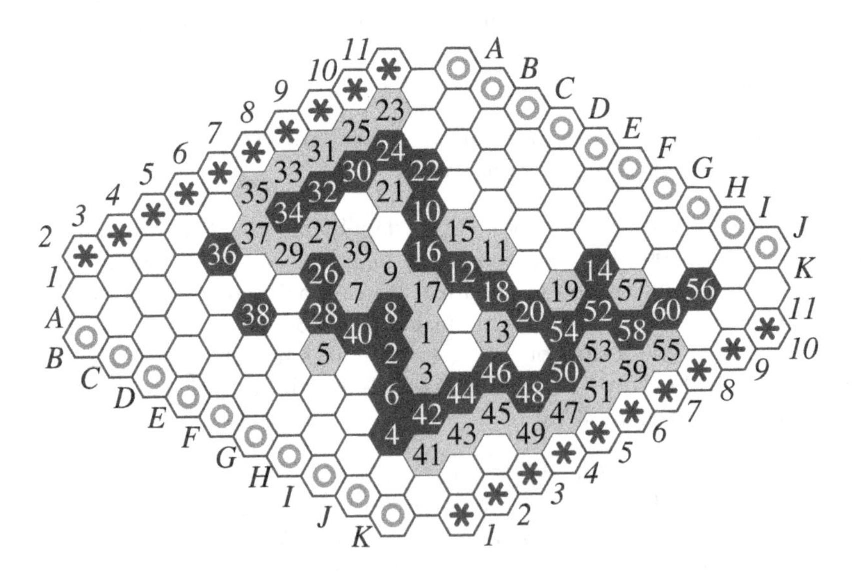

FIGURE 3.25: Winner of the best continuation contest. White erred with 17.f6 (i6 wins, this takes solver 150 seconds) and 39.d6 (i6 wins, 5 seconds). MoHex sees instantly that 40.f4 wins.

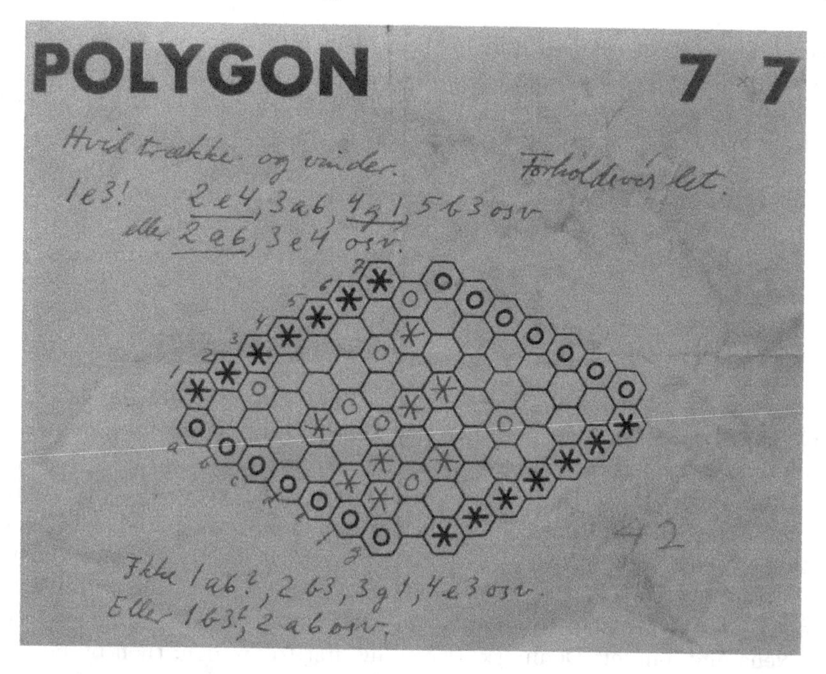

FIGURE 3.26: Puzzle 42 draft from the Hein archive. *White to play and win. Relatively easy. 1e3!* ...is in Lindhard's handwriting, '42' is in Hein's. This puzzle was expected in Politiken on April 17 1942 but never appeared. Puzzle 43 — with Solution 42 — appeared May 12 1942. Courtesy Hugo Hein.

In a later letter to Martin Gardner, Hein mentions that after Germany invaded Denmark, he "*had to underground in installments*". The Polygon columns gradually disappear.

First, errors creep in. The diagrams for Puzzle 36 and Solution 35 — published March 24, presumably as a late replacement for the best-opening contest winner column, which was delayed "*due to the large number of entries*" — are mistakenly interchanged. Puzzle 38 and Solution 38 have a missing marker whose location can be deduced from the solution commentary. The column with Puzzle 42 and Solution 41, promised for April 17 — the same day Hein's *Consolation Grook* is published — never appears. [In the Hein archive, we found a draft for only one puzzle: Puzzle 42! The puzzle is composed by Lindhard. We will have more to say about this in the next chapter. But we have no idea why this puzzle did not appear as promised in Politiken.] Puzzle 43 with Solution 42 appears four weeks later. From this point — mid-April — columns appear sporadically, often after the date promised by the previous column. Puzzle 47 has a marker missing, although it is back in place in Solution 47. And Solution 45 is incorrect: White has no winning move in Puzzle 45.

Did interest in Polygon fade? Was Hein preoccupied with other matters? Until 1937, when he first married, Hein lived with his mother in Rungsted.

Hein's first marriage ended, and in July 1942 Hein married Gerda 'Nena' Cronheim. Their son Jan was born January 9 1943. Nena was Jewish and Hein was a target of Nazi ridicule. In 1942 Tuborg printed a beer coaster with this Hein grook:

Klar de trange	Get through the hard times
Tider stolt:	strong and bold:
Hold Hodet klart	Keep your head clear
og Oellet koldt.	and your beer cold.
KUMBEL	(trans. BT)

In response, on January 6 the Nazi newspaper *Faederlandet* published an article that hoped for Hein to suffer the same fate as the businessman Carl Salomonsen, who had committed suicide on December 21 1942.

Danish resistance to the German occupation grew. From 1940, Germany had left the Danish government in place, but Hein anticipated a turn for the worse and made plans to leave. Erik Scavenius, Danish prime minister and married to Hein's mother's sister, requests Hein's emigration permit from Werner Best, the German commander-in-chief in Denmark. In summer 1943, Piet, Nena and Jan left Denmark, travelling first to Sweden, then by boat to the USA, and finally to Argentina, where Nena's parents were living.

The last Polygon column appeared on Wednesday August 11. There was no mention of Solution 49, and it never appeared. On August 29 1943, after the Danish government had refused to crack down on civil unrest, Germany declared martial law. Antisemitism was not strong in Denmark, and the Danish authorities had protected security for the well-integrated Jewish population. But this protection now disappeared.

In September, Niels Bohr — whose mother is Jewish, and who had been helping Jewish physicists leave Germany — fled Denmark with his family. By the end of the year he and Aage were working on the atomic bomb project in Los Alamos.

Starting in October those Danish Jews who had not escaped were arrested and transported to the Theresienstadt concentration camp. The news that this would happen was leaked, and most of the Jewish population — 7056 persons, including 1236 children, 1301 who were half-Jewish (such as Niels Bohr) and 686 non-Jewish spouses (such as Piet Hein) — escaped to Sweden, where they were granted immediate asylum.[8]

After the war Piet, Nena, Jan, and newborn Anders returned to Denmark. But Nena and the children soon went back to Argentina and Piet and Nena divorced. In 1947 Hein married Anne Cathrina (nee Krøyer Pedersen) and spent time in New York as a Politiken correspondent.

For a time, Polygon was forgotten.

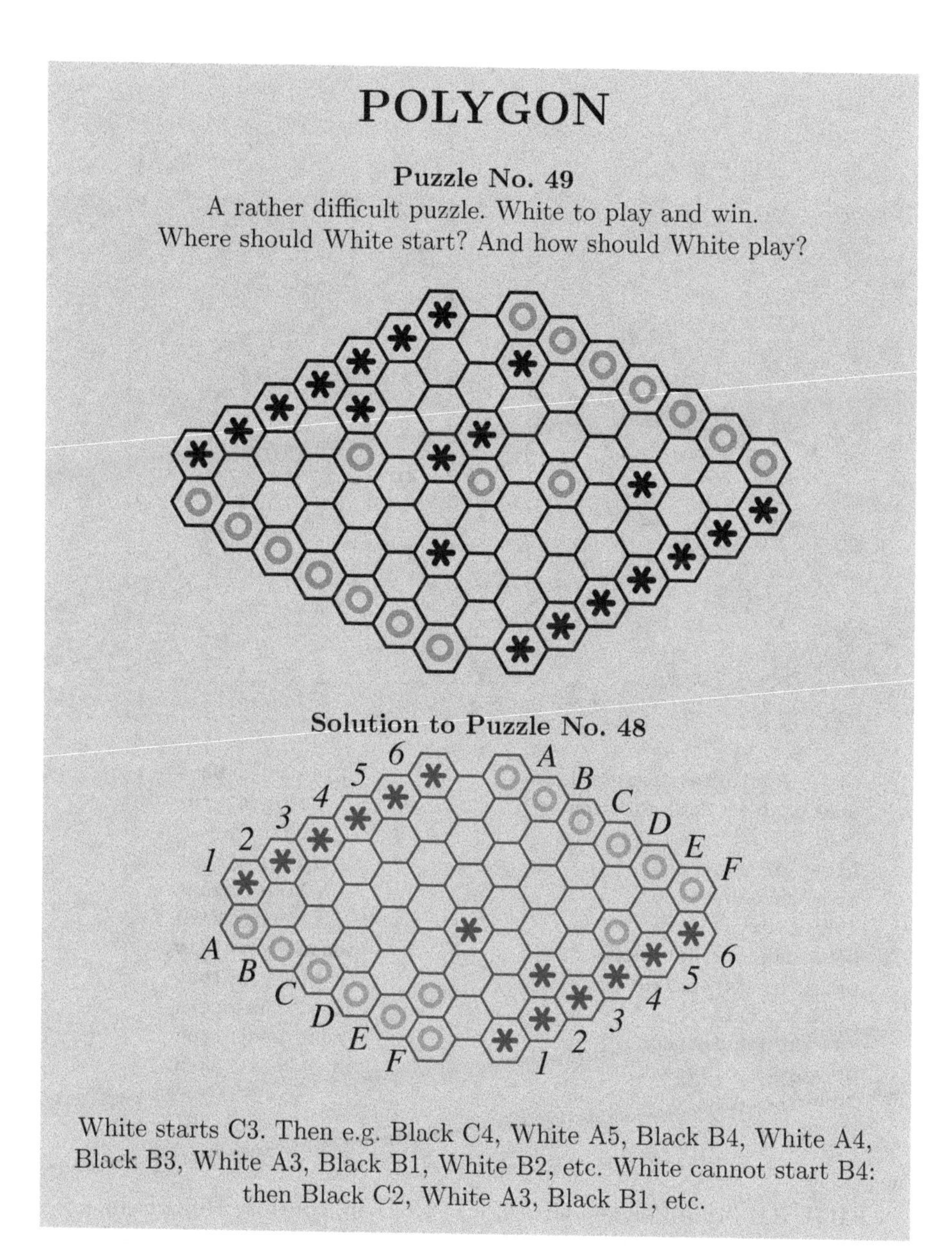

POLYGON

Puzzle No. 49

A rather difficult puzzle. White to play and win.
Where should White start? And how should White play?

Solution to Puzzle No. 48

White starts C3. Then e.g. Black C4, White A5, Black B4, White A4, Black B3, White A3, Black B1, White B2, etc. White cannot start B4: then Black C2, White A3, Black B1, etc.

FIGURE 3.27: Translated reproduction of the last Politiken Polygon column, August 11 1943. Original column © Politiken.

```
Livet Betragtet som Spil              Life as a Game
Gruk ved et Spil Polygon              A Polygon gruk

Livet er ganske                       Life is almost
some saadan et Spil                     like a game
let eller svaert                        easy - hard
some man gor det til,                decide your aim,
og: naar de enkleste                with the simplest
Regler er laert,                     rules, you start
saa er det lettest                    most easy then
at gore det svaert.                  to make it hard.

KUMBEL                                   (trans. BT)
```

FIGURE 3.28: Polygon Grook. © Piet Hein, courtesy Hugo Hein.

FIGURE 3.29: Piet Hein en route to Argentina, 1943-1944. © Hugo Hein.

Notes

[1]Polygon salon manuscript, page 8 [33].

[2][33, 51]

[3]1942.12.27 Politiken, page 4.

[4][33]

[5]Polygon salon manuscript, page 7 [33].

[6]Lindhard notes on Polygon terminology [49].

[7]Private communication. `www.anker.tiedemann.dk`.

[8]464 Danish Jews were arrested — either during the escape or because they did not flee — and imprisoned in Theresienstadt, where about 50 died. In April 1945 the survivors were transferred to the Swedish Red Cross, led by Folke Bernadotte, and transported back to Denmark in buses that had been painted white.

Chapter 4

Polygon puzzlist

Talent instantly recognizes genius.

Arthur Conan Doyle

Here is a game with commentary and a puzzle. I will send more puzzles and openings as soon as possible.

Jens Lindhard[1]

It is quite a three-pipe problem and I beg that you won't speak to me for fifty minutes.

Arthur Conan Doyle

4.1 Mystery of missing drafts

As our book-writing neared its end, one question nagged us: where were the rough drafts of the Politiken Polygon puzzles? The puzzles are exquisitely crafted: with only a handful of exceptions, each has exactly one winning move. As far as we know, the only way to compose such puzzles is by trial and error. So, in our research, we had hoped to discover either a new puzzle-composition method or rough drafts of puzzles. But we had found neither.

FIGURE 4.1: Jens Lindhard (shooting) playing field hockey. © History of Science Archives, Center for Science Studies, Aarhus University, Denmark (HSA-CSS-AUD).

Then, in September 2017, we uncovered Hein's call for puzzles, shattering our assumption that Hein had authored the Politiken puzzles. But if not Hein, then who? A short article in Politiken on January 26 gives a clue:

> *Monday February 1 at 8pm: Polygon-evening. In response to the spreading Polygon-mania, the game's creator Piet Hein, together with expert player Jens Lindhard, will demonstrate the game's strategy and finer points. They will use a giant game board to play members of the audience and answer questions. Tickets available at "Polygon" ticket office at the Town Hall Square today from 9am to 6pm. Free subscriber admission by handing in Coupon 1 from the subscription receipt. Cloakroom payment only 25* øre.[2]

At that point we knew little of Lindhard. Born March 26 1922, his father was a professor at the Veterinarian University in Copenhagen. Like Hein, Jens attended Copenhagen's *Metropolitanskolen* high school, graduating in 1940, 16 years after Hein. Lindhard was a strong chess player, and in 1942 a student at Copenhagen University. Later he worked closely with Niels Bohr. He was physics professor at Aarhus University from 1956 until his retirement in 1992, and president of the Royal Danish Academy of Sciences and Letters from 1981 to 1988. He retired from Aarhus in 1992 at age 70 but continued to come to the university each day.[3] After his unexpected death on October 17, 1997, his office — its blackboard still covered with physics problems — was cleared and his papers packed away.

In October 2017, hoping to find Polygon paraphernalia among Lindhard's papers, Bjarne Toft went to Aarhus. There, in a basement storage room of the

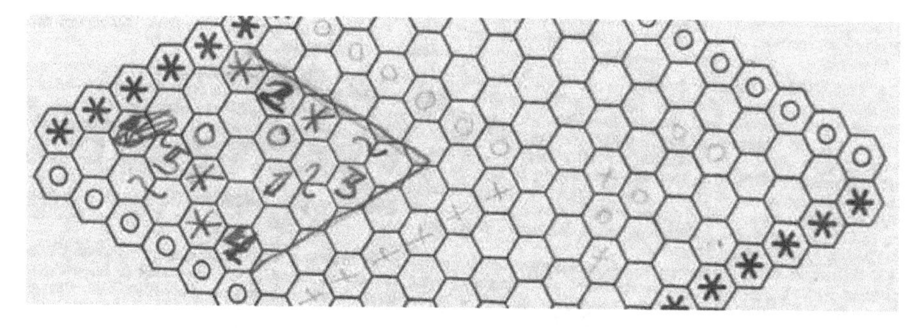

FIGURE 4.2: Politiken Puzzle 10 draft on call-for-puzzles sheet. © HSA-CSS-AUD.

university science museum, he found the final contents of Lindhard's office, packed into 116 archival boxes.

Lindhard had never married, and the boxes contained not only research papers but also personal items, including correspondence, photos and chess documents.[4] After hours of searching, Toft opened Box 115 — 'Unsorted Material' — to find a large envelope. And inside the envelope — eureka! — a treasure trove of Polygon material: letters from Hein, notes on terminology, diagrams for a winning first-player 6×6 strategy, Politiken contest entries ... and puzzle drafts. Finally, we had found evidence of Polygon puzzle construction! And we had confirmed the identity of the composer of (more than half of) the Politiken Polygon puzzles.

4.2 Puzzle drafts

The Lindhard archive fills in many of the historical gaps in the story of Polygon. In particular, it shows that the Polygon project was a two-man partnership, with Hein the expositor and Lindhard the game expert. Lindhard was Polygon's primary puzzlist, composing at least 27 of the 49 Politiken puzzles.[5] To date, we have seen only one Politiken puzzle draft — Puzzle 18 — in hand-writing that is not Lindhard's.[6] We suspect that Lindhard composed all but at most a handful[7] of the Politiken puzzles.

The puzzle drafts shown in this section are from the Lindhard archive: as expected, some are unfinished sketches. Unfortunately, Lindhard left no clues as to how he constructed puzzles: we guess it was a trial and error process informed by his Polygon- and chess-playing experience. We present finished versions of all Lindhard puzzle drafts in Appendix B.

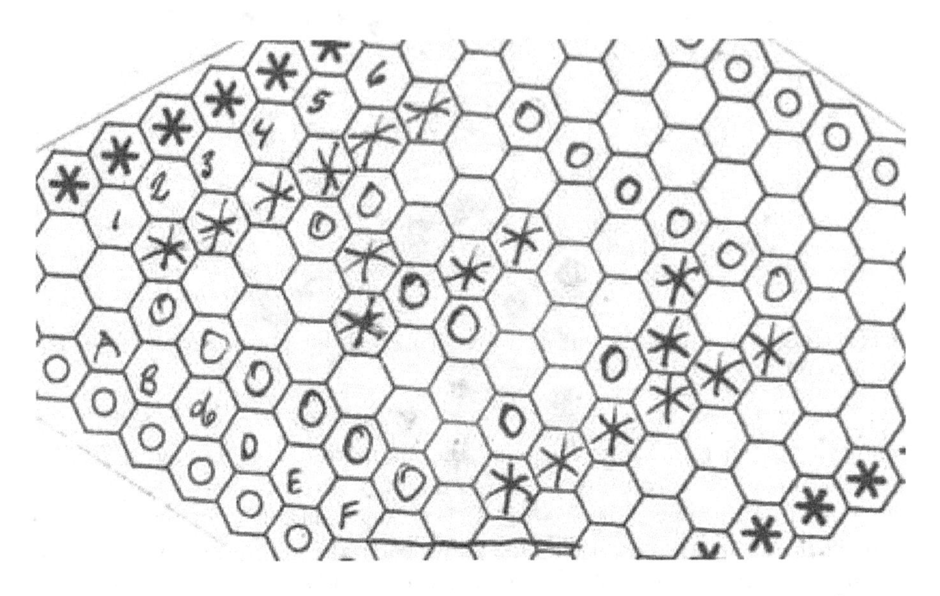

FIGURE 4.3: Politiken Puzzle 18 draft with name and address of Emil Christensen on back. © HSA-CSS-AUD.

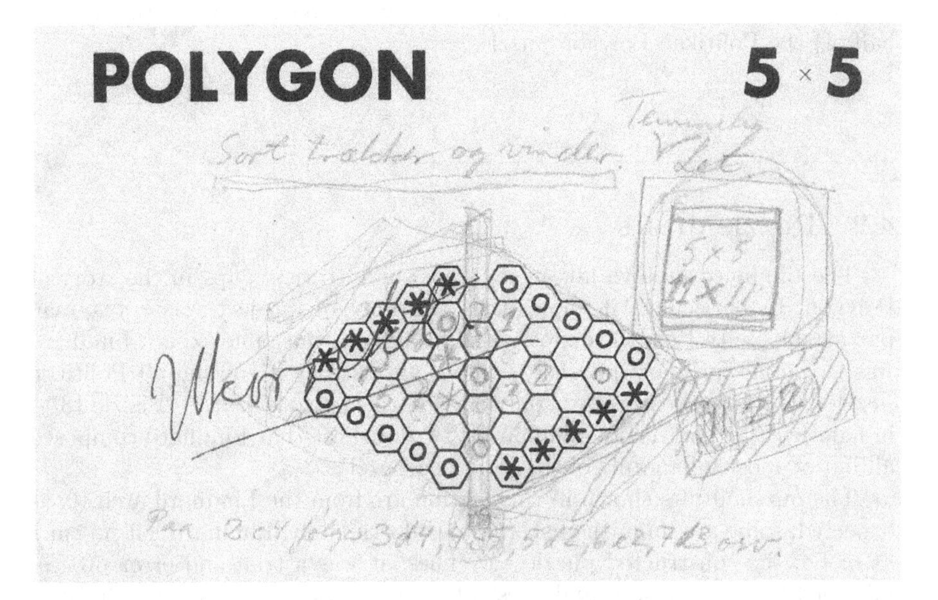

FIGURE 4.4: A puzzle draft, labelled *Ukorrect* (incorrect), perhaps because Black has three winning moves, not just one. © HSA-CSS-AUD.

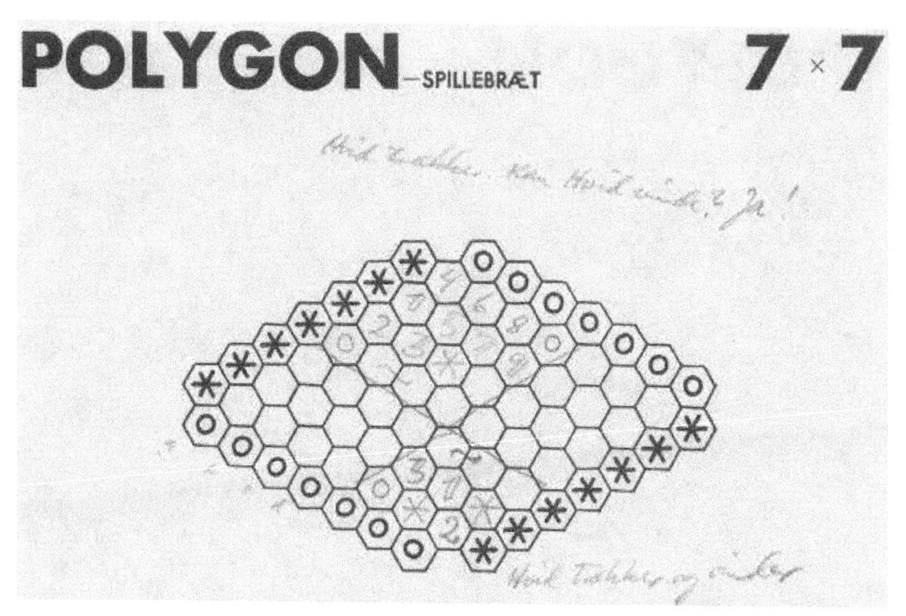

FIGURE 4.5: An unpublished draft (top) and Politiken Puzzle 4. © HSA-CSS-AUD.

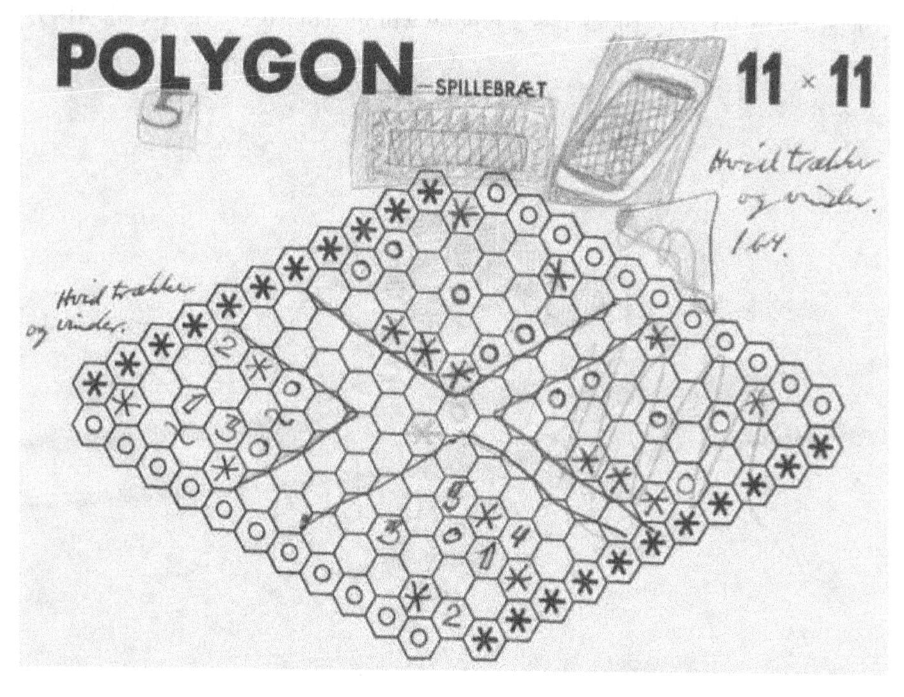

FIGURE 4.6: (clockwise from left) Drafts of Politiken Puzzles 9, 19, a variation, and unpublished. © HSA-CSS-AUD.

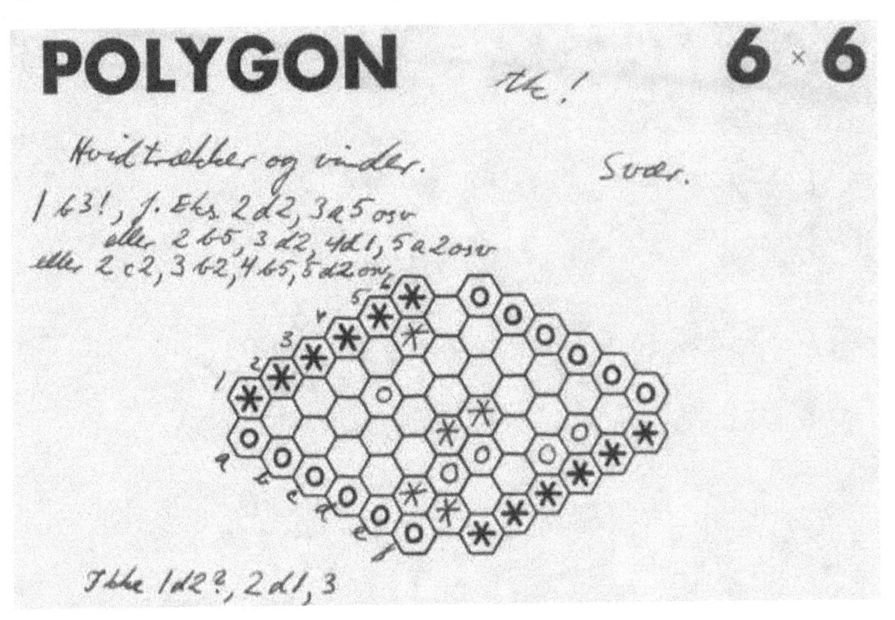

FIGURE 4.7: An unpublished puzzle draft. The proposed solution assumes no stone at b3 but is incorrect: White cannot win. Lindhard presumably added the stone at b3, now Black has a unique winning move. © HSA-CSS-AUD.

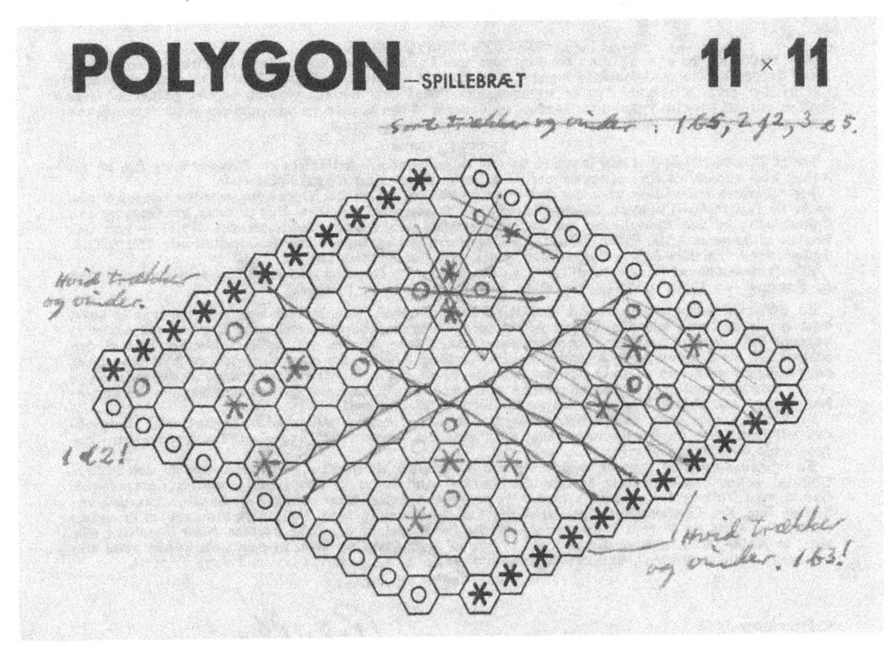

FIGURE 4.8: (clockwise from bottom) Drafts of Politiken Puzzles 2, 15, a reject, and variation of Puzzle 2. © HSA-CSS-AUD.

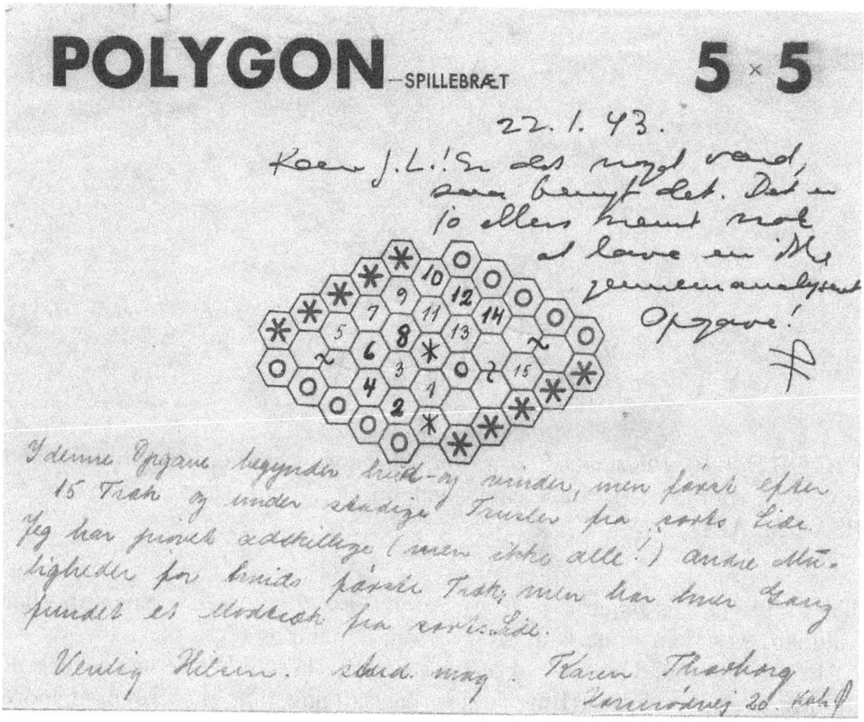

FIGURE 4.9: The Thorborg puzzle. At top, Hein's note to Lindhard. © HSA-CSS-AUD.

4.3 Thorborg puzzle

Some of the Lindhard archive's letters from Hein report reader reaction to the Politiken columns. Another has a puzzle and analysis by Karen Margrethe Thorborg (1921-2013):[8]

> *White starts and wins, but only after 15 moves, and under constant threat from Black. I tried very many (but not all!) other possibilities for White's first move, but in each case I found a winning Black reply.*
>
> *Kind regards, stud. mag.*[9] *Karen Thorborg*
> *Horserødvej 20, Copenhagen 0*

Hein forwarded the puzzle to Lindhard, adding a note:

> *22.1.43. Dear JL!* *If worthwhile, use this puzzle. But of course it is easy enough to construct an unanalyzed puzzle!* *PH*

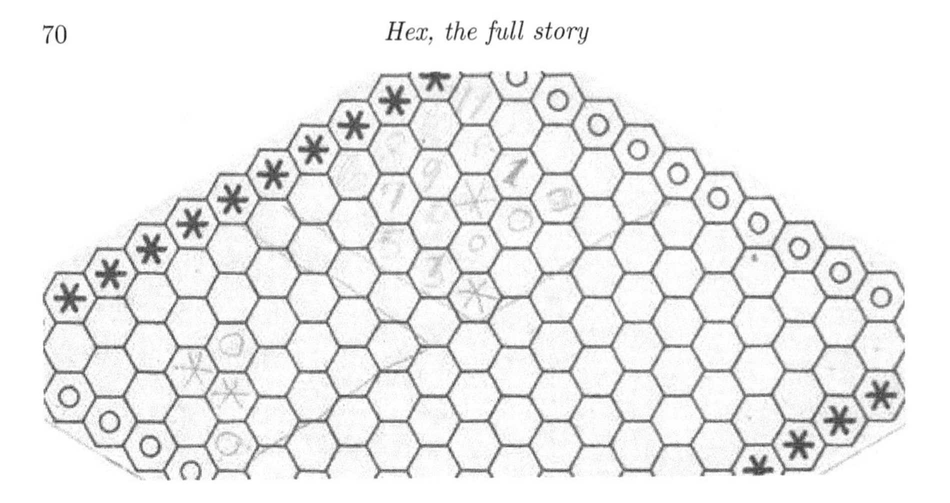

FIGURE 4.10: Puzzle draft based on Thorborg's puzzle. Black to play and win, with solution (top) and equivalent puzzle, White to play and win. © HSA-CSS-AUD.

The nature of this request — *if worthwhile, use this puzzle* — suggests that Lindhard was Hein's puzzle-analyst as well as composer.

In her solution, Thorborg observes correctly that all White moves, except the proposed move, lose. But she errs in claiming that the proposed move wins: Black has a winning reply. Lindhard presumably saw that this is the only winning reply, as he drafted the resulting 'Black to play' puzzle, together with the equivalent position in Hein's preferred 'White to play' form.

4.4 Unpublished Polygon booklet

Hein was busy in January 1943. His first child was born, and the Politiken columns had started. On January 13, as the Polygon columns shifted from appearing daily to twice weekly, Hein sent Lindhard an envelope. Large enough to hold blank 11×11 puzzle sheets, the envelope was rejected by the post office and returned to Hein.

Hein had good reason to supply Lindhard with puzzle sheets: in addition to the Politiken columns, Hein was planning a Polygon instructional booklet. An undated letter in the Lindhard archive reveals that — like the columns — the booklet was a joint venture with Lindhard.

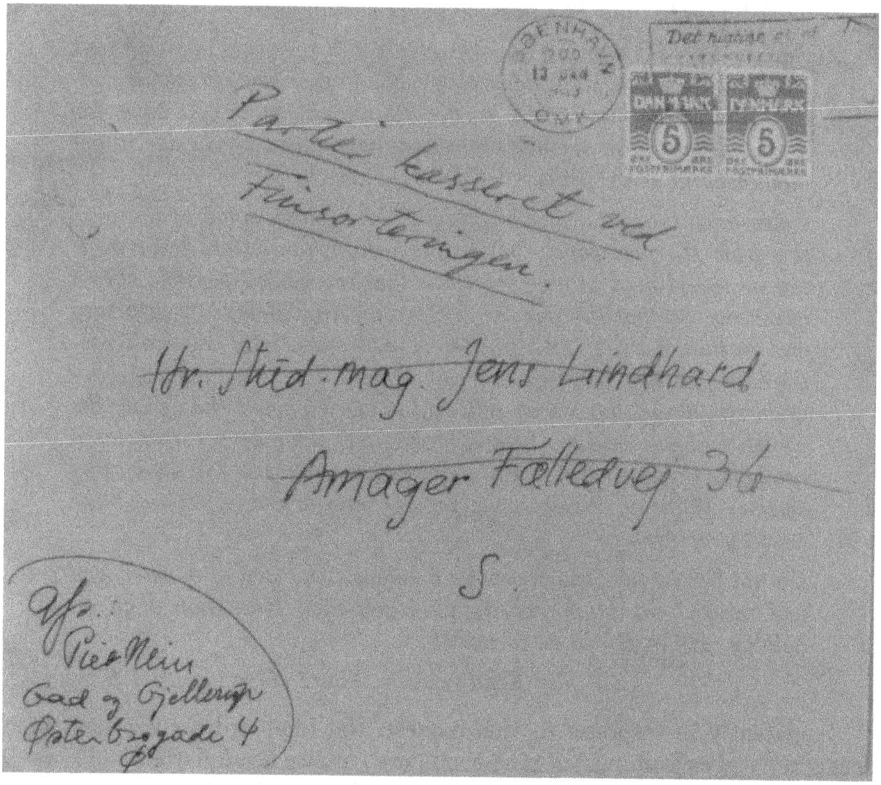

FIGURE 4.11: Envelope from Hein to Lindhard January 13, 1943 — perhaps with blank puzzle sheets — returned undelivered. Courtesy Hugo Hein.

FIGURE 4.15: Polygon booklet mock-up cover. *Instructions for Polygon.*

FORORD

Her har De Polygons eneste Regel, klart og anskueligt udtrykt, samt en Vejledning, som forærer Begynderen de første Erfaringer i Spillet og gir den Viderekomne et Grundlag for fortsat Udvikling af sin Spillefærdighed.

Februar 1943. Piet Hein.

FIGURE 4.16: Polygon booklet mock-up foreword. *Here is Polygon's only rule, expressed clearly and lucidly, together with instructions that give the beginner a taste of the game and the experienced player a foundation for further development. February 1943. Piet Hein.* Courtesy Hugo Hein.

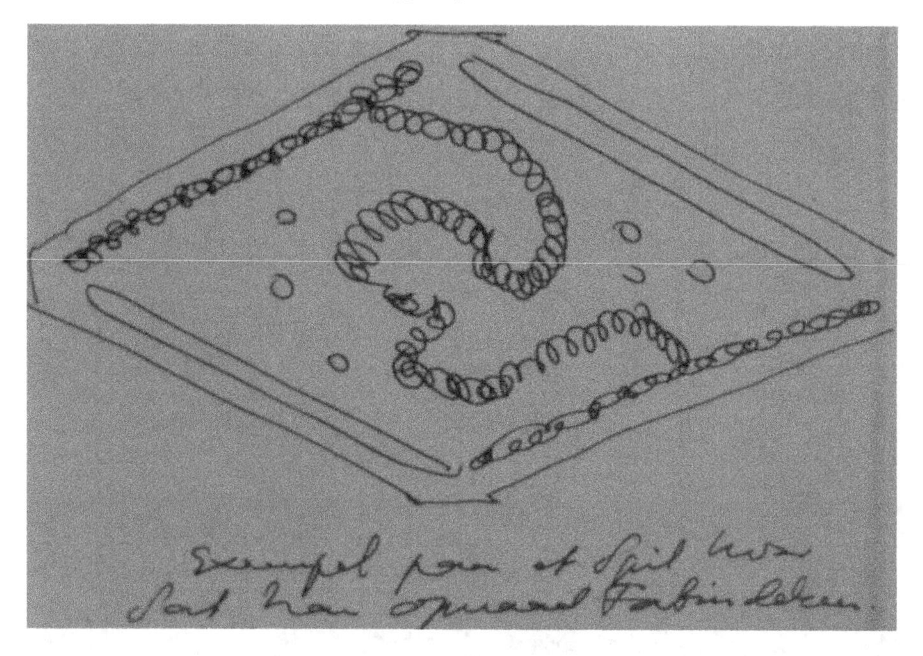

FIGURE 4.17: A page from the Polygon booklet mock-up. *An example of a game where Black joins her two fronts.* Courtesy Hugo Hein.

Hein and Lindhard seem to have liked a similar opening we call Tornehave Variant, leading to the position — page 10 of the salon manuscript — discussed at the Politiken salons. The Hein archive has an undated Lindhard letter with this variant opening and analysis:

> *Here is a game with commentary and a puzzle.*[10] *I will send more puzzles and openings as soon as possible.*
>
> *White starts in the center and Black answers as usual. White continues at h4 [the Tornehave Variant] - a fine move - and Black is not afraid of a small rubbing shoulders (RS), which Black stops in the right moment, namely with 8.g8. Then White plays h8 and when Black answers at h7, White plays j6 and has thus, in a fine way, formed a police headquarters. Now Black plays i3, a kind of double threat, because it may be of use in RS with i6-i5-i4, but if White stops this RS — which White in fact does with move 13 — then Black starts, as one can see, another RS. White interrupts this new RS with move 19.*
>
> *Black then takes the opportunity to start a distracting maneuver on the other front. With 23.i7 White prevents Black from connecting h7 to Black's right front.*
>
> *After this little diversion Black feels — in some sense correctly — fortunate to continue maneuvers near the left front. How-*

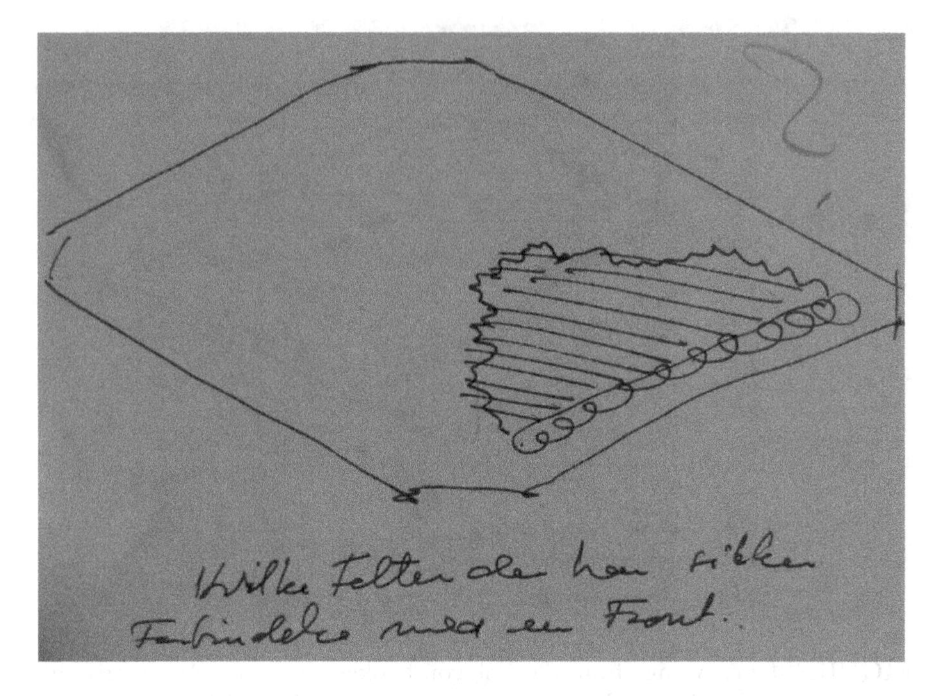

FIGURE 4.18: A page from the Polygon booklet mock-up. *Critical fields that can be safely connected to a front.* Courtesy Hugo Hein.

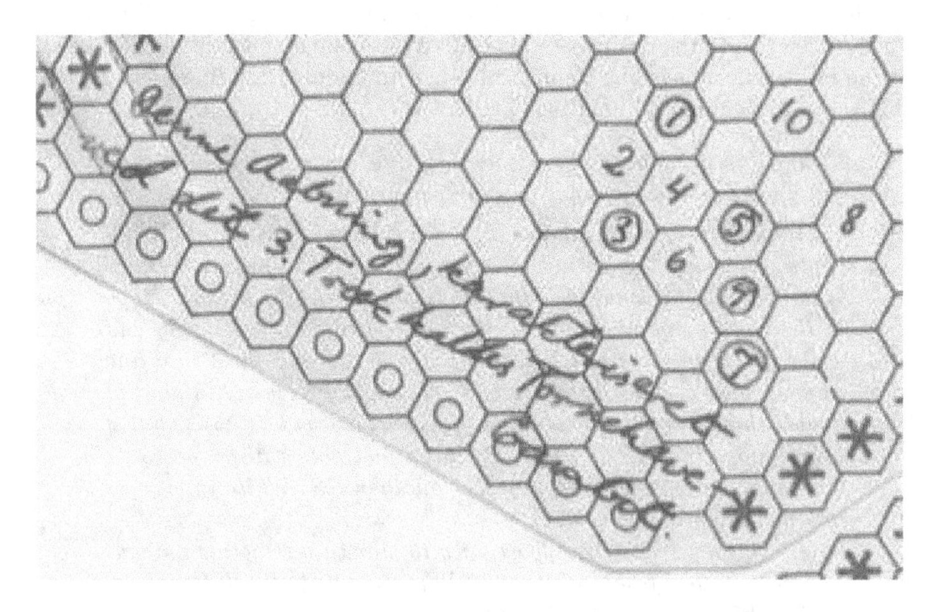

FIGURE 4.19: The Torenhave Gambit opening, characterized by move 3. With move 4 Black splits the two White groups. © HSA-CSS-AUD.

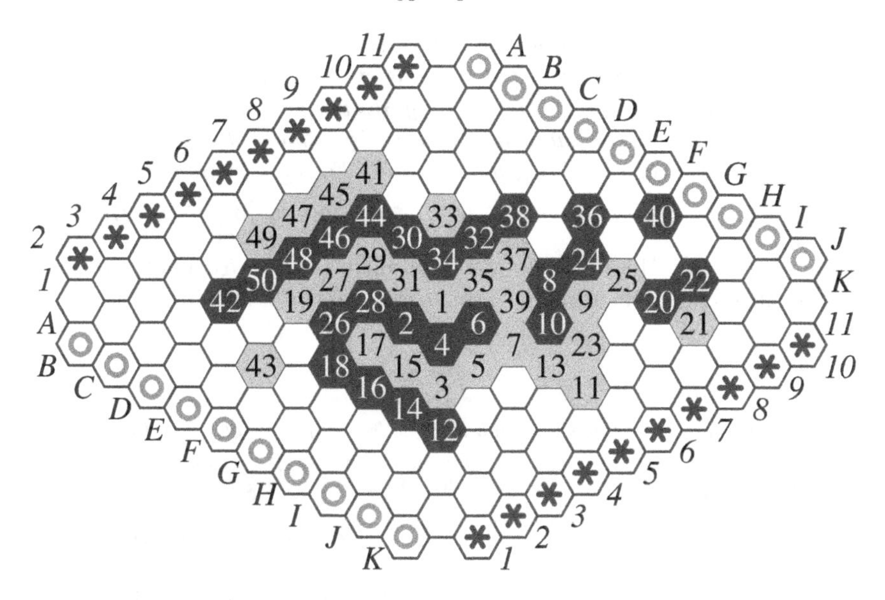

FIGURE 4.20: A Tornehave-opening game analyzed by Lindhard.

> *ever, one might have wished that White would have shown a little more originality, perhaps playing 29.c7. [MoHex also prefers this, but 29.c7 30.e2 31.c3 32.d1 wins for Black: 29.c6 looks better for White.] As it now stands, with 30.d7 Black can start the forcing sequence 31.36 ... 35.f7. [MoHex finds Black winning from at least move 32.] But this is futile, as Black plays the very strong 36.f10. Now 39.g7 is forced, whereafter with 40.g11 Black decisively joins [d7] to the right side. Black gives White the death blow with move 42, a well known double threat. The rest is silence.*
>
> *Kind regards, Jens Lindhard.*

4.6 Hein-Lindhard game

On Wednesday, February 17, 1943 Hein wrote Lindhard twice. The first note includes a letter from a reader who claims to have found a 'universal system' that guarantees a win.

> *Dear Jens Lindhard!*
>
> *Enclosed is an unusual human letter — from a Universal Methodist with this logic: "My system + a good player wins, my system +*

a bad player loses, hence my system is correct." As a system his examples have no interest, but perhaps it might interest you to check whether his corrections yield significantly stronger play. — I would like to get this back and likewise if you have others of that type. I assume that I will get a problem by Friday morning?

Kind regards, Piet Hein

Unfortunately among Hein's papers we found no letter, so we cannot comment on whether this reader's method had any merit. Later that evening, Hein wrote a second note:

Dear Jens Lindhard!

Thank you for the beautiful problem and the game tonight! Enclosed are 5×5, 6×6 and 7×7 boards; it is my impression that you miss having them.

Kind regards, Piet Hein

Hein's remark "*you miss having them*" might be in reference to the earlier undelivered envelope.

The second note suggests that, after having posted the first note, Lindhard met Hein, gave him a new puzzle — presumably Puzzle 29, which appeared on Saturday February 20 — and played a game, with Hein as opponent or spectator.

The Hein archive includes one game labelled 'Lindhard game', perhaps the game Hein refers to in the note. Figure 4.21 shows this fascinating game: Lindhard comes from behind to win. Hein, who publicly professed to not being an expert, plays well until move 11, which is nullified by move 12.[11] Almost anywhere else would have been better: for example, MoHex likes 11.h3. Hein kept the game close until move 21. Figure 4.22 shows the situation after move 20: can you find a winning move 21?

4.7 Perplexing Puzzle 45

In March 1943, as errors creep into the Politiken columns, Hein is late in his payment for Lindhard's puzzles. On March 14, Hein writes Lindhard on an unused puzzle sheet:

Dear Jens Lindhard!

Thank you for the most recent two letters with a problem in each! I have only now understood that you need Polygon pads, enclosed are 5 from me. More will be sent later from the publisher.

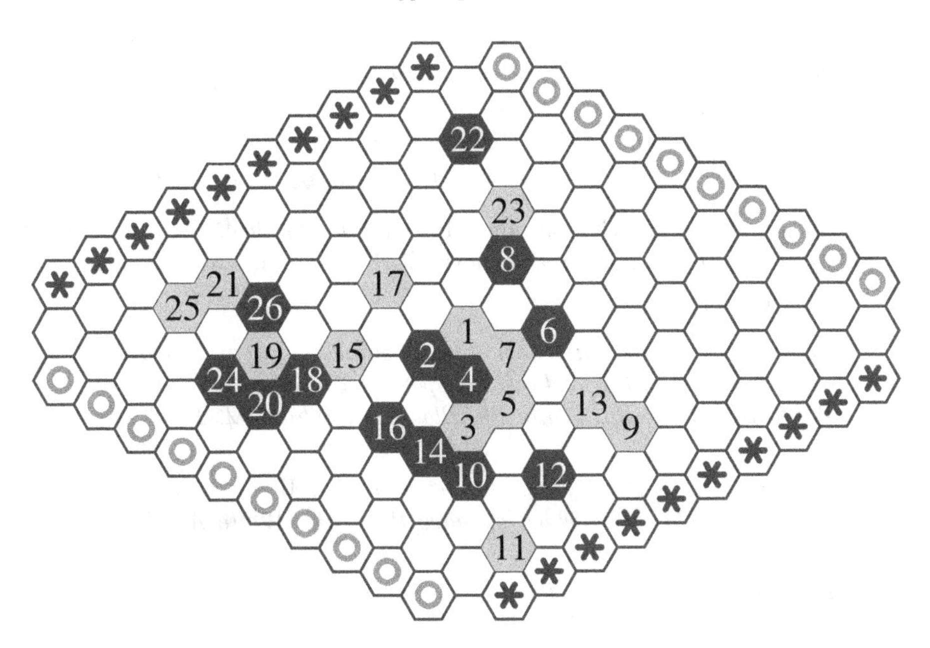

FIGURE 4.21: Hein-Lindhard game. Hein (White) plays a Tornehave opening but blunders on move 21 and resigns after move 26.

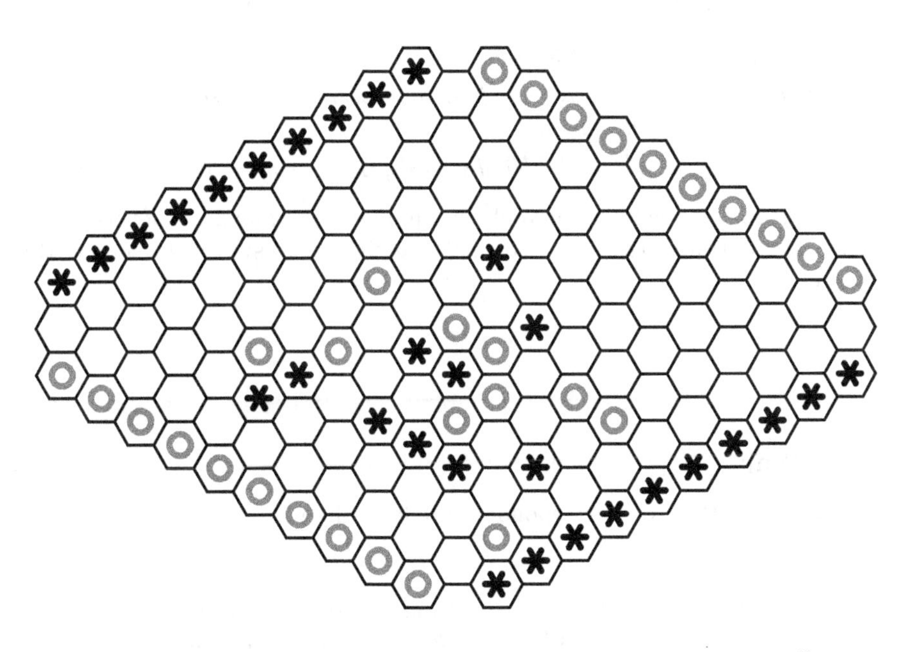

FIGURE 4.22: Hein-Lindhard game after move 20, White to play. Can you find a winning move? There are several.

I should have sent you money on the 1st, sorry that this has been delayed, it should arrive now. Up to which number have you already been paid? Please phone tomorrow concerning (1) this and (2) if you could look at the now-ended contest's best entry?

Kind regards, yours, Piet Hein

As mentioned earlier, the solution in Politiken to Puzzle 45 is wrong: White has no winning move. And a reader noticed! H.J. Hansen wrote to Hein:

Mr. Piet Hein.

Re Polygon Puzzle 45: I do not understand the mentioned solution (as shown in the enclosed newspaper cutting), it is as follows: White B6, Black B5, White F6, Black D3, White D4, Black C3, ...

Here I would as Black reply C2 instead of C3. White then has to reply B1 or B2, after which we have Black B3, White A4, Black B4.

I would be grateful for a short instruction concerning this question. Excuse me for the trouble!

Yours truly, H.J. Hansen

Hein apparently appended "?" to the letter and forwarded it to Lindhard, accompanied by this note:

11.8.43. Dear Jens Lindhard!

Is he right? Or what are the first countermoves? Be so kind as to find them and let me know!

Kind regards, Yours sincerely, Piet Hein

Unfortunately we found no reply by Lindhard, nor any draft of this puzzle. What do you think? Is Hansen correct? We give the answer at the end of this chapter.

4.8 War

By spring, Polygon correspondence reflects the war:

Dear Piet Hein!

I have been asked, if there are instructions for Polygon in languages other than Danish. It is from people who would like to send Polygon games to prisoners of war.[12]

Kind regards, Jens Lindhard

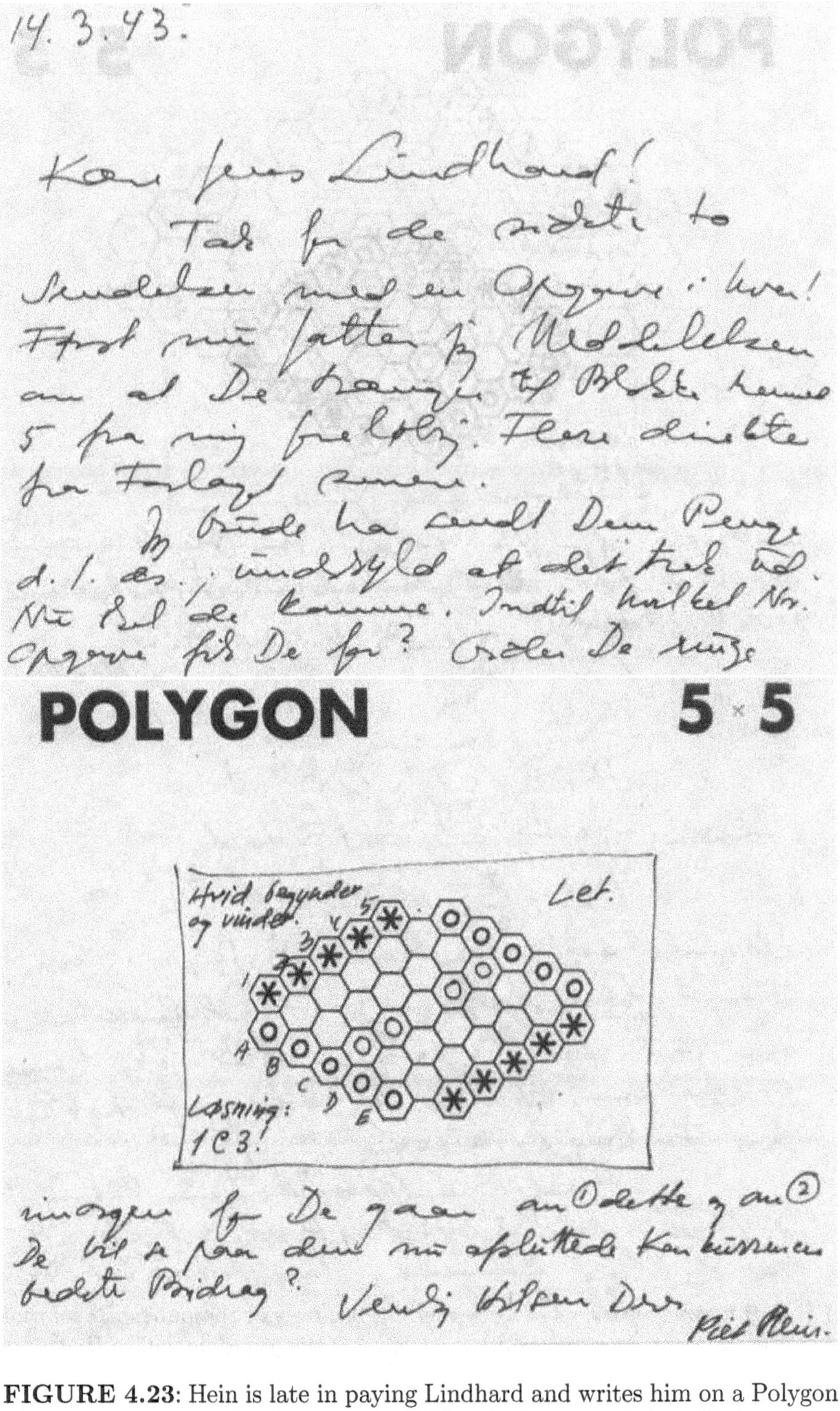
14.3.43.
POLYGON 5×5
Hvid begynder og vinder.
Let.
1
2
3
4
5
A
B
C
D
E
Løsning: 1C3!
Piet Hein.

FIGURE 4.23: Hein is late in paying Lindhard and writes him on a Polygon puzzle sheet. Notice Hein's gag puzzle: White's shortest win is 1.C3, but every White move wins. © HSA-CSS-AUD.

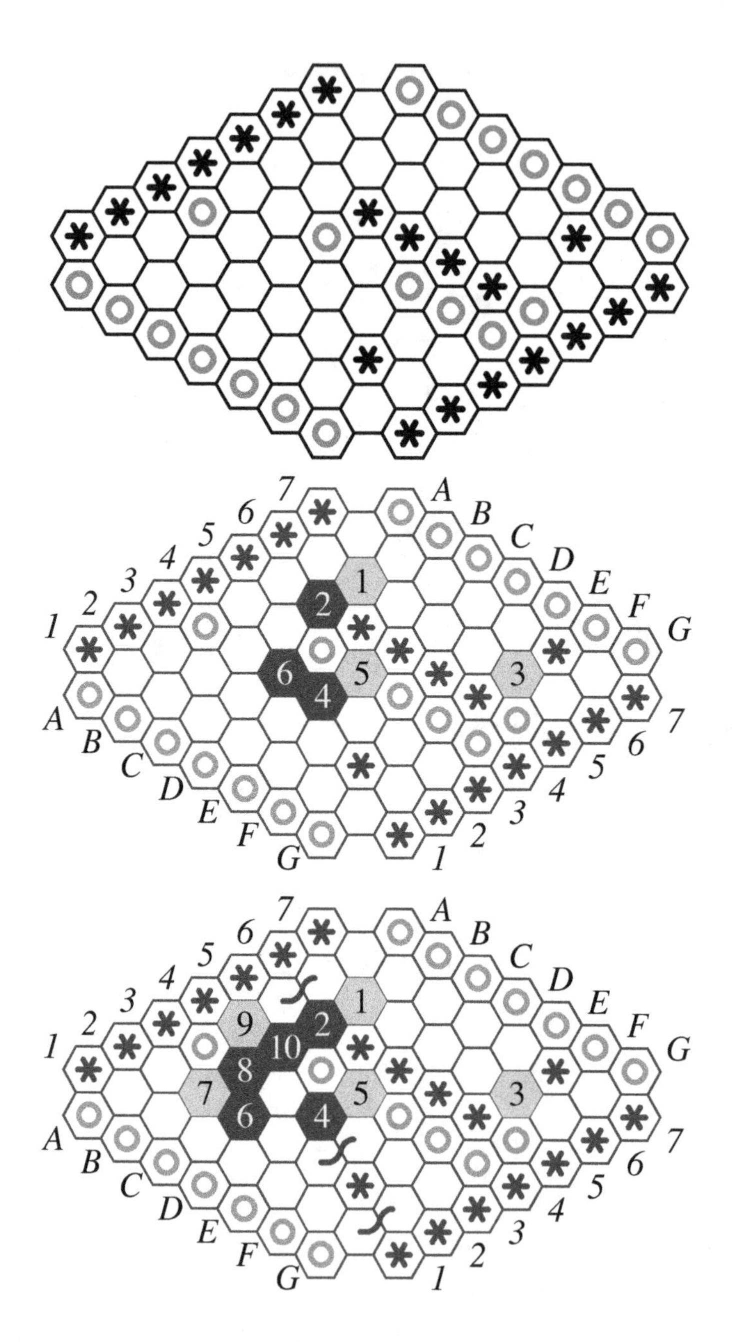

FIGURE 4.24: Puzzle 45 (top), the initial moves of the published solution (middle), and — move 6 onwards — Hansen's suggested deviation (bottom).

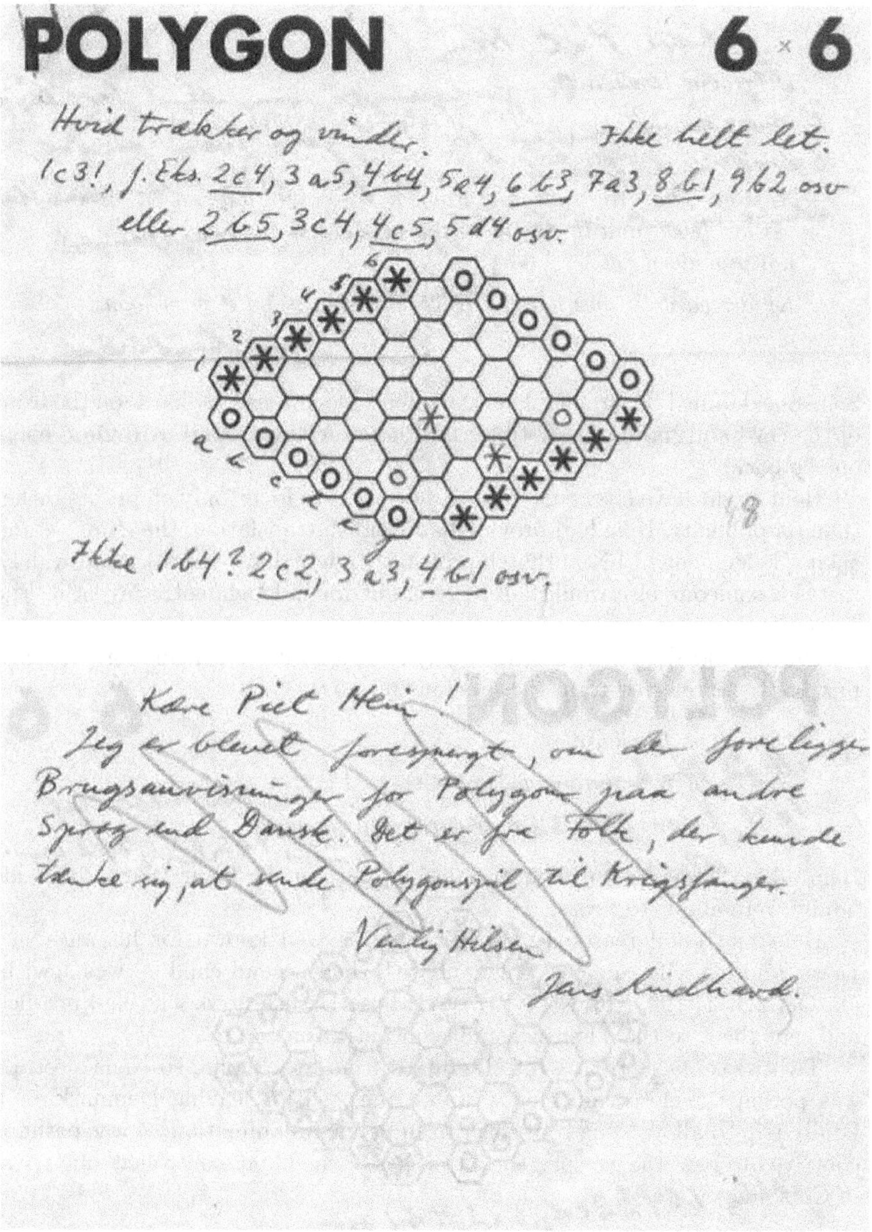

POLYGON 6 × 6

FIGURE 4.25: Draft of Politiken Puzzle 48, with a note on the back from Lindhard to Hein about prisoners of war. © HSA-CSS-AUD.

The message has been crossed out, perhaps by Hein who later returned the card to Lindhard. A card on May 14 is presumably Hein's reply to Lindhard's query:

> *Dear Jens Lindhard!!*
>
> *Thank you for the puzzle!*[13] *It is nice with these puzzles with so few pieces — which are unique.*
>
> *Answer: That [translation of Polygon into other languages?] is not yet available, and in consideration of protection abroad one should not provide it yet.*
>
> *Kind regards! I owe you for April and shall send it very soon.*
>
> *Yours, Piet Hein*

Someone doodled 'Dear Jens Jens Dear Jens, Dear Jens Lindhard' on the front of the card and 'no no many times no Phone Friday around 9-10 Øbro 6566' on the back.

Hein might have later regretted this decision to focus only on profits rather than on publicity. If he had provided an English translation, the game — and acknowledgement of his authorship of it — might have spread more widely. Instead, someone else would later get credit for independently inventing the game.

On May 30 Hein sent Lindhard payment for puzzles, including the April puzzles 38-41 referred to in the previous message:

> *Dear Jens Lindhard!*
>
> *Here is your honorarium for the seven Puzzles 38-44.*
>
> *Kind regards and see you, Yours, Piet Hein*

Hein added 'see you' to his letter closing: perhaps he knew that he and his family were about to flee.

Hein had good reason to leave: Hein was well known for his anti-Nazi views, and his wife Nena — pregnant with their second child — was Jewish. On October 1, 1943 the Nazis arrested those Danish Jews who had not fled and sent them to the Theresienstadt concentration camp.

Two weeks later Hein wrote Lindhard from Stockholm, so Nena's escape was perhaps a close call. Hein's stated purpose for leaving Denmark — a family trip to allow Nena to recover from pelvic inflammation — was perhaps contrived to pass the censors: the letter was opened and 'controlled' on arrival in Copenhagen:

> *Hotel Bellevue, Birger Jarls Gatan 14, Stockholm*
>
> *October 15, 1943*
>
> *Dear Jens Lindhard!*
>
> *We are on our way to my wife's home country Argentina. As I*

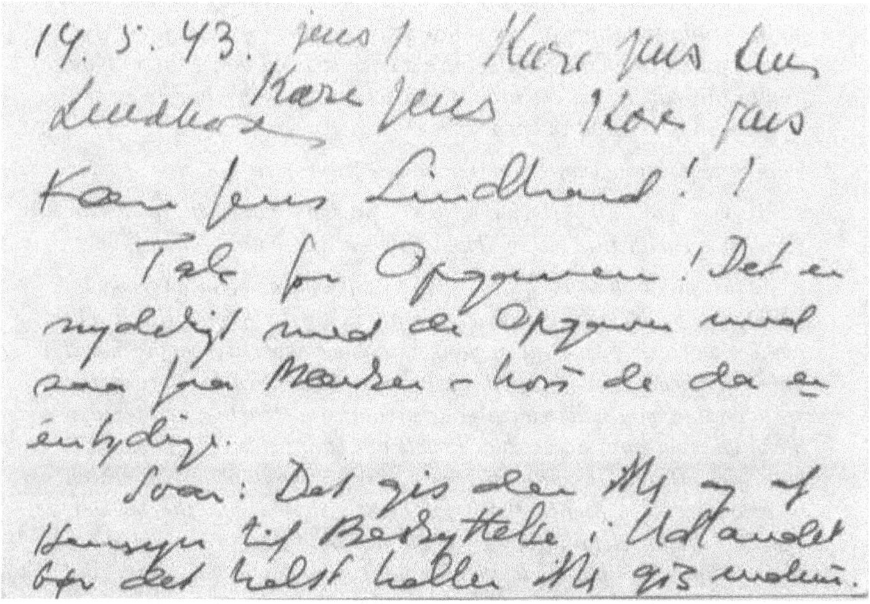

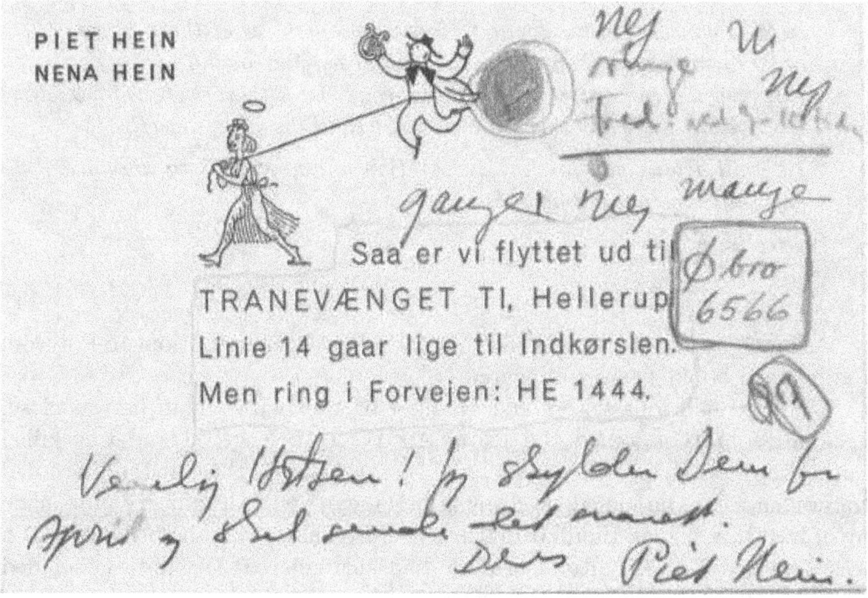

PIET HEIN
NENA HEIN

Saa er vi flyttet ud til
TRANEVÆNGET T1, Hellerup
Linie 14 gaar lige til Indkørslen.
Men ring i Forvejen: HE 1444.

FIGURE 4.26: Hein's reply to Lindhard — *Thank you for the puzzle ...* — is written on a printed change of address card: *Piet Hein, Nena Hein: We have moved to Tranevænget T1, Hellerup. Line 14 goes straight to our driveway. But phone ahead: HE 1444.* © HSA-CSS-AUD.

think you know, my wife has been very ill because of the cold climate – inflammation of the renal pelvis – and we have for 1 1/2 years wanted to follow the doctor's advice to head down there. Now finally the papers are in order, and we must leave now to arrive before our next child is born.

This is to explain why the letter comes from here.

1. I owe you 40 crowns. Please present yourself to editor Stephensen with this letter. Then you will get your money.

2. At the same place they are ready to publish, as soon as possible, a Polygon booklet, to introduce beginners to the game. But I will really not be able to convey your immense expertise to the usual unlearned beginners, that I think I am well qualified to represent. I do hope that you will realize your dream, and together with Steph make your wisdom accessible. It will not be much work. Discuss it with him. He can arrange things with you about payment, which of course will be higher than if I myself should edit the booklet, even if I as the originator of the game would like to have a share in it. I have brought your notes on the game with me. They are hopefully easy to write down again. Include this inconvenience in your discussions of an honorarium!

3. Will you do me the favor to inform me here what the numbers of the incorrect newspaper puzzles, either by your or my fault? You can read the newspapers at their office. I think there were 2 or 3. Do not tell me what was wrong, that I shall discover myself.

Be well! Thank you for fine cooperation and hopefully we shall see each other sometime again!

Greetings from my wife,

Yours sincerely, Piet Hein

As far as we know, this was the end of Lindhard's contributions to Polygon: the planned booklet never appeared.

Lindhard had four sisters and an older brother with whom he was close. Lindhard's brother joined the Royal Air Force in England and was killed during the Normandy invasion in 1944. Near the end of the war Lindhard fled to Sweden where he joined the Danish brigade, a Swedish military unit made up of Danish refugees. Lindhard returned to Denmark with the brigade, which was deployed in 1945 after Germany had surrendered. His duties included patrolling the Danish-German border.

With Niels Bohr in America, Lindhard made other arrangements to complete his M.Sc. The Swedish physicist Oskar Klein had frequently visited Bohr's institute and agreed to act as Lindhard's supervisor. And so Lindhard obtained an M.Sc. in Stockholm in 1945, but under the auspices of the University of Copenhagen.

Jeg spillede Hockey og der stødte jeg for første gang
på Aage Bohr, der spillede i en anden klub.
Vi begyndte også at studere ved universitetet samtidig
og jeg tror at Aage forsøgte at overtale mig til at studere
fysik og ikke matematik, men det var nu ikke nødvendigt.

FIGURE 4.27: Lindhard's retirement speech. *I played hockey ...* © HSA-CSS-AUD.

FIGURE 4.28: Lindhard archive chess documents. © HSA-CSS-AUD.

In 1992, on the occasion of his retirement, Lindhard reflected on time spent with the Bohrs: with Aage when they were both junior scientists in Niels' institute and with Niels as a research assistant:[14]

> *I played [field] hockey and there I first met Aage Bohr,*[15] *who played in a different club.*[16] *We also started our university studies the same year, and I think Aage tried to convince me to study physics instead of mathematics, but that was not needed.*
>
> *...I sit with Niels Bohr in front of a blackboard full of formulas. We are solving a difficult problem and forming our thoughts for a paper on the charge of fission fragments. After smoking a long time in silence, Bohr takes the pipe out of his mouth and says, "Now I have it! We shall start the sentence with the word 'Notwithstanding'!"*
>
> *This word is written down and the pipes are lit again. The new situation is considered and thought through for an even longer time, whereupon Bohr says: "I am extremely happy with how well our work is progressing today."*

FIGURE 4.29: Lindhard in Aarhus. © HSA-CSS-AUD.

4.9 Solutions

The complete set of Lindhard puzzles is shown in Appendix B. Here are solutions to three other puzzles from this chapter.

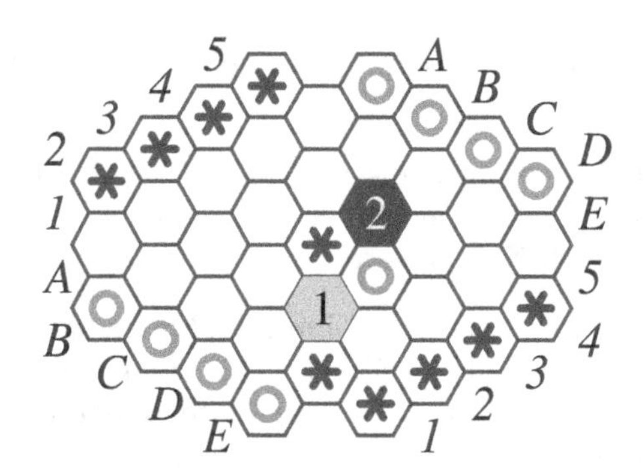

FIGURE 4.30: Thorborg's puzzle. Black wins: after White 1.d2, Black 2.c4! If 1.x for x in {a5-e5, b4-e4, b3, e2-e3} then 2.d2! For any other 1.y, 2.c4!

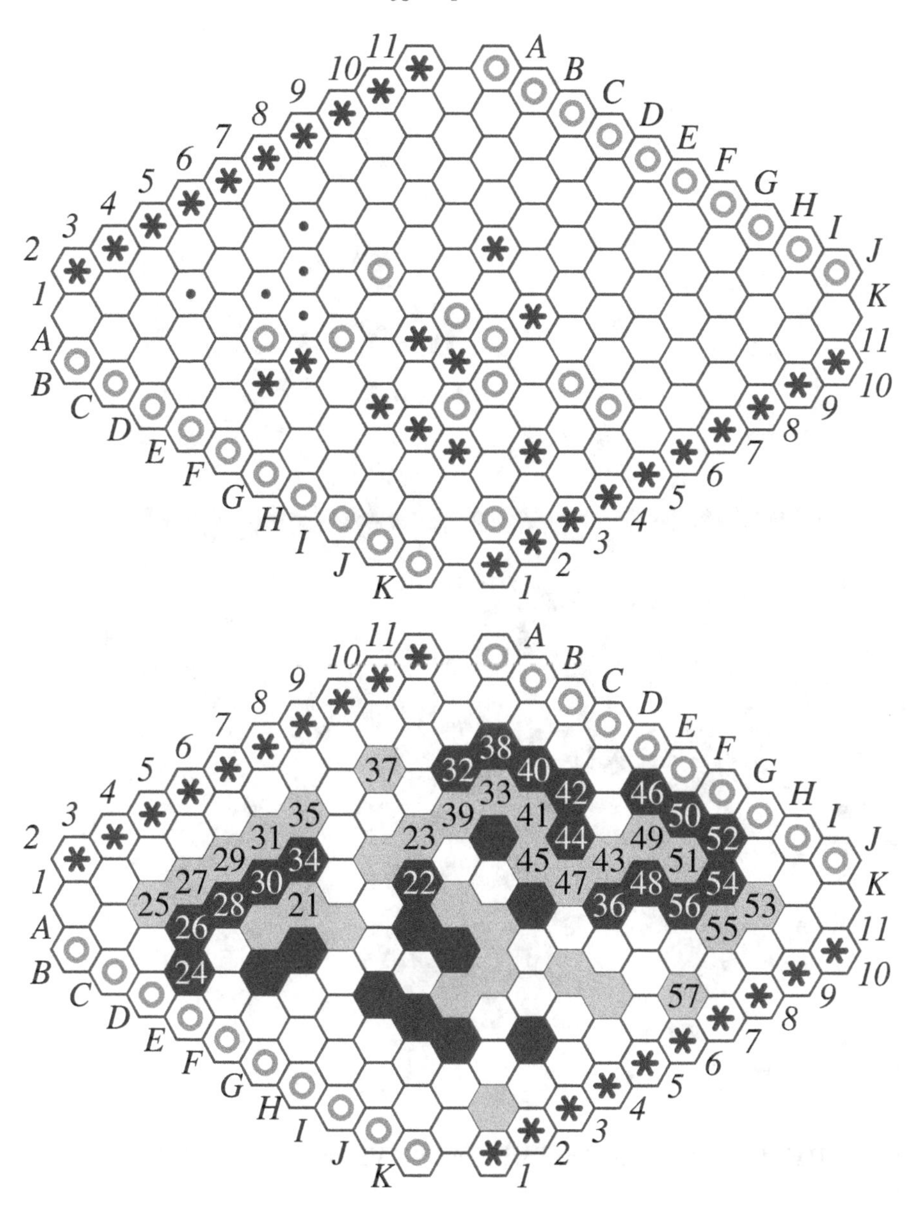

FIGURE 4.31: Hein-Lindhard game: White wins at each dot (top). A typical winning continuation from 21.d4 (bottom).

FIGURE 4.32: Jens Lindhard with sisters Vibeke, Jytte and Annelise. © HSA-CSS-AUD.

FIGURE 4.33: Lindhard, perhaps in Switzerland in the late 1940s. © HSA-CSS-AUD.

Notes

[1]From an undated letter to Hein with analysis of a Tornehave game [49].

[2]A Danish *krone* (crown) has 100 *øre*. In 1942, 5 Danish crowns was about 1 US dollar.

[3]Bjarne Toft studied mathematics and physics at Aarhus University in the 1960s, but never had Lindhard as a teacher.

[4]One newspaper clipping reports Lindhard's promotion in 1940 to Danish Chess Master: *In Class 1 at the Danish Chess Union Easter tournament in Randers 1940, the three young Copenhageners Aage Hau, Kjeld Hansen and Jens Lindhardt [sic] were the big sensation, completely outclassing their provincial rivals. Of the 52 players in this class, these three — with points 7, 6 and 5.5 — were the only ones to obtain the title "Danish Master" and advance to Danish Master Class.*

[5]The Lindhard archive has drafts in Lindhard's writing of Puzzles 2, 4-7, 9-11, 13-14, 16-17, 19-24, 48, and a note from Hein for payment for Puzzles 38-44. The Hein archive has a draft in Lindhard's writing of Puzzle 42. Hein also sent Lindhard a reader's enquiry to the solution of Puzzle 45, so Lindhard might have composed that puzzle as well.

[6]The writing of the Puzzle 18 draft matches neither Hein's nor Lindhard's. The name on the back — Emil Christensen — is presumably that of the composer.

[7]Besides 18, the early puzzles not found in the Lindhard archive — 1, 3, 8, 12, 15 — could well have come from other people.

[8]Born 1921, Karen Margrethe Thorborg studied at Copenhagen University and taught high school at Aurehøj Gymnasium in Copenhagen. She died 2013, age 91.

[9]Thorborg's title *stud. mag.* is that of a university student who has not yet graduated. At that time in Denmark, there were no undergraduate degrees: the first degree was the master's, which a student typically earned after five or six years of study.

[10]Unfortunately the letter is undated, and we do not know which puzzle Lindhard included with his Tornehave Variant game commentary.

[11]We will see why in later chapters: move 12 effectively captures eight cells for Black, including the cell of move 11.

[12]Unfortunately we have no further information concerning this request.

[13]This message seems to reply to Lindhard's query re prisoners of war, written on the back of the draft for Puzzle 48.

[14]Lindhard retirement speech, History of Science Archive, Center for Science Studies, Aarhus University, Denmark.

[15]Field hockey was apparently popular in the Bohr family: Aage's older brother Ernest played for the Danish field hockey team at the 1948 London Olympics.

[16]Aage also attended a different school from Lindhard, the *Sortedam Gymnasium* in Copenhagen. In 1892, Hanna Adler (1859-1947) — Niels Bohr's mother's sister — and a fellow student were the first Danish women to obtain master's degrees in science. In 1893 Adler founded *H. Adlers Fællesskole*, the first Danish co-educational school, based on ideas she had learned on a study tour of the United States. She led the school until 1929. The school, which later became *Sortedam Gymnasium*, still exists under the name *Sortedamskolen, H. Adlers Fællesskole*. Adler, who was Jewish, did not flee to Sweden in 1943 like the rest of her family, and at age 84 she was arrested by the Nazis and imprisoned. After much protest, including from many former pupils of her school, Adler was eventually released from prison, rather than sent to Germany, and given a pass that allowed her to move freely in (but not to leave) Copenhagen, the only such permission given to a Jew in Denmark during the war.

Chapter 5

Rebirth

What one can invent, another can discover.

Arthur Conan Doyle

It's a matter of connecting topology and game theory.

John Nash[1]

5.1 New game in Fine Hall

Six years after disappearing from Politiken, Hex mysteriously reappeared in Princeton. One morning in late winter 1949, perhaps March, John Nash bumped into David Gale on the grounds of the graduate college at Princeton University. Nash was a first-year Ph.D. math student; Gale, who would receive his Ph.D. later that year, was at the time a Henry B. Fine math instructor. In a 1995 interview with Sylvia Nasar, Gale recalls Nash's excitement:[2]

> *Gale! I have an example of a game with perfect information. There's no luck, just pure strategy. I can prove that the first player always wins, but I have no idea what his strategy will be. If the first player loses at this game, it's because he's made a mistake, but nobody knows what the perfect strategy is.*
>
> *Assume that two squares [of a checkerboard] are adjacent if they are next to each other in a horizontal or vertical row, but also on the positive diagonal.*

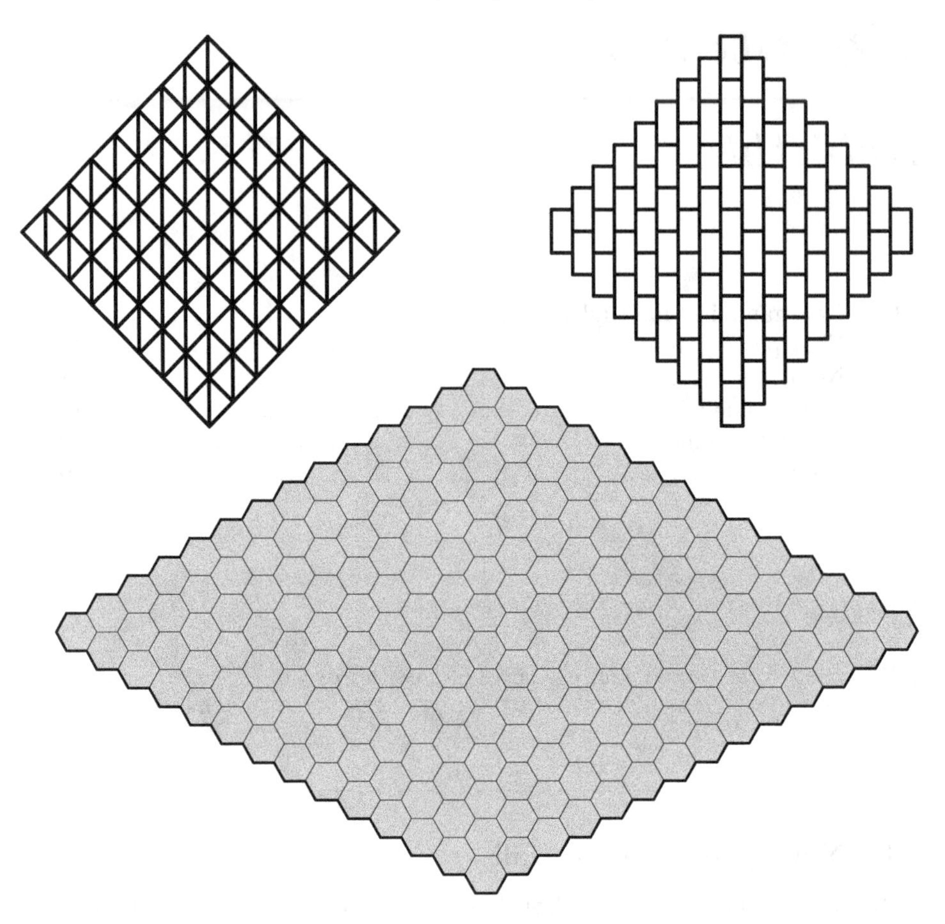

FIGURE 5.1: Gale converted Nash's description of the game board (top left) to its dual (right) and then to the more natural hexagonal grid (bottom).

Gale further described his reaction:[3]

> *I was quite intrigued and thought of the hexagonal format, and the same day I built a [14x14] Hex board which I gave to the Fine Hall common room. The game became quite popular there and [the board] was still being used, somewhat the worse for wear, when I revisited Princeton 4 or 5 years later.*

The common room in Fine Hall was where graduate students in the Princeton Mathematics Department gathered to relax and play games, such as Go and Kriegspiel. It is perhaps also where Nash revealed some properties of the game.

FIGURE 5.2: John Nash, circa 1948.

5.2 First player wins

One of the mysteries of Hex is that there exists a winning strategy for the first player, even though — except for smaller boards — a particular winning strategy is not known. Hein mentions this at the Parenthesis talk and in the Politiken columns, and gives a proof in a 1957 letter to Martin Gardner.[4]

> *You tell me that you have a rather lengthy but quite elegant proof that the first player has a sure win. You may be interested in a rather short one too (just to have one of each kind): Regard the second player B as the first[,] with the peculiarity that the first player A who is now the second has one mark in advance – already there on the board. Now if the second player is sure to (be able to) win, that mark won't prevent him from winning (as it can only connect his areas and bar the opponent's connection) so the premis [sic] that the second player wins leads to the conlution [sic] that the first player wins; indirect proof.*

Here is Nash's proof in his own words, also from 1957:[5]

> *(1) When the board is filled one or the other of the players will have connected but not both.*
>
> *(2) Either the first player or the second will have a winning strategy.*
>
> *(3) Suppose the second player could force a win.*

FIGURE 5.3: David Gale, perhaps in the 1950s. © Katharine Gale.

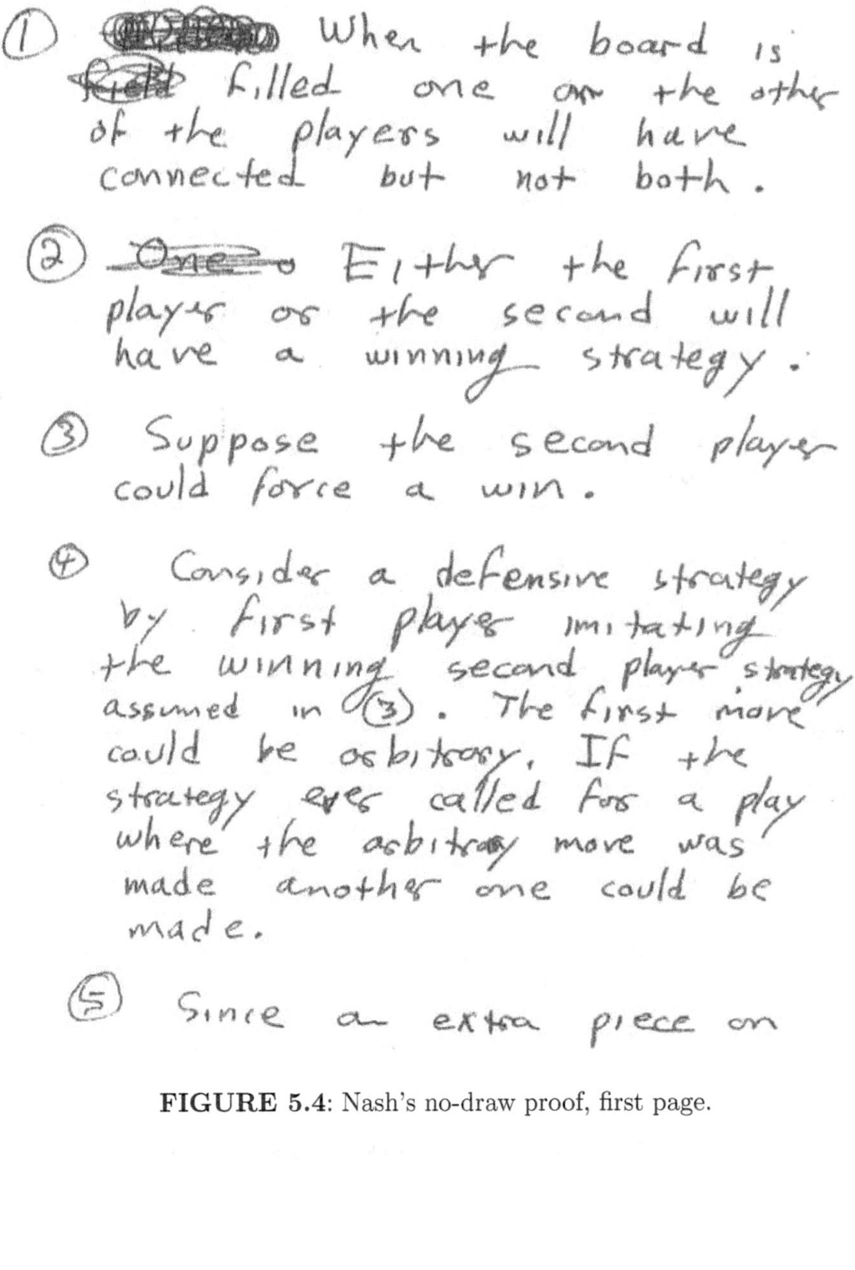

(1) When the board is ~~field~~ filled one or the other of the players will have connected but not both.

(2) ~~One~~ Either the first player or the second will have a winning strategy.

(3) Suppose the second player could force a win.

(4) Consider a defensive strategy by first player imitating the winning second player strategy assumed in (3). The first move could be arbitrary. If the strategy ever called for a play where the arbitrary move was made another one could be made.

(5) Since an extra piece on

FIGURE 5.4: Nash's no-draw proof, first page.

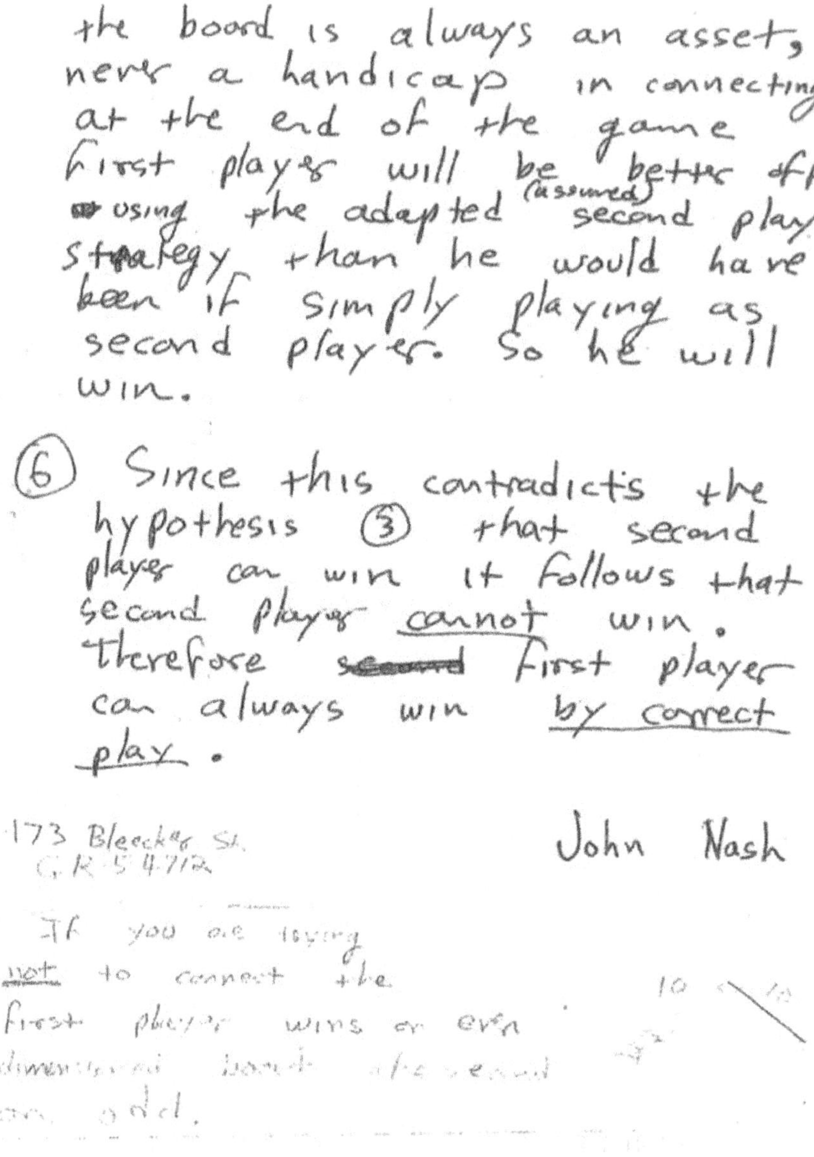

the board is always an asset,
never a handicap in connecting,
at the end of the game
first player will be (assumed) better off
using the adapted second player
strategy than he would have
been if simply playing as
second player. So he will
win.

(6) Since this contradicts the
hypothesis (3) that second
player can win it follows that
second player <u>cannot</u> win.
therefore ~~second~~ first player
can always win <u>by correct
play</u>.

173 Bleecker St.
GR 5 4712

John Nash

If you are trying
<u>not</u> to connect the
first player wins on even
dimensional board the second
on odd.

10 ⟍ 10

FIGURE 5.5: Nash's no-draw proof, second page.

(4) Consider a defensive strategy by [the] first player imitating the winning second player strategy assumed in (3). The first move could be arbitrary. If the strategy ever called for a play where the arbitrary move was made another one could be made.

(5) Since an extra piece on the board is always an asset, never a handicap in connecting, at the end of the game [the] first player will be better off using the adapted (assumed) second player strategy than he would have been if simply playing as second player. So he will win.

(6) Since this contradicts the hypothesis (3) that [the] second player can win it follows that second player cannot win. Therefore [the] first player can always win by correct play.

5.3 No draws

Nash was a genius, so it is not a surprise that his proof takes for granted properties that might not seem obvious to others.

In the first part of (4), Nash uses this property: for a given Hex position, if a player has a winning strategy, then the player also has a winning strategy if the position — but not the player to move next — is changed by putting one of that player's stones in an empty cell. As Nash then hints, it suffices to modify the original strategy. But what if the required arbitrary move cannot be made, because at that point all the cells are occupied? Can this occur? And if it does, does the player still win?

In (2), Nash makes this assumption: no game of Hex can end in a draw. Hein's explanation of this property, from the first Politiken column, is that wherever three cells meet it is impossible for each player to block the other, so globally — on the whole board — it is impossible for each player to block the other. Hein's argument holds for any board where, at each point where three or more cells meet, the number of cells that meet there is odd. But his argument, while valid, is not a rigorous mathematical proof.

The first such proof appears in 1961 in the book *Symbols, Signals and Noise* by John R. Pierce[6]. Pierce, then an electronics scientist at Bell Labs, was a colleague of Claude Shannon and had supervised the construction of the first transistor. Pierce's book was one of several in which he explained technology to a general audience. He coined the term 'transistor', and wrote science fiction under the pseudonym J. J. Coupling.

Here, we outline Pierce's proof, also given by Gale in the American Math Monthly in 1979[7]. The 1969 college textbook *Excursions into Mathematics* by Anatole Beck, Michael N. Bleicher and Donald W. Crowe[8] has another proof.

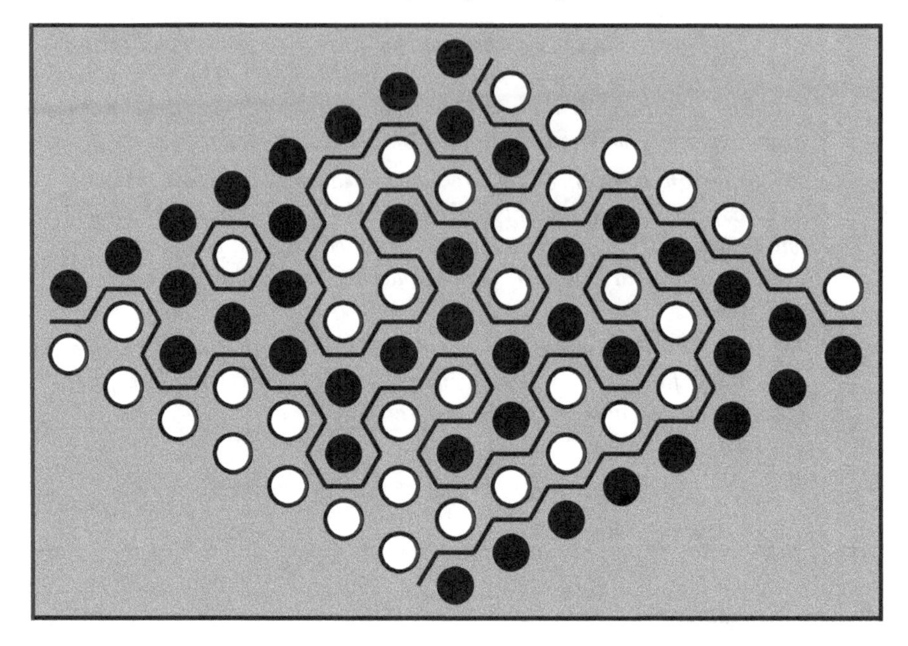

FIGURE 5.6: From Pierce's proof that Hex has no draws: put a segment between adjacent stones of different colours.

Consider a full Hex position: each of the upper left and lower right borders has a row of black stones, each of the other two borders has a row of white stones, and each interior cell has a stone. We saw in an earlier chapter that at most one player joins their two sides. Here we want to show that at least one player joins their two sides.

On the filled Hex board, draw a line segment along each cell edge that separates stones of different colour. See Figure 5.6. Start at the bottom corner, between the adjacent white and black stones, and form a continuous path by following the separating segments. For example, in the diagram, after the first segment the path meets a white stone and so turns, keeping the black stone touching the first segment also touching the next segment.

Each segment ends where three stones with collectively two colours meet. Whenever one segment arrives at such a point, exactly one segment leaves. So the path terminates only when it exits the board, so at one of the board's four corners. The path starts with white on the left and black on the right: each turn preserves this property, so the path also ends this way. The path cannot end at the top corner, since the last segment would then have white on the right and black on the left. So the path ends either at the left corner or the right corner.

In the former case — shown in Figure 5.7 — the path that starts at the bottom ends at the left corner, so the chain of black stones immediately to the right of this path joins the two black borders and black wins. In the latter case, the chain of white stones immediately to the left of this path joins the two white borders and white wins. This completes the proof.

Mathematicians often seek to simplify proofs. Here is a modification of Pierce's proof that avoids any reference to orientation properties (right side, left side) of the plane.

As before, start with a full Hex board, with black stones on the upper left and lower right rows and white stones on the upper right and lower left rows. Now recolour as red all white stones that are either on the lower left row or that reach this row by a white path. Recolour also as red all black stones that are either on the lower right row or that reach this row by a black path. There are three cases.

In the first case, the upper right row is red, so there is a white path joining the lower left and upper right rows, so we are done.

In the second case, the upper left row is red, so there is a black path joining the lower right and upper left rows, so again we are done.

In the third case, the upper right row is still white and the upper left row still black. We claim that this case cannot occur.

Start from the top corner, between the adjacent black and white stones, and follow a path of segments that separate black and white stones. This path does not stop on the edge of the board: the bottom rows are red, so the only place on the edge with a black stone touching a white stone is at the top corner, and the path cannot double back on itself. So the path stops where three stones with colours black, white, red all meet. But this is not possible: if the red cell was originally black (respectively white) then the black (resp. white) cell should be red. So the third case cannot occur. This ends the proof.

That three cells coloured black, white, red must meet in the third case above follows directly — without needing the continuous path argument — from a property known as Sperner's Simplex Lemma. In fact, the statement that a full Hex board has either a black path joining the black sides or a white path joining the white sides is equivalent to the two-dimensional case of this lemma. In his 1979 paper Gale explains that this Hex property is also equivalent to a discrete version of another famous property, namely Brouwer's Fixed Point Theorem. This theorem is familiar to Nash, who uses it in his Ph.D. thesis on non-cooperative games, for which he receives the Nobel Memorial Prize in Economics in 1996. So it follows that Sperner's Simplex Lemma for two dimensions and a discrete version of Brouwer's Fixed Point Theorem are equivalent[9].

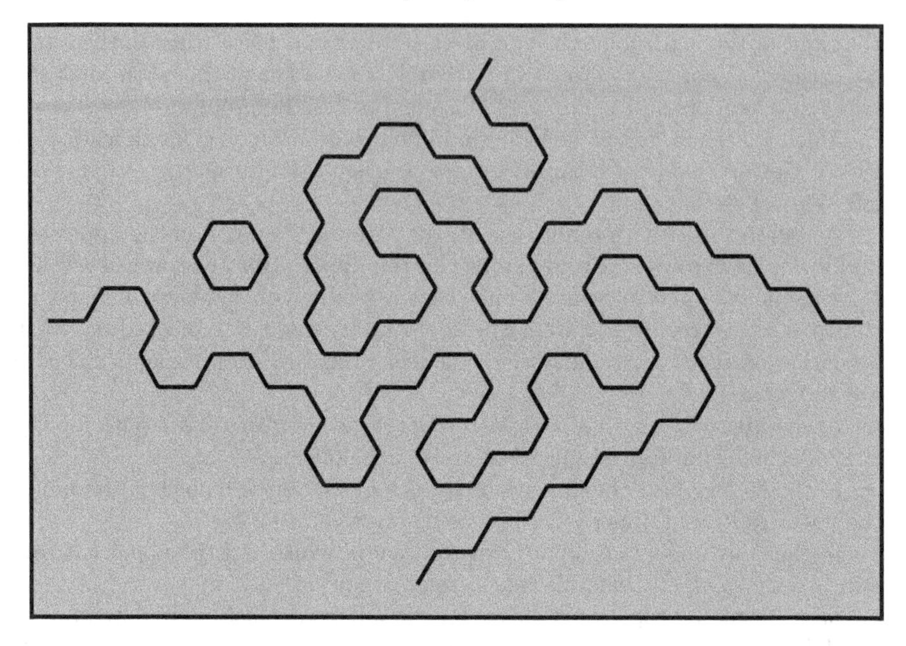

FIGURE 5.7: From Pierce's proof. The path that starts at the bottom must end at the right or left. Here, it ends at the left.

5.4 Longer side wins

The proof that the first player wins Hex assumes that the second player can steal the first player's strategy. This requires that the board looks the same — or symmetric — to the second player as it does to the first player. This happens if the board is regular, namely has the same number of rows as columns.

But on irregular boards, whoever owns the longer sides — and so has the shorter distance to cross the board — can win even when playing second, by following this pairing strategy, apparently due to Nash:

Partition an $n \times (n+1)$ board into two triangular parts each with n cells on each side. Label the parts in mirror fashion, as shown in Figure 5.8. If playing first, play anywhere; if not, wait. Then, in response to each opponent move, reply in the cell with the same label on the other part of the board.

Again, Nash's argument omits some details: here, how does this strategy ensure that longer-side-player wins? We leave the answer as an exercise for the reader. Hint: try to construct a winning path for the longer-side player that does not repeat a letter. Or, more simply, use the five princes problem from Chapter 1.

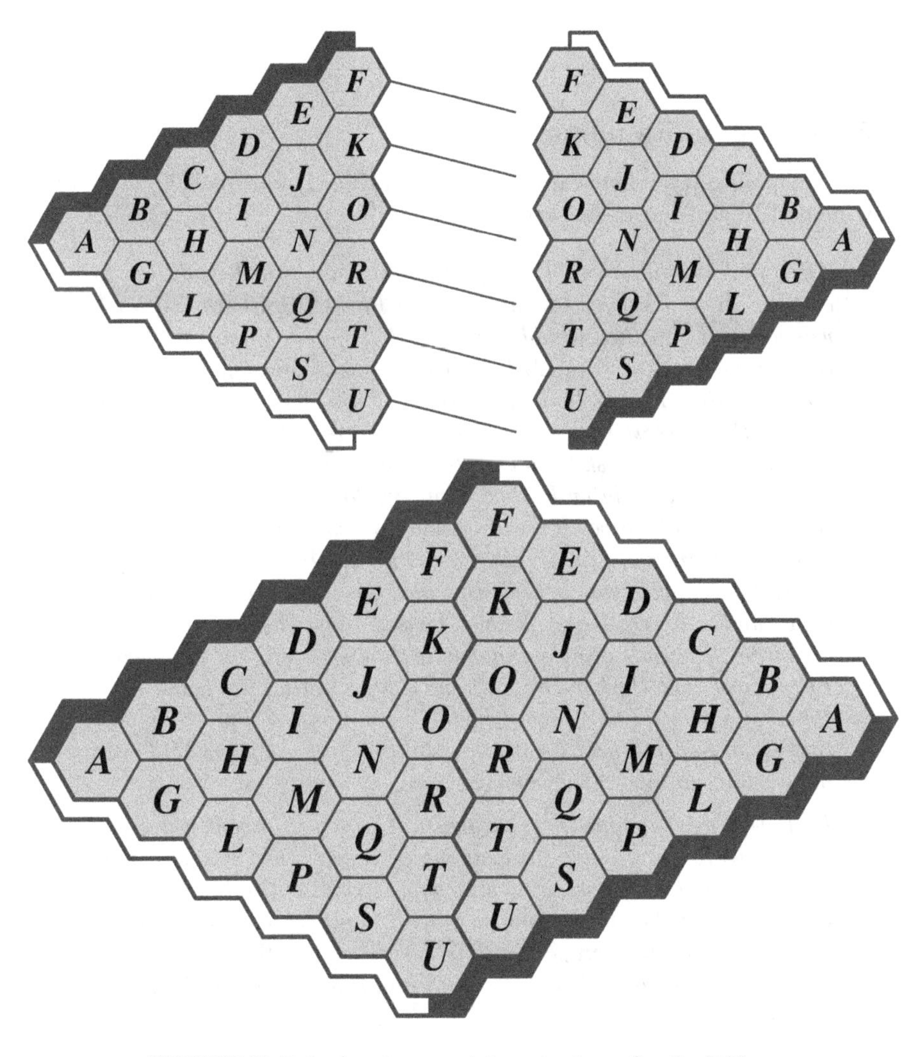

FIGURE 5.8: A mirror pairing strategy for 6×7 Hex.

The reader might have noticed that there is another way to partition the $n \times (n+1)$ board into two symmetric parts, as shown in Figure 5.9. This also gives a winning pairing strategy for the second player. Again, we leave a proof that this strategy wins as an exercise.

5.5 Hex gets its name

In an email to Ryan Hayward and Jack van Rijswijck in 2000, Gale recalled how he and Nash tried to market their game:[10]

> *After making the board, I thought maybe Nash and I could try to cash in on it. I suggested sending it to Parker Brothers but Nash was worried that they would steal it. I did show it to various game companies in New York, Milton-Bradley, was one, but they were not interested. I think I may have sent a description of the game to Parker Brothers and I seem to recall they were not interested. They said games of pure skill don't sell. People prefer dice or spinner games. So n years later Parker Brothers does come out with Hex. It was too small to be interesting, maybe 8x8. By that time I was at Brown University and Nash was at MIT. One morning, maybe 1958, I get a call from Nash who had seen the Parker Brothers game and believed I had double crossed him, cashing in on his invention. He was not aware that Piet Hein scooped him by 7 years. I explained that Parker Brothers must have gotten the game from Hein, and I think he believed me. (It was not too long after that that Nash had his first breakdown, he disappeared from MIT and for a while no one knew where he was.)*
>
> *I believe Nash, a game theorist, was not trying to invent a popular game, but rather to give an example of a game which was a first player win but whose winning stategy was (and still is) not known. (I later invented a simpler game, called Chomp by Martin Gardner, which also has this property. It is described in "Winning Ways" which I'm sure you are familiar with and also in one of Gardner's columns.)*

In fact, Parker Brothers had copyrighted 11×11 Hex in 1950. In the November 13 and December 4 issues of Life magazine, an advertisement for their famous games — Monopoly, Clue, Sorry, Make-A-Million — also mentions in small font *Hex the Zig Zag game of block and counter-block \$2.00.* In a 1957 letter to Gardner, Parker Brothers Development Manager Le Roy Howard explained:

FIGURE 5.9: Another mirror pairing strategy for 6×7 Hex.

FIGURE 5.10: Parker Brothers 1950 copyright design for Hex.

> *This game was originated by Mr. Piet Hein ... We published HEX some years ago, but the sale has not been up to expectations.*

Executive Vice President Edward P. Parker elaborated:

> *Like yourself, I personally had a high regard for the game HEX. I played it a great deal and found that it involved the use of considerable skill and judgment. The general public, however, apparently missed the whole idea of the game and its sale was very disappointing; so much so that we could not continue it.*

So, for a short time around 1950, Hex was commercially available, but sold poorly. It was played a bit in the academic world, but not well known. That would change in 1957.

Notes

[1]December 1999 phone conversation, Nash and Hayward and Jack van Rijswijck.
[2]1995.09.20 Nasar-Gale interview [54].
[3]1957.04.10 letter Gale-Gardner [23], 2000.02.08 email Gale-Hayward/Van Rijswijck.
[4]1957.03.23 letter Hein-Gardner [33].
[5]Undated 1957 letter Nash-Gardner [23].
[6][61]
[7][20]
[8][5]
[9][70]
[10]2000.02.08 email Gale-Hayward/Van Rijswijck.

Chapter 6

Games and machines

I believe inspiration is contagious.

Amy Purdy

6.1 Contagion

Graduate students in Princeton were pushing the frontiers of many branches of mathematics, including topology and what would later be called artificial intelligence. The appearance of Gale's board sparked a creative flurry that yielded other topological games and some game-playing machines. Of Hex and Fine Hall, Nash commented[1]

> *[Hex] became moderately popular, it became one of the games that was played in the common room and the board was around there. It was not as popular as go or chess, but it was played occasionally. I was really the strongest player in [Hex], but not in the other games.*

The Fine Hall Hex players — including Nash, Gale, John Milnor, John Tate, Robert O. Winder — are today well known for their mathematics. Winder had a successful career at RCA Laboratories. Nash won the Nobel Memorial Prize and Abel Prize. Gale won the John von Neumann Theory Prize. Tate won the Wolf Prize and Abel Prize, and Milnor won the Fields Medal, Wolf Prize and Abel Prize.

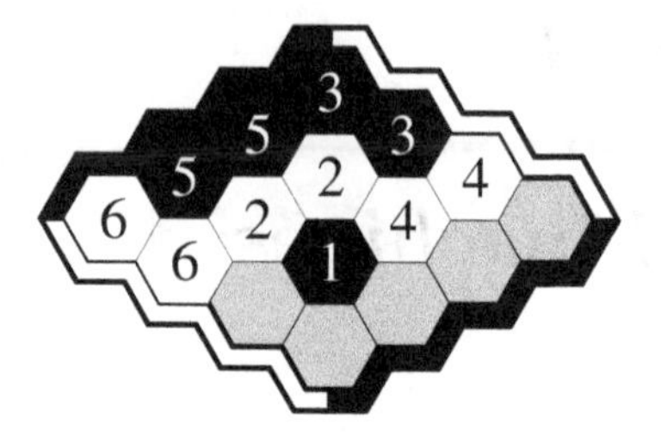

FIGURE 6.1: 1-2-2 Hex: the 2nd-player wins this game. Can you find a winning 1st-player strategy? Move 1 does not win.

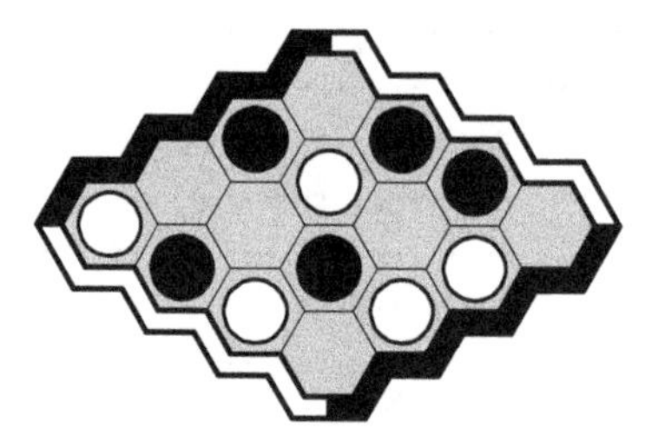

FIGURE 6.2: A Rex puzzle by Evans: White to play, who wins?

Claude Shannon — mathematician, electrical engineer, and a bit of a jokester — also knew this group. In 1940, he was a member of the Institute for Advanced Study in Princeton, a 2-mile walk from Fine Hall. He then worked at Bell Telephone Laboratories in Murray Hill, 40 miles away, until joining the MIT faculty in 1956.

6.2 1-2-2 Hex

The Fine Hall group created several Hex variations. Tate suggested 1-2-2 Hex: the first player places one stone; thereafter, on a turn, a player places two stones — or one, if only one empty cell remains.[2] See Figure 6.1. Tate's alternation rule mitigates the first-player advantage: on the 2×2 board the second player wins; on the 3×3 board the first player wins, but has only one winning first move; on the 4×4 board, the first player wins, but with only two symmetric winning first moves. Can you find these moves? We give a solution at the end of the chapter.

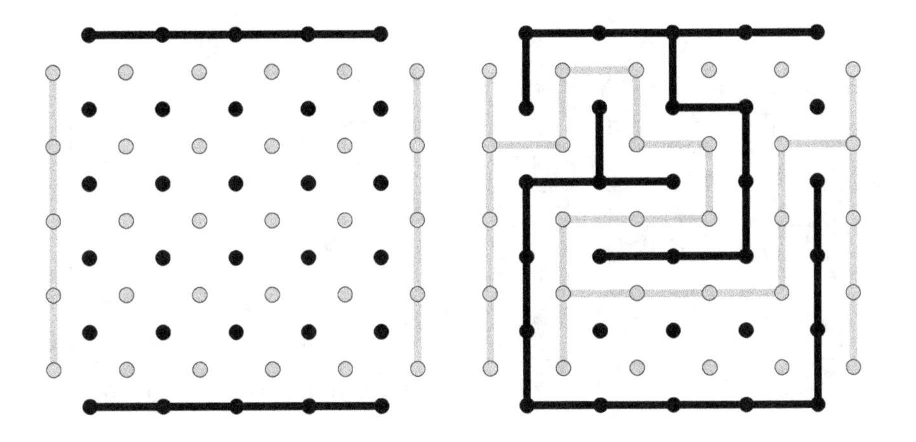

FIGURE 6.3: Gale's game, now called Bridg-it: (left) an empty board, (right) White wins.

6.3 Rex

Another variation is Reverse Hex, or Rex: whoever joins their two sides *loses*. For the $n \times n$ board, Robert Winder showed by strategy stealing that the first (resp. second) player wins Rex when n is even (odd)[3]. Figure 6.2 shows a Rex puzzle composed in 1974 by Ronald J. Evans while an instructor at the University of Wisconsin-Madison.[4]

6.4 Gale's game, or Bridg-it

Shortly after building the Fine Hall Hex board, Gale invented the game now called Bridg-it.[5] See Figure 6.3. One player has black borders and links, the other has white borders and links. On alternate turns, a player joins two of their pegs with a link. To win, a player must join her or his two borders.

In a 1957 letter to Gardner, Hein refers to a version of Hex in which the board is covered with octagons and squares instead of hexagons.[6] as in Figure 6.4 Later, after Gardner describes Gale's game to him, Hein explains how Bridg-it can be played on such a board:[7]

> *The game Prof. Gale proposes is really topologically an instance of Hex with one of the other [board] patterns with 3 polygons meeting in the corners which I mentioned to you, in [fact] the pattern of oc-*

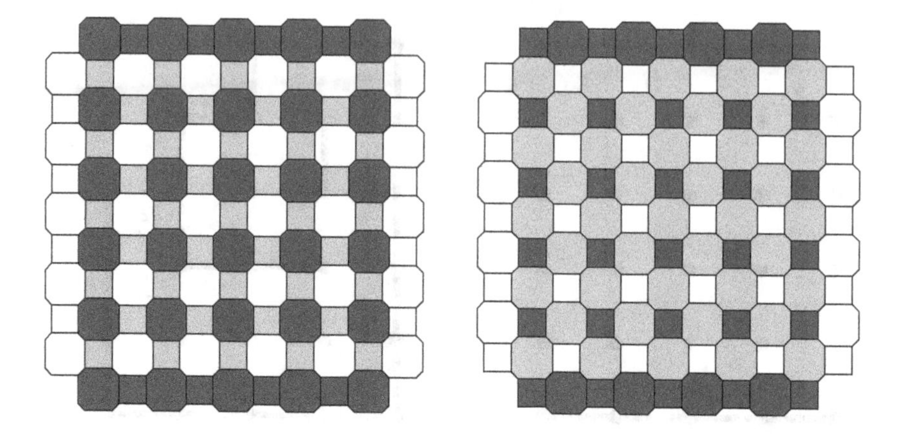

FIGURE 6.4: (left) Hein's formulation of the empty Bridg-it board. (right) An equivalent formulation. For each, play on the grey cells.

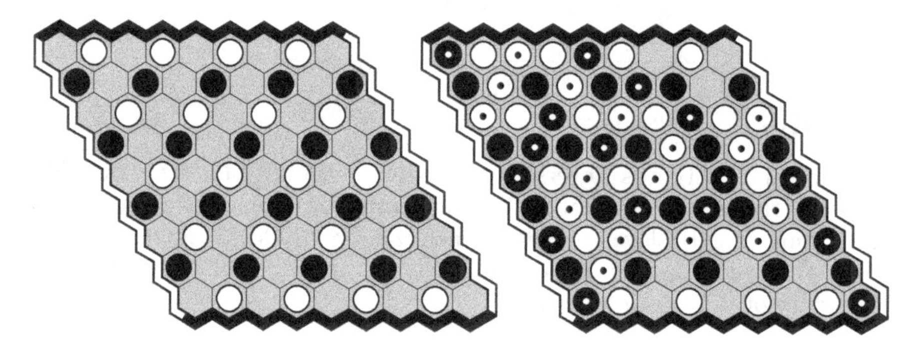

FIGURE 6.5: Playing Bridg-it on a Hex board: (left) the empty board (right) the finished game from Figure 6.3.

> *tagons and squares, where the octagons are given a priori to Black and Red(White) alternately in a chess pattern and the squares are left to choose among — as you'll easily see.*

We have not seen this noted elsewhere, so we mention it here: Bridg-it is also a special case of Hex as played on the usual Hex board. As shown in Figure 6.5, any $m \times n$ Bridg-it position corresponds to a $2m-1 \times 2n-1$ Hex position. To simulate the effect of the Bridg-it pegs, add stones to the empty Hex board as shown. Then, whenever a link is played in the Bridg-it game, add the corresponding Hex stone to the Hex position. Each Bridg-it position and corresponding Hex position are topologically equivalent.

It follows that a Bridg-it game cannot end in a draw: if a Bridg-it position has no empty cells, the corresponding Hex position has no empty cells, so one

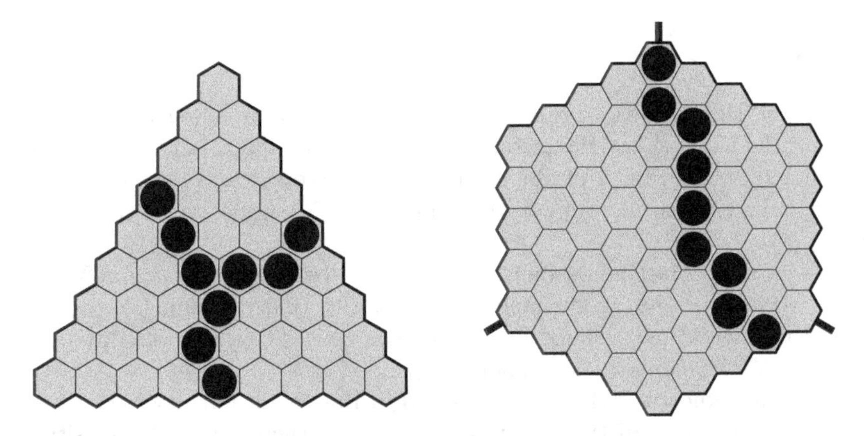

FIGURE 6.6: (left) A winning set on the usual-shape Y board. (right) A winning set on an hexagonal-shape Y board: the short lines mark the three sides' extremities.

player wins the Hex position, and the same player wins the corresponding Bridg-it position.

Also, the empty Bridg-it board looks the same for each player, and adding an extra link to a position is never disadvantageous for the player with that link. So the strategy-stealing argument holds, and in Bridg-it — as in Hex — the first player has a winning strategy.

Gale later invented the game of Chomp, which can be played on a chocolate bar. The board is a rectangular bar composed of squares. On a turn, a player breaks off any rectangular piece that includes a bottom-right square, i.e. with no square below it and no square to the right of it. The loser is whoever takes the piece that includes the top-left square.

On the 1x1 board the first player loses, but on any larger board the first player wins: there is a proof using strategy stealing. Hint: argue by contradiction, and consider what happens if the first move takes only the bottom-right square.

Gale's interest in games continued. From 1991 through 1997, Gale wrote the Mathematical Entertainments column for the Mathematical Intelligencer. His work on mathematical economics includes — jointly with Lloyd Shapley, who won a Nobel Memorial Prize in Economics — a solution to the stable marriage problem.

6.5 Y

Another Hex variation, called Triangle by Milnor, is now called Y. The rules are the same as for Hex, except that the board shape is triangle rather than diamond. To win, a player must join all three sides. As shown in Figure 6.6, a winning stone configuration can have a 'Y' shape, hence the game's name.

In a letter to Martin Gardner in 1957, Gale described a version of Y that can be played on a hexagon-shaped board. Notice that a winning Y configuration can have a linear shape if it includes a corner cell. Gale credited this version of Y to Shannon and Milnor independently.[8]

We do not know whether Y was played in Fine Hall, but it was certainly played in Ann Arbor. In 1953 Charles Titus, an instructor at the University of Michigan, introduced Hex to fellow researcher Craige Schensted. Fascinated by the notion of a connection game, and independently of the Fine Hall group, they invented other games, including (triangular) Y and a game similar to Bridg-it.[9] They also created the tile-arranging game Kaliko, originally called Psyche-Paths. See Figure 6.7. With production assistance from Stepen Titus, Kaliko — still available from Kadon Games[10] — made the cover of Games Magazine in June 2000.

Y is closely related to Hex, as shown by three properties:

1) As shown in Figure 6.8, Y generalizes Hex. For any Hex game starting from any Hex position, there is a corresponding Y game starting from a particular position, such that the two games have isomorphic game trees (so, from corresponding positions, the set of available moves is in 1-1 correspondence, and corresponding terminal positions have the same winner). To find the Y position corresponding to a given Hex position, start with the empty Y board, add black and white stones as shown in the figure, and then add stones in the same position as on the Hex board.

Y is strictly more general than Hex: there are Y games that do not correspond to any Hex game. The empty Y board with 2 cells on each side gives such a game. We leave proving this as an exercise for the reader. Hint: show that there is no Hex position with exactly 3 empty cells, where each player wins by occupying any 2 of the 3.

2) Y cannot end in a draw. Here is a proof due to Craige Schensted, using what he calls a *median algebra*. We want to show that a Y position with all cells filled has a winning configuration for exactly one player.[11]

Consider P_k, a Y position with k cells on each side. If $k = 1$ then the stone there touches all three sides and we are done.

If $k \geq 2$, perform 'Y-reduction' as shown in Figures 6.9 and 6.10: replace P_k with a position P_{k-1} on a board with sides one shorter than of P_k, as follows. First, orient P_k and label its cells as shown in Figure 6.9.

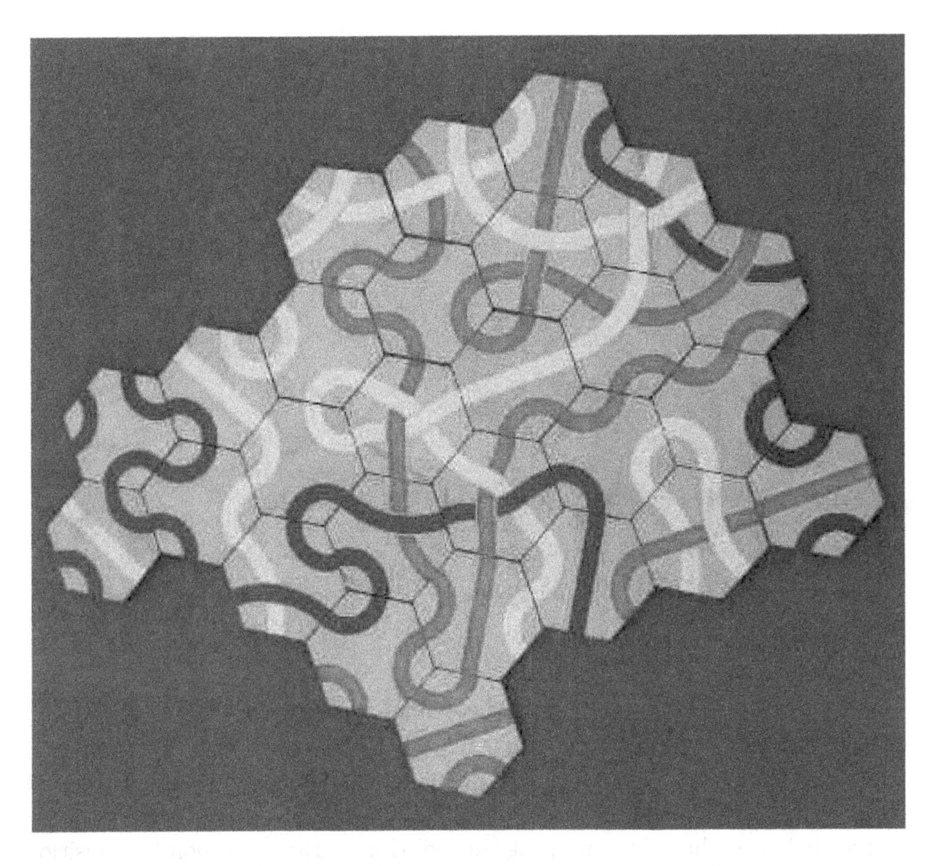

FIGURE 6.7: Kaliko, a tile-arranging game. Photo © Kadon Games.

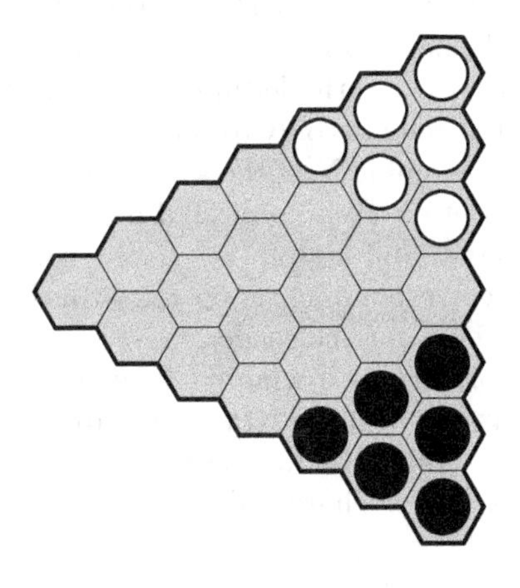

FIGURE 6.8: Y generalizes Hex.

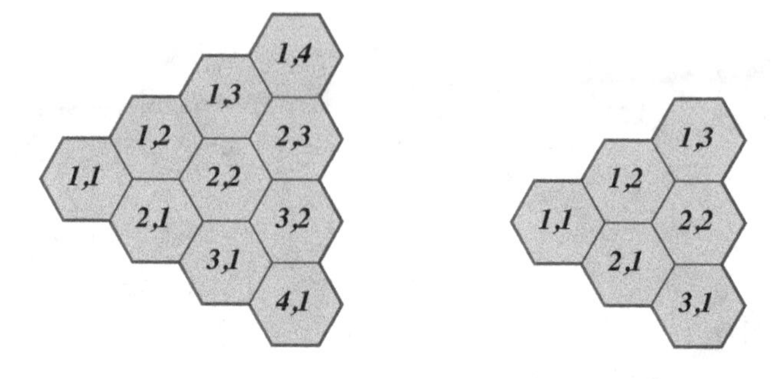

FIGURE 6.9: Y reduction: left triple (1,1 1,2 2,1) maps to right cell 1,1, etc.

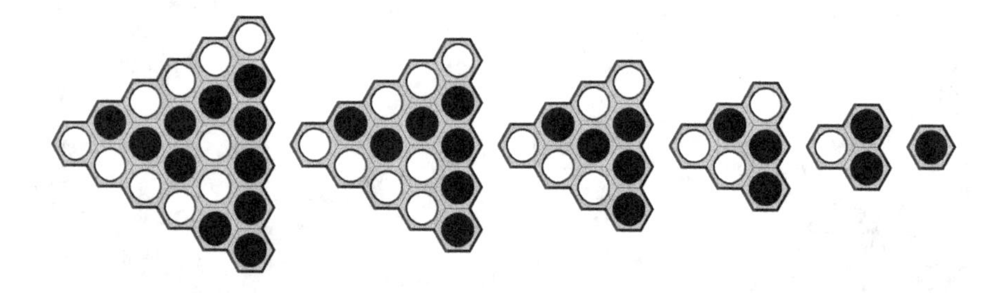

FIGURE 6.10: Y reduction: the winner is the same for each board.

Next, for each triple of pairwise touching cells of P_k oriented the same as P_k — so, with one cell on the left and two cells on the right — put a stone whose colour is that of the majority in the triple in the corresponding position of P_{k-1}. So, each triangle $\{(x, y), (x+1, y), (x, y+1)\}$ of P_{k-1} corresponds to cell (x, y) of P_k.

Figure 6.10 shows P_6 reduced to P_5 and then P_4 and so on down to P_1, the leftmost triple of P_6 has 2 white stones and 1 black, so majority white, so the corresponding cell in P_5 — the leftmost cell — is white.

To finish the proof, it remains only to show that there is a winning Y on P_k if and only if there is a winning Y of the same colour on P_{k-1}. There are two things to prove: a cell set on P_k is connected if and only if the corresponding set on P_{k-1} is connected, and a stone set on P_k touches a particular board side if and only if the corresponding set on P_{k-1} touches the same board side. We leave these as exercises for the reader.

3) In Y, the first player has a winning strategy. This follows by strategy stealing, since there are no draws, extra stones are not disadvantageous, and the starting position is symmetric for the two players.

To Schensted, Hex and Y share a flaw that motivates the design of new boards:[12]

> *A difficulty with [the Y and Hex boards] is that plays in the center of the board are very strong while plays in the acute-angled corners*

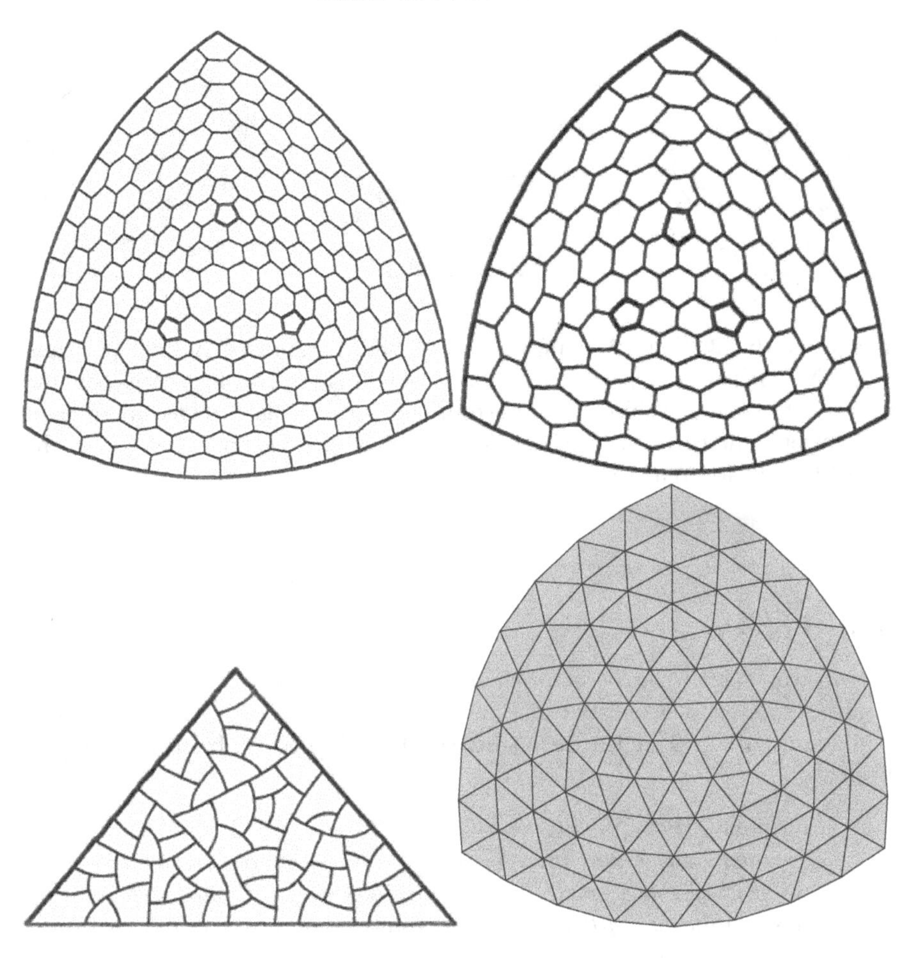

FIGURE 6.11: Mudcrack Y boards, (top) from 1972-4 Ann Arbor tournaments, (bottom right) in dual form. Drawings © Craige Schensted.

> *are very weak. Thus the first player can obtain an overwhelmingly strong position by playing [first] in the center. In 1969 we came to realize the importance of what might be called the Mudcrack Principle, namely that great freedom is possible in the design of the boards.*

Mudcrack Y has a 3-sided board and the same rules as Y, but cells can have different shapes as in Figure 6.11. So the board — any planar drawing — can look like a patch of dry cracked mud.

Robert Hochberg, Colin McDiarmid and Michael Saks used Sperner's Simplex Lemma to prove a theorem implying that Mudcrack Y has no draws,[13] so strategy stealing implies that the first player wins.

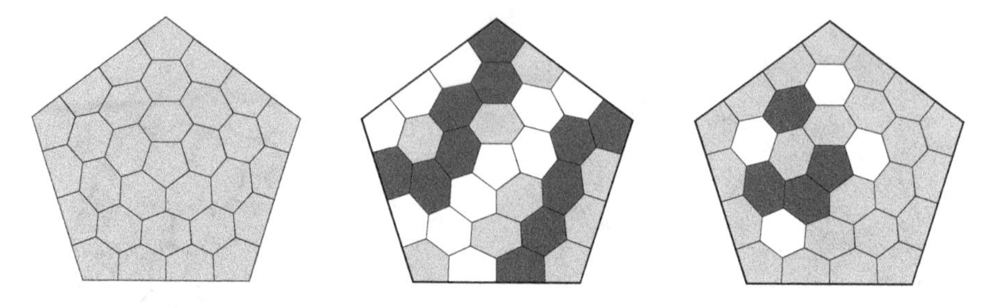

FIGURE 6.12: (left) An empty Poly-Y board. (middle) A final position. Black Y touches left, upper-left and upper-right sides, so Black owns top and upper-left corners; White owns bottom-left corner; Black owns upper-right and lower-right corners: Black wins 4-1. (right) Puzzle. Black to play, who wins? Repeat for White.

6.6 Poly-Y

Motivated by Irene Verona Schensted — a physicist, Y fan, Go fan and Craige's spouse — Schensted and Titus sought a game with the complexity and nuance of Go. This led them to create Poly-Y in 1970. Irene's review was positive:[14]

> *I am now a Poly-Y-phile as well as a Go-phile. ... Poly-Y seems ... subtler and deeper [than Y]* **without** *being more complex tactically. Poly-Y has much the same 'feel' as Go, with many related but simultaneous battles raging separately across the board.*

Poly-Y is played on any mudcrack board with an odd number — at least five — of sides. A player owns a corner if they have a Y — contiguous cell set touching at least three sides – that touches the two sides of that corner. The winner is whoever owns more corners, as shown in Figure 6.12.

Using the Mudcrack Y no-draw property, Schensted and Titus showed that a completely filled Poly-Y board has each corner owned by exactly one player.[15] So Poly-Y has no draws, and the first player has a winning strategy.

The Schensteds founded a game company, Neo Press, that in 1975 published Titus and Schensted's *Mudcrack Y and Poly-Y*. This beautiful book – designed, lettered, and drawn by Schensted — is still available from Kadon Games at `www.gamepuzzles.com`. Schensted later changed his name to Ea Ea[16] and created Poly-Y successors Star and *Star, also still available from Kadon.

6.7 Switches

After the war, technology shifted from special purpose machines to general purpose computers, and scientists starting designing machines that could think. Shannon, known for his work on information theory, also worked on computational intelligence. In 1949 he built an electric chess computer, and in 1950 he wrote *Programming a Computer for Playing Chess*, which includes a search algorithm for any two-person game. In 1956, John McCarthy — who together with Marvin Minsky had worked as a graduate student under Shannon's supervision in the summer of 1952 — introduced the term "artificial intelligence" to describe this research field.[17]

An electrical switch "makes" (completes) or "breaks" (disconnects) a circuit. Shannon's 1937 MSc thesis — *A Symbolic Analysis of Relay and Switching Circuits* — laid the foundation for the use of Boolean algebra in computer design.[18] It was presumably Hex that inspired the invention of his eponymous switching game.

The Shannon switching game is played on a graph with two special nodes. One player is Cut, the other Short. On her turn, Cut breaks (or erases) any unmarked link. On his turn, Short marks any unbroken link. The game ends either when the two special nodes are joined by a path of marked links — so Short wins — or the special nodes have been separated into different components — so Cut wins.

Shannon's Birdcage game is the Shannon switching game played on the graph in Figure 6.13 (left) with the two special nodes at top and bottom. We have seen Birdcage before: it is topologically identical to Bridg-it. The Birdcage position in the figure corresponds to the earlier Bridg-it position.

In 1951 Shannon built the machine shown in Figure 6.14 to play Birdcage. The machine models the game as a flow process. Imagine trying to force electrons across a circuit. As the opponent blocks you by cutting connections, the remaining links will have more current flowing through them. Shannon's machine has a resistor switch for each graph link. A resistor opens (breaks) or is shorted (connects) when respectively either Cut or Short plays at the corresponding link. To move, the machine picks the resistor through which the most current flows. In 1958, not knowing of Shannon's machine, engineers W.A. Davidson and V.C. Lafferty at the Illinois Institute of Technology built a similar machine.

Shannon's machine plays well against humans. After hundreds of games, when given the first move the machine lost only twice, perhaps due to circuit failure or operator error. When given the second move the machine usually lost, unless the human made a weak move.[19] As part of his 2011 doctoral studies, Thomas Fischer showed a clever opponent can lead Shannon's machine into a position where, after starting first, the machine eventually makes a wrong move and loses.[20]

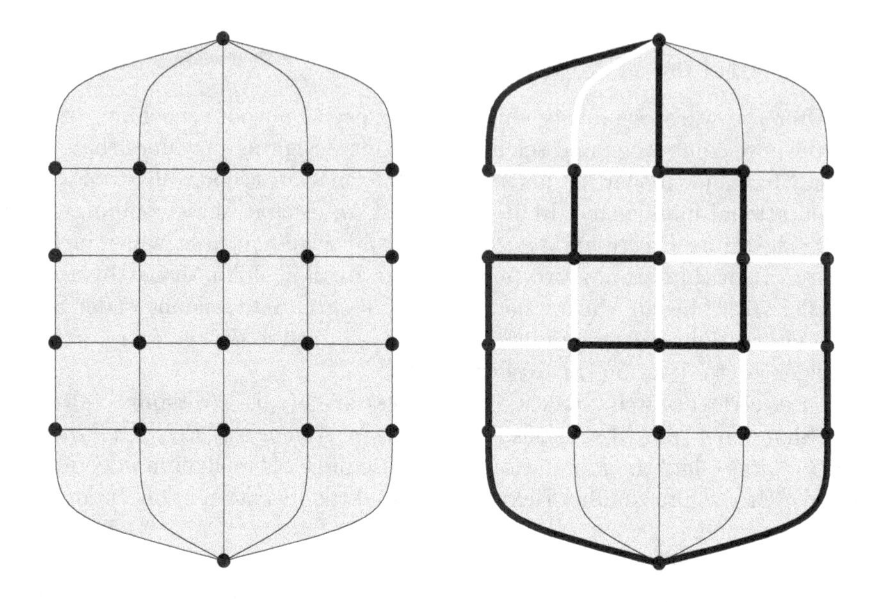

FIGURE 6.13: (left) Birdcage is the Shannon switching game on this graph. (right) A finished game: Cut (White) wins.

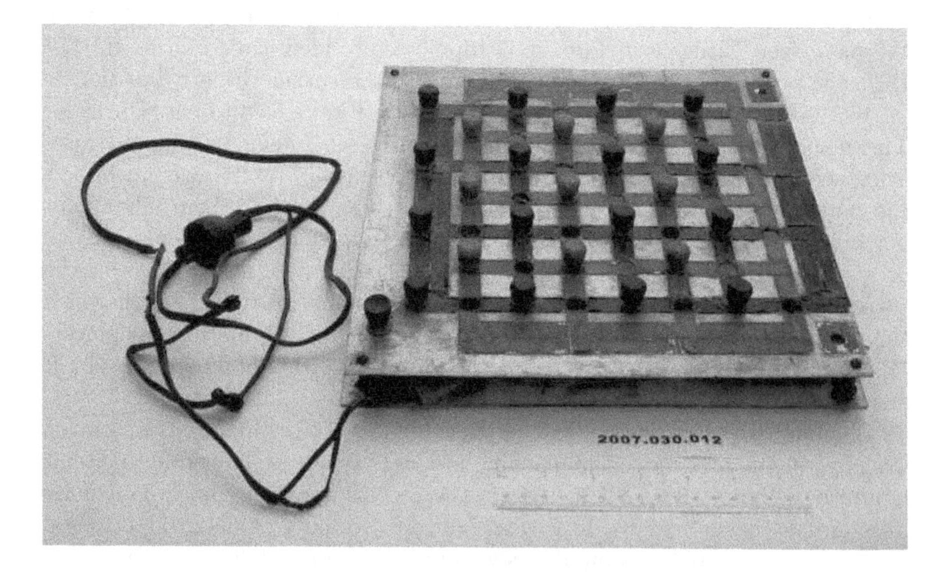

FIGURE 6.14: Shannon's Birdcage machine. © MIT Museum.

Prior to 1953, using a different electrical model, Shannon and E.F. Moore built a Hex-playing machine:[21]

> *A two-dimensional potential field is set up corresponding to the playing board, with white pieces as positive charges and black pieces as negative charges. The top and bottom of the board are negative and the two sides positive. The move to be made corresponds to a certain specified saddle point in this field. ... With first move, the machine won about 70 per cent of its games against human opponents. It frequently surprised its designers by choosing odd-looking moves which, on analysis, proved sound. We normally think of computers as expert at long involved calculations and poor in generalized value judgments. Paradoxically, the positional judgment of this machine was good; its chief weakness was in end-game combinatorial play. It is also curious that [this] Hex-player reversed the usual computing procedure in that it solved a basically digital problem by an [analog] machine.*[22]

Shannon described his machines in a letter to Gardner:[23]

> *I did indeed build two [Hex machines]: the first, an honest machine built in collaboration with E. F. Moore as described briefly in my "Computers and Automata" paper and the second, playing on a 7 x 8 board.*
>
> *The [latter] machine played second by the symmetry strategy and had the short distance to go. The machine was built largely as a joke. It was especially [effective] against people who knew the ordinary form of Hex and knew that playing first they should, with proper play, be able to win. In most cases, people would play many games against the machine without realizing the inequality of the board, even when many broad hints were dropped.*
>
> *One feature of the construction was the use of thermistors to slow down the operation of the machine. It took from one to eight seconds for the machine to make its move. This gave the impression of the machine figuring out various properties. ... Not more than two or three people noticed the inequality without hints or suggestion.*
>
> *... I might add that I also built some other game playing machines for a game I invented, called "bird cage" or the "switching" game. This game has many of the same properties as Hex, but lends itself to easier mechanization.*

In 1952, after a conversation with Shannon, Nash wrote a Rand Corporation technical report in which he described Shannon's machines, noting that Shannon's group at Bell Labs had found a class of games based on self-dual

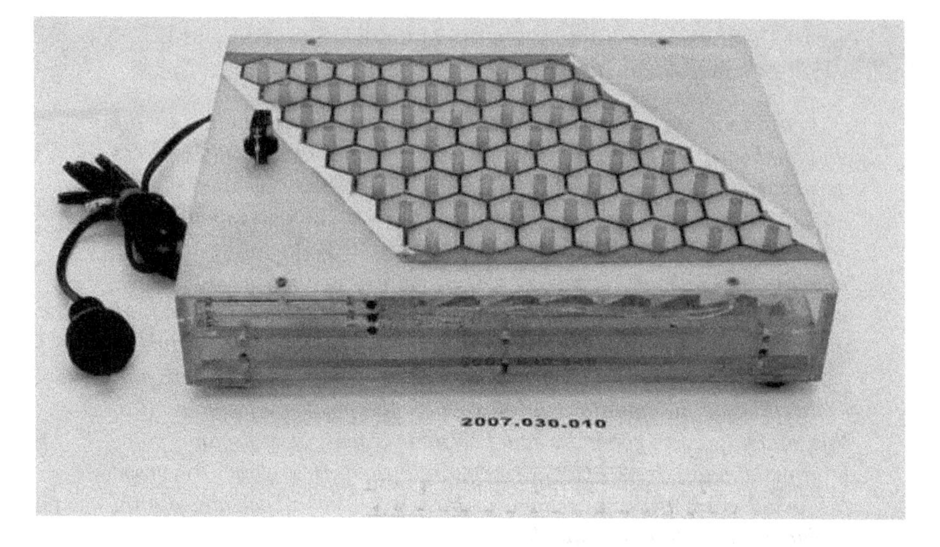

FIGURE 6.15: Shannon's unbeatable gag machine. © MIT Museum.

graphs.[24] Birdcage is such a game. Take the Birdcage graph, add a link between the two special nodes, and draw the dual graph: in Figure 6.16 the dual has white nodes and dashed links. Notice that — after adding the extra link — the Birdcage graph is isomorphic to its dual. Also notice that cutting (respectively shorting) a Birdcage graph link corresponds to shorting (respectively cutting) a Birdcage dual link. So again we see that Bridg-it — or Birdcage — is symmetric for the two players.

To close this section, we mention two generalizations of Shannon switching. One is Shannon node-switching, also called Shannon vertex-switching, a game played on a graph. The graph has two special nodes. The players are Cut and Short. On a turn, Short marks any unmarked node. On a turn, Cut deletes any unmarked node. Short wins by joining the special nodes, Cut wins by separating them. Figure 6.17 shows how Hex is just Shannon node-switching played on the appropriate graph.

In their paper *Biased Positional Games*,[25] Vašek Chvátal and Paul Erdös coined the term *maker-breaker* for a more general game, played on a set of points, each initially uncoloured. Each winning subset is specified. On a turn, a player colours any uncoloured point. Maker wins by colouring all points of a winning subset, otherwise Breaker wins. This game generalizes Shannon node-switching and all Y versions we have mentioned. For example, Figure 6.18 shows 3×3 Hex as a maker-breaker game.[26]

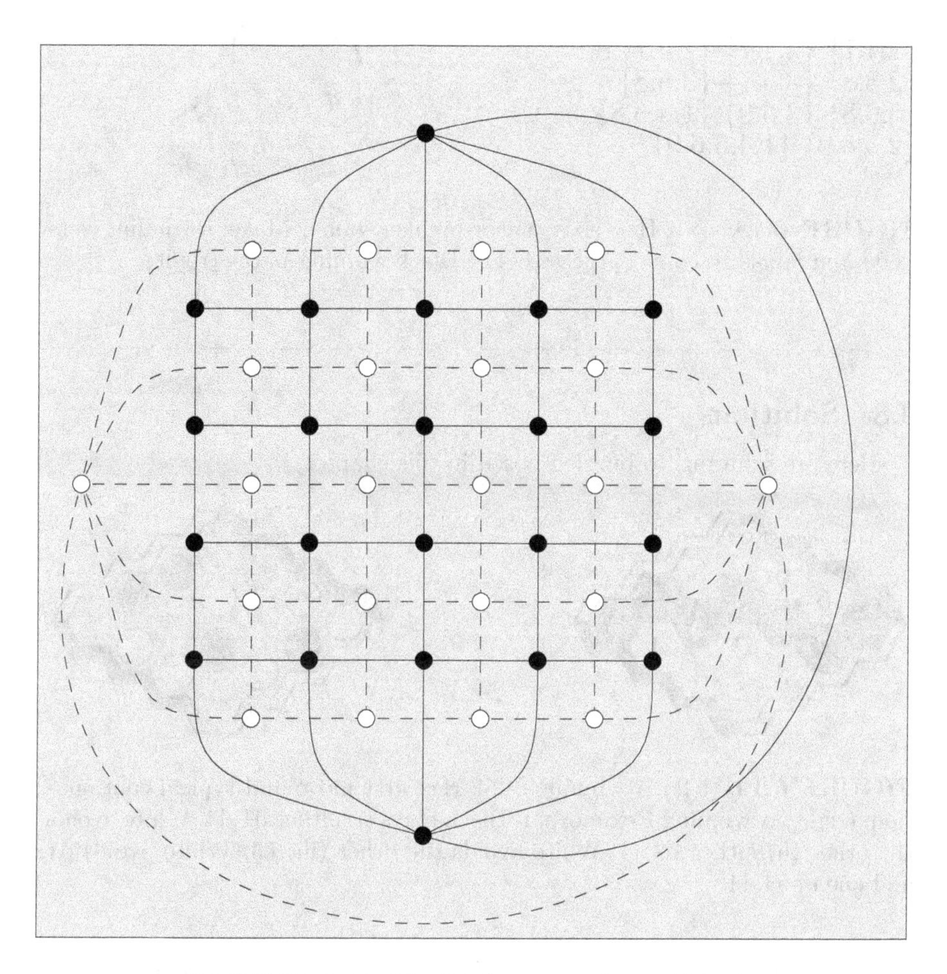

FIGURE 6.16: The Birdcage graph is self-dual.

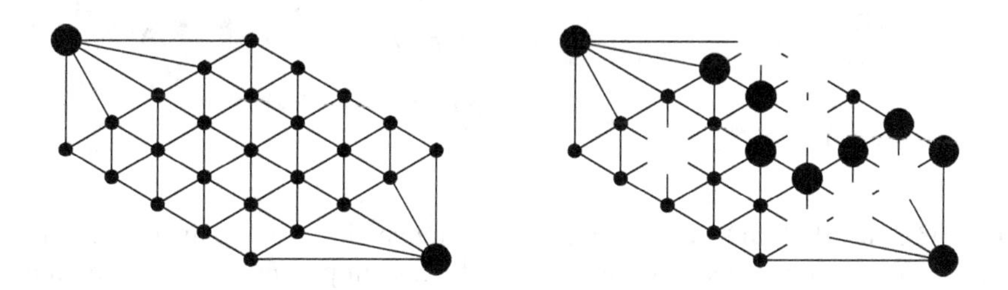

FIGURE 6.17: 5×5 Hex as Shannon node-switching. (left) A starting network, (right) Short (Black) wins.

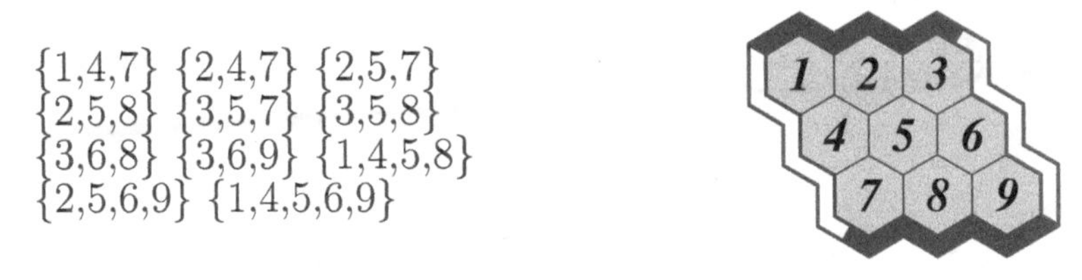

FIGURE 6.18: 3×3 Hex as a maker-breaker game: Maker's winning sets (left) and labels to convert these sets to Black winning paths (right).

6.8 Solutions

Here are solutions to puzzled posed in this chapter.

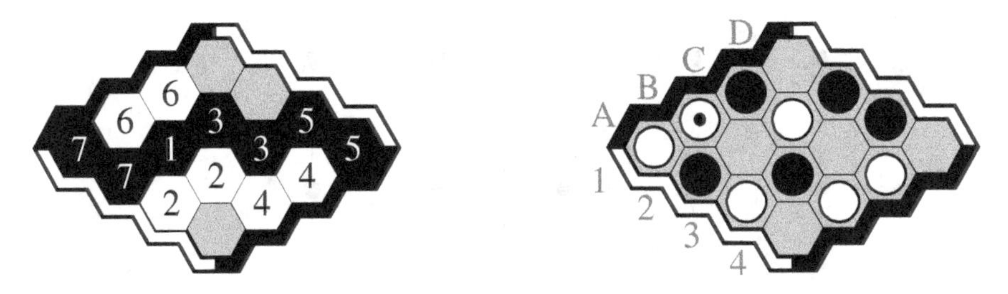

FIGURE 6.19: (left) A winning 1-2-2 Hex first move and typical continuation. (right) A winning Rex move. If Black plays (i) either d1,d4, White avoids the other (ii) either b2,a4, White avoids the other (iii) c3, White avoids d1 and one of a4,d4.

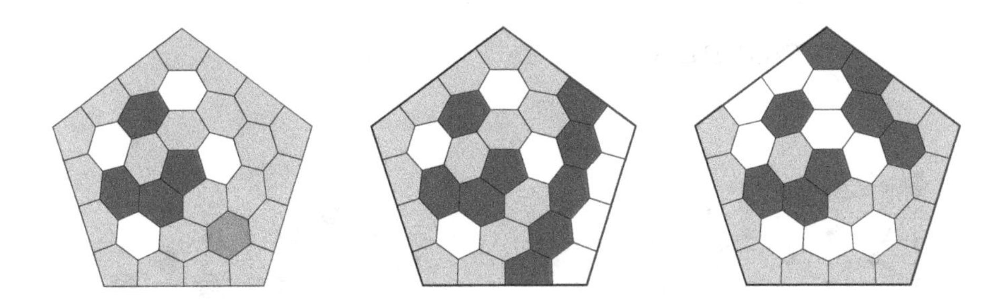

FIGURE 6.20: (left) A winning Poly-Y move, for each player. (middle) Assume Black plays next: now Black can attach to upper left, left and bottom, and also one of the remaining two sides, as shown. (right) Assume White plays next: from continuation, with Black to play, White can attach to each side except upper right and win.

Notes

[1]1999.12.03 phone conversation: Nash, Hayward and Jack van Rijswijck.

[2]1957.04.06 letter Gardner-Hein [23].

[3]1957.08.19 letter Winder-Gardner [23]. See if you can find a proof. There are two cases, depending on the parity of n: the proof is simpler when n is even.

[4][17, 26]

[5]1957.04.10 letter Gale-Gardner [23].

[6]1957.10.04 letter Hein-Gardner [33].

[7]1957.12.05 letter Hein-Gardner [33].

[8]1957.04.10 letter Gale-Gardner [23].

[9][64]

[10][43]

[11]American game designer Steven Meyers independently observed a related Y property. Take a Y position with all cells occupied, and consider the three smaller-board positions obtained by removing one of the three outermost rows: whoever wins the original position is the same as whoever wins at least two of the three smaller-board positions. Meyers invented the game BoxOff, featured in the August 2013 edition of GAMES magazine [53]. Cameron Browne and Frederic Maire analyzed BoxOff with Monte Carlo methods [11].

[12][64] pp 182-4.

[13]In Appendix G of Browne's *Connection Games* [10], Chris Hartman explains how the theorem of Hochberg et al. [41] implies no draws in Mudcrack Y.

[14]Footnote on the history of Mudcrack Y and Poly-Y [64] p 185.

[15][64] p 4.

[16]Ea Ea is named after the Sumerian god Enki.

[17]Minsky and McCarthy later founded MIT's Computer Science and Artificial Intelligence Laboratory.

[18][67]

[19][25], see chapter titled *Recreational Topology.*

[20][19], Figure 5.6 page 94.

[21]We found no images or further details of this machine. See next endnote.

[22]We found no images or further details of this machine. In a letter 1957.03.22 to Gardner, Shannon writes: *The actual hex machine that [Moore] and I built is at Murray Hill and you might be able to contact Moore there and learn of the details and see it in operation, although, it may be out of commission by now. Our only model was a crude red board layout.*[23]

[23]1958.09.18 letter Shannon-Gardner [23].

[24][56]

[25][14] Chvátal and Erdös worked on this paper mostly on a sightseeing bus ride at a conference in Qualicum Beach in Canada. As with his doctoral student Hayward, Chvátal learned Hex from Claude Berge.

[26]For the maker-breaker game, a natural question is whether the obvious greedy strategy — pick a move that hits the most remaining winsets — always wins. The answer is no: we leave finding counterexamples as an exercise. As we see in the next chapter, Hex (and so also the maker-breaker game) is computationally hard: if there is an efficient method to solve arbitrary Hex positions, then there is an efficient method to solve thousands of other problems that, so far, seem intractable.

Chapter 7

Hex goes global

It is something of an occasion these days when someone invents a game that is both new and interesting.

Martin Gardner

Fame is a vapor, popularity an accident.

Horace Greely

There is always more mystery.

Anais Nin

7.1 Mathematical games

By the mid-1950s Parker Brothers had stopped selling Hex and Gale and Nash had stopped trying to market it. In 1952 Nash wrote a report on the game for the RAND Corporation, but the article was classified secret and not publicly distributed.[1] Only a small number of academics had heard of the game. But that changed in 1957, thanks to the curious science writer Martin Gardner (1914-2010).

In 1956, having just finished writing the book *Mathematics, Magic and Mystery*, Gardner submitted an article on hexaflexagons — triangle-covered Möbius strips with curious folding properties — to the popular science periodical *Scientific American*. Editor Gerry Piel liked it and asked, is there enough similar material to make a monthly column? Gardner said yes, Piel agreed, and Gardner then scrambled to find out if there actually *was* enough material. Thus began Gardner's famous *Mathematical Games* column, which ran monthly for 25 years.[2]

We do not know how Gardner first heard of Hex, but by January 1957 he was tracking down everything he could on the game. He started with Parker Brothers, who replied promptly:[3]

FIGURE 7.1: Martin Gardner. Courtesy James Gardner.

> *This will acknowledge with thanks your letter of January 13th regarding the game of HEX.*
>
> *This game was originated by Mr. Piet Hein, who can be reached, we believe, through Mr. J. Luming at 667 Fifth Avenue, New York at the George [sic] Jensen Company, Inc.*
>
> *We published HEX some years ago, but the sale has not been up to expectations. Fortunately, we have a few copies left, one of which we are sending to you under separate cover - retail price was $2.00.*

On January 18 Gardner then wrote Hein, care of Politiken in Copenhagen:[4]

> *Since the December issue, I have been doing a monthly column on recreational mathematics for the magazine Scientific American. In some future issue I plan to discuss games based on topology, including your own interesting game Hex. Parker Brothers were kind enough to give me your address.*
>
> *I would like very much to have from you, if possible, some details about the historical background of your game – when you invented it, circumstances that led to its invention, and so on. Anything, in fact, that would be of interest to the readers of the magazine. Also, a bit of biographical data about yourself – your present work, any background you may have had in mathematics, and so on.*
>
> *I was sorry to hear that Parkers no longer were issuing the game. Perhaps my discussion of it will stimulate interest and lead to a reprinting!*
>
> *Do you know if any analyses of the game has [sic] appeared in any mathematical journal?*

This query is the start of a long and prolific correspondence between Gardner and Hein: in 1957 alone they exchanged 30 letters. In his *Scientific American* columns in 1958, 1969 and 1972 Gardner discussed Hein's Soma cube. In Chapter 1 we quoted from Hein's reply to Gardner's query. Keen to trace the origins of Hex, Gardner wrote Hein again in March:[5]

> *I owe you an apology for delaying so long my reply to your intensely interesting letter. My chief excuse is that I have been waiting for replies to other letters I have sent out regarding Hex, and hoped that I would have more information to pass on to you. ...*
>
> *What I would most like to know at this point is whether Hex is your own invention, or whether you came across it as an older game of unknown origin, and passed it on [to] the Parkers in the form in which they put it out. I could not tell from your letter which of these two possibilities was true. There are all kinds of myths floating about concerning Hex. One physicist told me that he had heard it had been invented by a group of mathematicians*

at Princeton, who played it on a bathroom floor and originally called the game "John".[6] *A Canadian mathematician wrote that he had heard a similar story, only it was at Harvard. And a third mathematician assures me that it is an ancient European game of unknown origin! What is the true account? ...*

Five days later Gardner wrote Shannon:[7]

By a startling coincidence, I ran into John Nash himself, in the NYU mathematics library. He's currently at NYU on leave of absence from MIT. I persuaded him to come to my apartment (a few blocks from his own place of residence) the other evening, and we spent several hours discussing the history and other aspect[s] of Hex.

Nash had a Sloan fellowship, a 3-year research grant that included one year without teaching. He was based at the Institute for Advanced Study, but living in New York City.[8] In April Gardner wrote Hein more about his meeting with Nash:[9]

In 1949 Aage Bohr, son of Niels Bohr, showed him a pencil and paper game that Bohr said was played in Denmark. Nash thinks (but of this he was not sure) that the game Bohr showed him was played on the intersection points of the following type board:

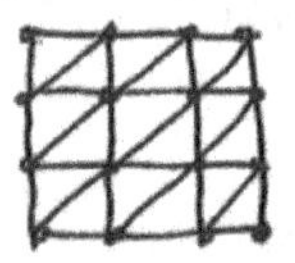

This would, of course, be topologically identical with the connectivity properties of a board of hexagonal cells. Nash says that he realized that the same game could be played on a board of hexagons.[10] *He explained this to a David Gale (now a prof. at Brown University) who constructed the first hexagonal board at Princeton. ...*

[Nash] says that in 1949 his chief interest in this game was that it provided such a neat illustration of the game for which you could prove a win for first player although the winning strategy is not known, and he claims that he himself was the first to devise this proof. It is identical with the proof you cited in your letter, except that it is made more rigorous and precise by breaking it down into a larger number of steps. ...

Nash gave me an analysis of a four by four board, proving that the first player wins only if he plays on the main diagonal; and an even simpler proof that he wins on a 5 by 5 board by playing in the center cell. ...

What clearly must have happened is this. Young [Aage] Bohr picked up the game from you, or from his father or someone at the Bohr Institute, but either did not know it came from you, or failed to mention this fact to Nash. ...

I never dreamed I would get involved in such complicated historical matters! Actually, I won't have too much space to go into the game's history, but what I do print I want to be as accurate as possible, as it may be the only documentation on the game's history for many years to come.

7.2 Hex history mystery

And so Gardner uncovered a history mystery: was Nash's invention of Hex truly independent of Hein's? The mystery deepened days later, when Gardner heard back from Gale:[11]

I did in fact receive your earlier letter, but time being what it is I never got around to answering. Here, however, is what I know on the subject. ...

Concerning the Princeton discovery, this occurred in the winter of 1948.[12] *John Nash, now Assistant Prof. of Math at M.I.T. originally conceived of the orthogonal checkerboard version you mentioned, to be played on a checkerboard. The hexagonal version was my idea and I constructed the first board. The game became quite popular with math and physics students at Princeton. The discovery by Nash and myself was completely independent and in fact without knowledge of the earlier Danish version. I tried for some time to sell it to various game companies, even had some preliminary correspondence with Parker Bros. No one wanted it. Games without a chance element wouldn't sell, everyone insisted. Nash and I were both amazed when the game came on the market. We investigated and got the information about Hein that you already know.*

On this letter, Gardner added an exclamation mark in the margin beside ... *a checkerboard.* Is this in response to the discrepancy between Gale's version of events and Nash's version: was it Nash or Gale who observed that this checkerboard game could be played on a hexagonal grid?

Meanwhile, Hein replied to Gardner:[13]

... I immediately called up Aage Bohr who is a friend of mine. He remembered having shown the game to people in the U.S. including,

he thought, John Nash, but he didn't remember whether in the triangular form or in the hexagonal.

I launched the game in a lecture I gave in the students union at the Atomic Institute here in Copenhagen (Niels Bohr's Institute) in the end of 1942.

It was usually played on the hexagonal pattern – of which they have a "mosaic" there of tiles like a bathroom floor, made by me for that purpose, and of which everybody there and at The Mathematical Institute, The Polytechnical Institute a.s.o. always had their pockets full.

...

Neither [Aage] nor I remember to have used the triangular pattern, but of course it is easier to draw than the hexagonal, and of course they are topologically equivalent ...

On April 18 Gardner replied to Hein, thanking him for the photocopy he had sent. Throughout his career, Gardner was respected for his healthy scepticism and as a stickler for evidence.[14]

I am grateful to you for the photostat, which arrived yesterday, because it puts into my hands a documentary proof of the date your game was given to the public, and also because I am pleased to have it for my collection of mathematical recreational material. As you will see, I have listed this article in the bibliography.

I think all the required data is now clear, and I'll just sidestep the whole Princeton business by referring to it briefly without saying anything at all about a possible independent invention. I finally heard from Prof. David Gale, who was associated with Nash at Princeton, and he gave me a still different version! He writes that Nash showed him the game in the triangular board form and that he (Gale) thought of making the board hexagonal (all this in 1949). Then he and Nash agreed to a fifty-fifty split if they could market the game, but they were unable to do so. When Parkers brought out Hex, Gale continues, he obtained from Parkers the information about you and the prior Danish game. Gale is under the impression that Nash thought of the triangular form independently of the Danish game, but Nash has confessed to me that it was shown to him by Aage Bohr. (He is not positive it was Bohr, since there were other Danish mathematicians around Princeton who might have shown it to him – but he is fairly sure it was Bohr.)

In July 1957 *Scientific American* published Gardner's column: *Concerning the game of Hex, which may be played on the tiles of the bathroom floor.* As promised, Gardner had omitted any reference to Nash re-inventing Hex, instead crediting Aage Bohr for introducing it in America:[15]

> *[Hex was] introduced 15 years ago in Niels Bohr's Institute for Theoretical Physics in Copenhagen. ... It swept the Scandinavian countries in the middle 1940s, and in 1949 was taken up by game theorists in the U.S. ... Hex was invented by Piet Hein, ... In 1949 Aage Bohr, son of Niels, introduced the game at the Institute for Advanced Study in Princeton, where it quickly captivated students of mathematics both at the Institute and Princeton University.*

This upset Nash, who contacted Gardner, who relented and appended this to his October column:[16]

> *In the discussion of the game of Hex, to which this department was devoted in July, it was not mentioned that the game was independently invented in 1948 by John Nash, then a student at Princeton University. The popularity of the game among Princeton students resulted from this version, which was developed and extensively analyzed before news reached Princeton of the earlier Danish invention.*

This in turn upset Hein, who wrote Gardner in October:[17]

> *I was greatly suprised to read the note at the end of your article in the October issue of Scientific American, page 138.*
>
> *What has caused it and how do you reconcile it with your letter to me of April 18, the 19th line, where you write "... Gale is under the impression that Nash thought of the triangular form independently of the Danish game, but Nash has confessed to me that it was shown to him by Aage Bohr. (He is not positive it was Bohr, since there were other Danish mathematicians around Princeton who might have shown it to him – but he is fairly sure it was Bohr.)"*
>
> *The note has the faults (1) of confusing Mr. Nash's activity for making Hex known, with any kind of proof of his independent invention of it, and (2) of presenting his independent invention of it as a kind of fact, though it can never be anything more than a hypothesis based on his own postulate. There were hundreds of Scandinavian students who knew Hex in America before 1948 many of which in Princeton (I have got the numbers from American Scandinavian Foundation) and hundreds of copies of Politiken (the Danish paper who published description of it and problems and games of it since 1942); I was on your continent from 1944, and I w e n t t o P r i n c e t o n i n 1 9 4 8[18] and showed it to a lot of scientists and students among which were Albert Einstein, who was greatly fascinated by it.[19]*
>
> *Furthermore there are many ways of developing the idea apart from the rhombic board consisting of hexagonic (or the topologically equivalent six-starred) elements.*

So it is overwhelmingly likely that Mr. Nash — possibly unconsciously — got inspired with the idea from these sources.

I must expect you to formulate a note in the November issue correcting the false impression, that the "fact" given in the note in the October issue is based on anything but Mr. Nash's unproved postulate in the face of all the facts mentioned above.

I should like to hear from you at your earliest convenience and see, in advance, the mentioned note for the November issue.

Gardner replied to Hein in November:[20]

> *Here's the full story about the note on Nash. When Nash came over to see me, and we talked at length about Hex, he told me that someone, probably Bohr, had shown him a version of the game, and I assumed, naturally, that this was the origin of his interest in the game. After my article appeared, he phoned me in great agitation. He had not made clear, he said, or at least I had not clearly understood, that he had invented and analyzed the game before he had seen your prior game. It is true, of course, that I have no way of knowing if this is really true. But it seems that Nash for many years has claimed this, and when my article strongly implied that he had merely picked up the game from Danish students, this painted him as a liar in the eyes of his colleagues. I checked with some of his associates at Princeton at the time, and they expressed their belief that Nash had in fact hit on the idea independently, though of course much later than you did. This was the opinion, for example, of Shannon, of Bell Telephone. Now I am sure you can see the dilemma this placed me in. Nash felt deeply wounded by the implication that for years he had been telling lies about his early work on the game. He is, I should add, a distinguished game theorist. His Ph.D. thesis at Princeton was on game theory and contains, I believe, a discussion of Hex.*[21] *A recent publication of John Wiley and Sons, titled Games and Decisions, by R. Duncan Luce and Howard Raiffa, is filled with references to Nash's important contributions to game theory. [He] is now a professor at Massachusetts Institute of Technology. So what was I to do? He demanded that I run the paragraph just as it appeared, I discussed it with the editor, and we decided that the charitable thing to do was to give Nash the benefit of the doubt. I, personally, believe that he is telling the truth as he remembers it, though you are certainly right in pointing out that he may have come across something in a casual moment that he later forgot about completely, and later imagined that the idea had come to him out of the blue, so to speak, when it really traced back to some Danish source at Princeton.*

My feelings about it are this. The fact that you invented the game before anyone else is undisputed. Any number of people can come along later and say that they thought of the same thing at some later date, but this means little and nobody really cares. I think it is obvious to anyone who reads the note in Scientific American that it is based on Nash's claim, and since it is something that can't be proved, one merely has to accept the man's word for it, or to feel sceptical of his claim. I don't think anything is served by saying in another note, words to the effect that after all we must realize that this is merely Nash's unproved assertion. Obviously there is no way Nash could prove his claim, and to say this is to express doubts about Nash's veracity. This would be an extremely unkind thing to do to a distinguished mathematician, and I can say without qualification that I know Scientific American would refuse to do this. When Nash phoned me, he was genuinely and deeply hurt by my article, and under the circumstances don't you think that the charitable thing to do is to let his claim stand? It detracts in no way from your credit as the inventor, and it gets Nash off the hook in regard to his statements over the past few years to his colleagues.

I am sure you can see how I would not want to become involved in a dispute over whether Nash did or did not think of Hex some years after you did. My article made perfectly clear that you were the inventor, and I am sure that readers who read the later note realize that Nash's claim is something that cannot possibly be proved. If I were to publish the note you suggest, Nash would feel wounded a second time and I think nothing whatever would be gained by it.

Two things occur to me. (1) that you drop a letter to Nash, at M.I.T., then form an opinion on the basis of his reply as to whether he is telling the truth. (2) if you think not, then write a letter to the editor of Scientific American, for publication in the magazine's letter columns, expressing your scepticism, and I will urge the magazine (which is the most I can do) to print it. My feeling is, however, that someone's claim to a later and independent invention is so minor and trivial a claim, that it is not worth the trouble of controversy.

I am deeply sorry if my note was offensive to you. I can only say that it was done in a spirit of charity toward a mathematician who felt strongly that he was being made out to be unreliable in the eyes of his colleagues. I think if you will exchange letters with Nash, you will find him a very humble sort of fellow. In fact, the chief reason I failed to understand him clearly when I spoke to him was his modesty and his reluctance to talk about himself.[22]

Do you know the Baroness who writes under the name of Isaak Dinesen?[23] *Her new book of short stories is receiving laudatory reviews, and I find her writing fascinating.*

Hein replied a week later:[24]

I greatly appreciate your frank answer to my question concerning Mr. Nash's postulated independent though later invention of Hex. Though of course the abundant communication between Denmark and America in particular Princeton and the strange fact that the second invention of the game had exactly the same form as the first (though the board could have all forms [and] the elements could be bricks of two squares each and mixed octagons and squares a.s.o.a.s.f. ...) makes it a priori extremely unlikely that the idea has arisen entirely independently of any fertilisation from the original source which may have come through a mere flash of a suggestion and may have been unconscious or have become unconscious.

...

The baroness you mention is my cousin and lives almost next door to me here in Rungsted. Isn't she a great imitator and manufacturer of synthetic decadence!

(As we mentioned earlier, covering the board with octagons and squares yields a game equivalent to Bridg-it. In 1942 Hein already knew that the only necessary requirement for no-draw connection game boards was that at most three cells should meet at a point. This is the same observation that occurred to Schensted and Titus when they generalized Y to Mudcrack Y.)

In November Gardner shared his final thoughts on this affair with Hein:[25]

Just between you and me, and off the record, I think you hit the nail on the head when you referred to a "flash of a suggestion" which came to Mr. Nash from a Danish source, and which he later forgot about. It seems the most likely explanation. And I am glad that I was able to give publicity to the full story of Hex's origin, because most mathematicians in this country knew nothing about it and simply supposed that it all came from Nash. In fact, a few correspondents that I heard from had been calling the game Nash, having picked it up from Princeton students who were calling it that. ...

In her biography *A Beautiful Mind*, based on interviews with Gale and others, Sylvia Nasar dates the initial Nash-Gale Hex conversation as late winter 1949, with Nash describing the game as played on a checkerboard, and Gale realizing its hexagonal form. In an interview with Nasar, Hans Weinberger recollects that Nash might have had the idea for Hex while at Carnegie

Institute in Pittsburgh, before coming to Princeton.[26] However, in a phone interview on December 2 1999 with Ryan Hayward and Jack van Rijswijck, Nash recalled inventing the game while at Princeton.

Aage Bohr was in Princeton as a member of the Institute for Advanced Study in the first half of 1948, and from January 1949 in New York City as a visiting fellow at Columbia University, seven miles north of New York University. So Bohr was in Princeton and New York before the Nash-Gale meeting at which Nash first told Gale about his new game.

In December 1999 Ryan Hayward and Jack van Risjwijck — who had just written an MSc. thesis on computer Hex at the University of Alberta[27] — phoned Nash, asking how he had invented the game. Nash replied:

> *Well, let's see, it was a long time ago, I don't exactly remember in steps, because it's a matter of connecting topology and game theory, so I don't exactly remember any steps towards that I was [in] an area where Go was being played a lot, and of course Go has a geometrical structure on the board, black and white stones, so probably the Go influenced me.*

To add to the mystery, Thomas Maarup — a student at the University of Southern Denmark, writing a thesis on the history of Hex[28] — contacted Nash, Gale and Bohr around 2004. Nash replied immediately, but said he had no clear memory about the matter, and put Maarup in contact with Harold Kuhn. Gale remembered well, and gave details consistent with his letter to Gardner in which he dates the meeting where Nash told him about Hex as winter 1948.[29] Aage Bohr had no clear recollection, and did not remember talking to anyone about games.

Hein presumably accepted Gardner's conclusion: perhaps a "flash of a suggestion" came to Nash from a Danish source, that Nash later forgot about. In any event, it seems he never contacted Nash nor *Scientific American* on this affair. Instead he moved on to other projects, for example, providing Gardner with information for the upcoming *Scientific American* column on the Soma cube.

Gardner's columns undoubtedly contributed to Hein's fame in America. In October 1966 — eight years after the first *Scientific American* column on the Soma cube — Life Magazine profiled Hein.[30] A month later, MIT Press issued the first English edition of Hein's grooks, and two months after that Esquire Magazine profiled Hein.[31] In 1966, Hein wrote Gardner:[32]

> *I shall never forget that I am your discovery, perhaps even creation.*

Gardner was in turn appreciative of Hein: *Undiluted Hocus-Pocus* — the title of Gardner's autobiography — is from a Hein grook.

With respect to the origins of Hex in America, nothing seems certain. How did Nash discover Hex? Was it at Princeton, or earlier in Pittsburgh? Was it inspired by thinking about ladders in Go or someone's forgotten mention of a connection game? Who was the mysterious Dane who explained Hein's game to him, and was the conversation only verbal, or was a board drawn? And when? And was the grid hexagonal or triangular? How did Parker Brothers hear of the game, and how did they track it to Hein?

According to Gale, Nash conceived the game as played on a checkerboard, i.e. on the dual of the usual hexagonal board. Nash or Gale or both decided on board size 14×14, unlike Hein's preferred 11×11. Gale built a 14×14 hexagonal grid board and left it in the Fine Hall common room, crediting Nash as the game's creator.

Nash's explanation of inventing Hex after watching Go games is plausible. Go is also a connection game, and it would not be surprising for Nash to wonder how to change the underlying rectangular grid so that mutual blocking cannot occur.

Regardless of whether Nash's invention of Hex was completely independent of Hein's, Nash deserves credit for making Hex known and for his theoretical insights on the game. Nash's theory and Gale's board started a Hex renaissance.

Notes

[1][56]
[2][73]
[3]1957.01.13 letter Gardner-Parker Bros, reply 1957.01.16 [23].
[4]1957.01.18 letter Gardner-Hein [23].
[5]1957.03.18 letter Gardner-Hein [23].
[6]Unfortunately we found no records detailing these 'myths' on the origin of Hex.
[7]1957.03.23 letter Gardner-Shannon [23].
[8][54] p 202.
[9]1957.04.06 letter Gardner-Hein [23].
[10]Here, with respect to conversation where Nash first told Gale of his game, Nash's version differs from Gale's: each recalls himself as realizing that the game could be played on a hexagonal grid.
[11]1957.04.10 letter Gale-Gardner [23].
[12]When interviewed by Sylvia Nasar many years later, Gale puts the time as late in the winter of 1949, so perhaps here Gale means *in the winter of 1948-49.*
[13]1957.04.12 letter Hein-Gardner [23]
[14]1957.04.18 letter Gardner-Hein [23].
[15][21]
[16][22]
[17]1957.10.30 letter Hein-Gardner [33].
[18]In this type-written letter, Hein inserted spaces here for emphasis.
[19]Unfortunately we have no evidence verifying this claim.
[20]1957.11.04 letter Gardner-Hein [23].
[21]Nash's PhD, *Non-cooperative games*, does not mention Hex [55], but his Rand Corporation technical report does [56].
[22]In her biography of Nash, Sylvia Nasar reports Nash as sometimes communicating awkwardly. In 1959, after experiencing delusions, Nash was involuntarily admitted to McLean Hospital, a psychiatric hospital affiliated with Harvard Medical School. At the start of her book, Nasar reports this conversation from May 1959 between Nash and Harvard professor George Mackay at the hospital [54]:

> *[Mackay:] ... How could you believe that you are being recruited by aliens from outer space to rule the world? ...*
>
> *[Nash:] Because the ideas I had about supernatural beings came to me the same way that my mathematical ideas did. So I took them seriously.*

Did Nash's inability in 1959 to distinguish between dreams and reality have anything to do with his earlier recollections of how, around 10 years earlier, the idea for Hex had first come to him? We will likely never know.
[23]Isaak Dinesen is the pseudonym of Karen Blixen, the best-selling Danish author of Out of Africa. The book Gardner refers to is probably Last Tales, published in 1957.
[24]1957.11.10 letter Hein-Gardner. This letter, incorrectly dated 1957.10.10, describes events occurring just before 1957.11.10. [33].
[25]1957.11.14 letter Gardner-Hein [23].
[26][54]
[27][71]
[28][51]
[29]In her Nash biography, based on interviews that include a 1995 dinner with Nash, Gale

and Kuhn in San Francisco, Nasar dates the meeting at which Nash first described Hex to Gale as late winter 1949, so before spring 1949.

[30] [40]

[31] [13]

[32] 1966.10.15 letter Hein-Gardner [33].

Chapter 8

Is Hex easy?

Problems worthy of attack
prove their worth by hitting back. Piet Hein

8.1 Bridg-it falls

In his October 1958 *Scientific American Mathematical Games* column — *Four mathematical diversions involving concepts of topology* — Martin Gardner mentioned David Gale's game.[1] Soon after, Hassenfeld Brothers Inc. (later Hasbro) published a version of the game, which they called Bridg-it. Gardner wrote Gale, asking whether he had heard this news. Gale replied by postcard:[2]

> *No, I didn't know about BRIDG-IT. Since you published my game in Sc.A. a number of people have written me about trying to sell it and I've told them to go ahead, but I didn't think anything had come of these efforts. I had long since given up wasting my energy on trying the sell the game after some futile attempts to sell HEX back in '48*[3] *(before Parker Bros. bought it from Hein).*
>
> *Anyway I'm glad you told me about BRIDG-IT. I'll keep an eye out for it, and in case it starts sweeping the country maybe I'll investigate further.*

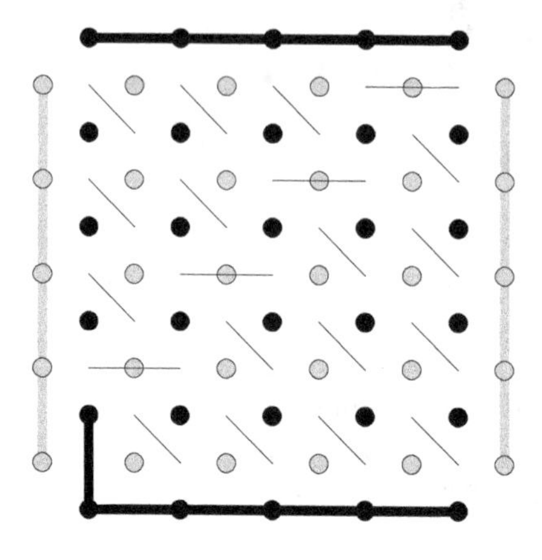

FIGURE 8.1: Gross's winning first-player Bridg-it strategy. After Black's first move (thick vertical line), follow the pairing: after each White move, play at the other end of the thin line. Can you prove that this strategy wins?

But Bridg-it did not sweep the country. A year after getting Gale's postcard, Gardner learned that Oliver Gross of the RAND Corporation had found a first-player winning strategy.

See Figure 8.1. Gardner wrote Gross, who replied promptly:[4]

> *It was not my intention of publishing a paper describing my solution of the game currently known as Bridg-it, although I appreciate Professor Gale's discretion on my behalf. Nor do I have any objection to your describing the strategy (and/or any other information you might glean from this letter) in your column.*
>
> *The winning strategy which I have found for player 1 of Bridg-it is probably best described by recourse to the enclosed figure ... I'm afraid you will be forced to admit that the strategy is indeed a "simple" one in almost any sense of the word.*
>
> ... *[It cannot] be said that [Bridg-it] is "solved" in the strictest sense of the word, since I know of no simply described strategy for the second player which will take advantage of player 1's first deviation from optimal play.* ...

Any game can be solved by brute force search: just examine all possible continuations of the game. When Gross says that Bridg-it is not solved, he means that he does not know an *efficient method* — in particular, better than brute force — that solves arbitrary Bridg-it positions, i.e. finds a winning move if there is one.

Around this time, computer users noticed that many problems fall into one of two categories: those for which there is an efficient algorithm, and

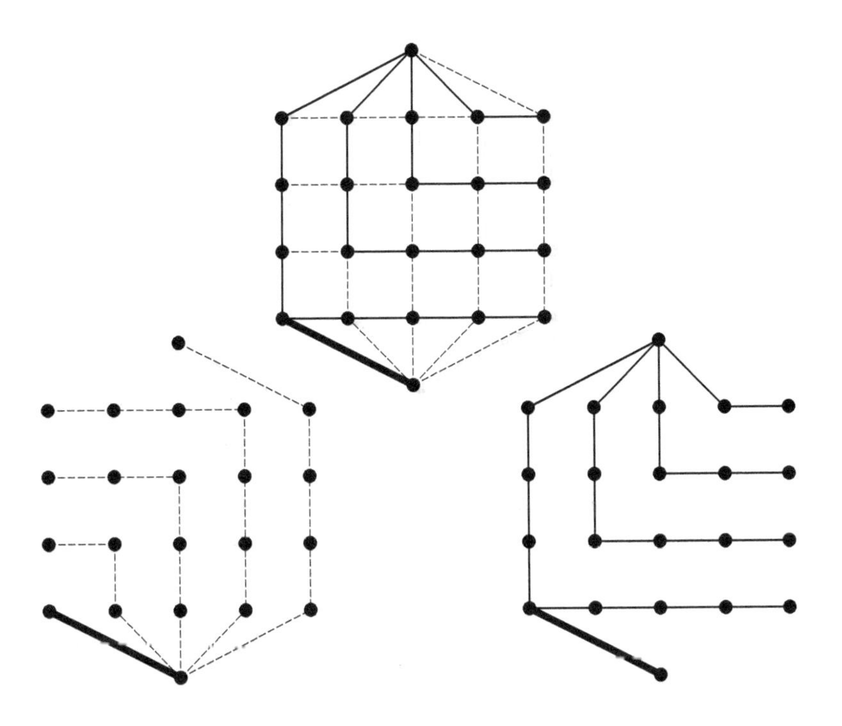

FIGURE 8.2: A winning first-player Bridg-it strategy (top). The diagram decomposes into two trees (bottom).

those that are virtually intractable. In the 1960s, the Canadian mathematician Jack Edmonds advocated using the word "good" or "efficient" to describe an algorithm whose runtime increases at most polynomially with problem size, e.g. if the problem size is doubled, the runtime increases by at most a constant factor.

Is there an efficient — in the sense of Edmonds — algorithm to solve Bridg-it? The answer is yes. We now illustrate such an algorithm. It does not yield a pairing strategy, such as Gross's: instead, it builds two special trees that can be used to play and win.

Figure 8.2 illustrates a first-player Bridg-it strategy returned by the algorithm. We will prove that the strategy is indeed winning. Here Bridg-it is in the form of a Shannon switching game: Short wants to join the top and bottom nodes, Cut wants to separate them. On respective turns, Short thickens any thin edge, and Cut erases any thin edge. We see that Short has made the same first move as in Gross's strategy. We have drawn the thin edges, solid and dashed, in a special way: the two edge sets are disjoint, and each edge set connects each pair of nodes (or group of nodes connected by thick edges) in tree-like fashion.

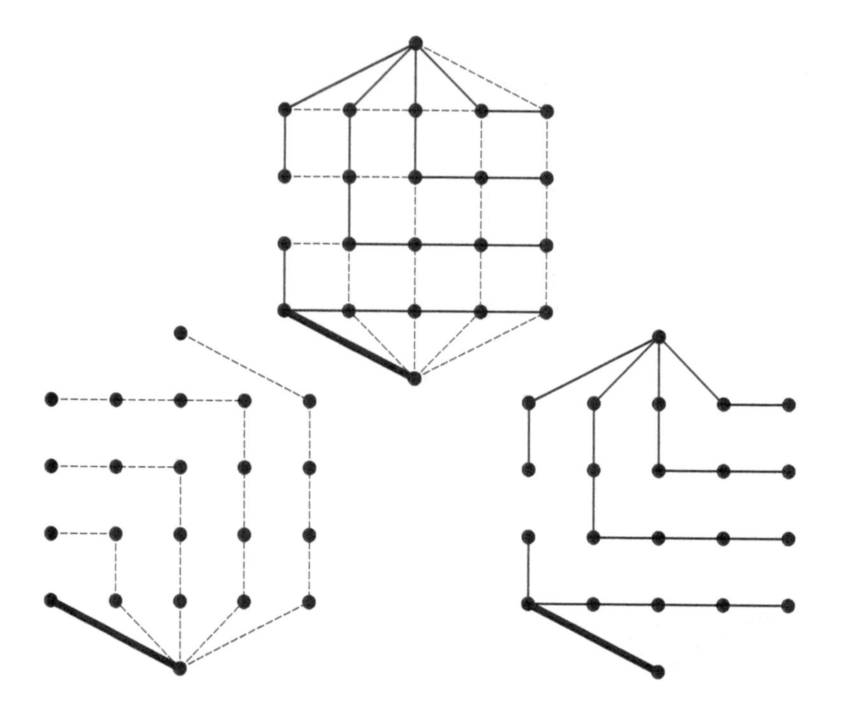

FIGURE 8.3: Cut's move has disconnected the solid-edge tree.

Here is how Short uses these two trees to win: after each Cut move, Short picks an edge of the other type (e.g. solid if the opponent cut dashed) that restores the node-group connectivity.

For example, in Figure 8.3 Cut has erased an edge of the solid-edge tree, so Short seeks an edge of the dashed-edge tree that will reconnect the solid-edge tree. Such an edge always exists: to prove this is an exercise in graph theory, on the properties of trees. In our example, Short has five such edges in the dashed-edge tree to choose from: each choice yields a winning strategy. Here, Short picks the rightmost such edge. See Figure 8.4. Now it is Cut's turn. Continuing in this way, Short always restores connectivity, so Short wins.

We leave it as an exercise to show that for the usual Bridg-it board — each side has 5 points and 4 links — every first move wins.

A "solution" of Bridg-it — an efficient algorithm to solve arbitrary Bridg-it positions — came soon after. In 1963 Alfred Lehman gave a characterization that holds for any Shannon switching game:[5] for a position with Cut to play, Short wins if and only if there are two edge-disjoint trees that span the same node set and each contain the two terminals. Edmonds then gave an efficient method that finds two spanning trees with minimum edge overlap.[6] Lehman's

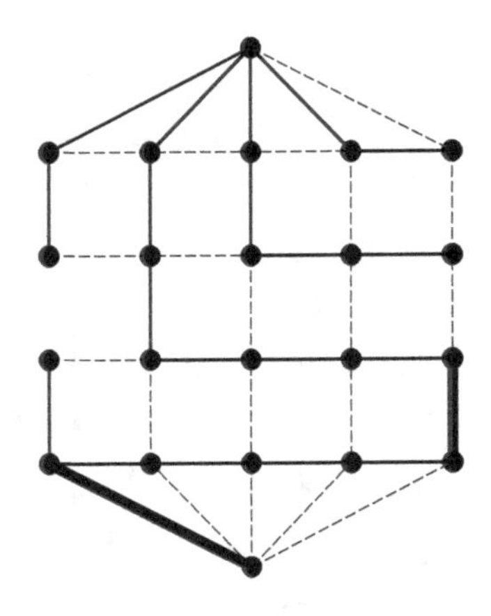

FIGURE 8.4: Short picks an edge that reconnects the solid-edge tree.

proof of his characterization uses matroids, which generalize properties of cycle-free edge sets in graphs.

In general, Lehman's characterization has a different form for positions with Short to play. But for Bridg-it, because the underlying graph is planar and the graph is self-dual, each player can contruct an equivalent Shannon switching game in which that player is Short, and so use the above characterization.

So Bridg-it is 'solved' because there is an efficient way to check for this characterization. We illustrate this in the next section.

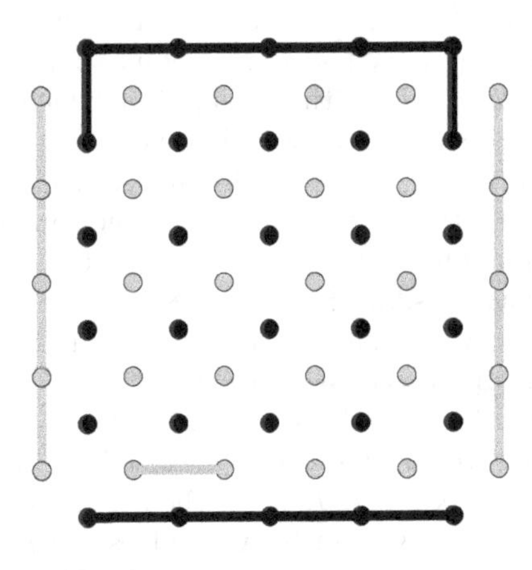

FIGURE 8.5: A Bridg-it puzzle. White to play: who wins?

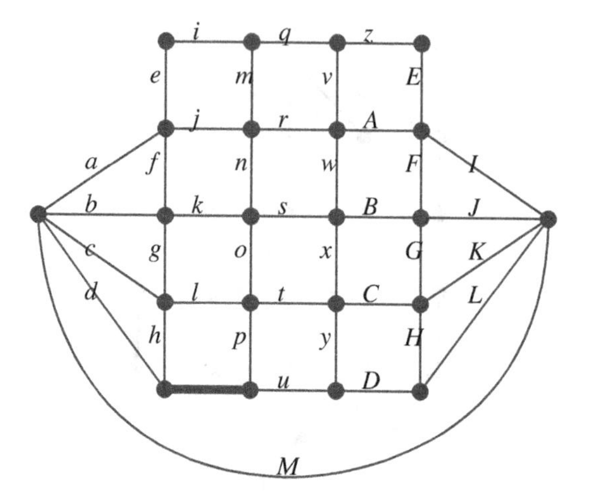

FIGURE 8.6: Represent the Bridg-it puzzle in Figure 8.5 as a Shannon switching game. Then add an edge — here M — that joins the two terminals.

8.2 Solving Bridg-it: an example

Let's solve the puzzle in Figure 8.5. White plays next, so construct the corresponding Shannon switching game with White as Short. Erase each Black move, thicken each White move, and contract each thickened edge, i.e. treat each shorted group as one node. Add an extra edge connecting the two terminals. (In puzzles where the opponent plays next, do not add this edge.) For puzzles where we play next as Short, we can win if and only if — after adding this edge — some subgraph has two edge-disjoint spanning trees, each joining the terminals. Cut must remove this edge, after which we can deduce what our next move as Short should be. See Figure 8.6.

Now search for the two trees. Recall that a tree is a graph that is connected and has no cycles. From the set of graph edges, pick any subset S_0 that forms a tree. From the remaining edges, take any subset S_1 that does not form a cycle and is as large as possible. See Figure 8.7

If S_1 is connected then it is also a tree and we are done. In some cases, as in our example, S_1 is disconnected, so we are not yet done.

In our example we have used all but three edges: F, G, H. Now we want to re-arrange our trees to include one of these edges. If we add F to S_1, we form a cycle in S_1 with edges I, J. If we then break the cycle by moving I from S_1 to S_0, we create a cycle in S_0 with edges a, j, r, A, M, I. But now we break this cycle by moving M from S_0 to S_1, and this does not create a cycle in S_1. So add F to S_1, move I to S_0, move M to S_1. See Figure 8.8.

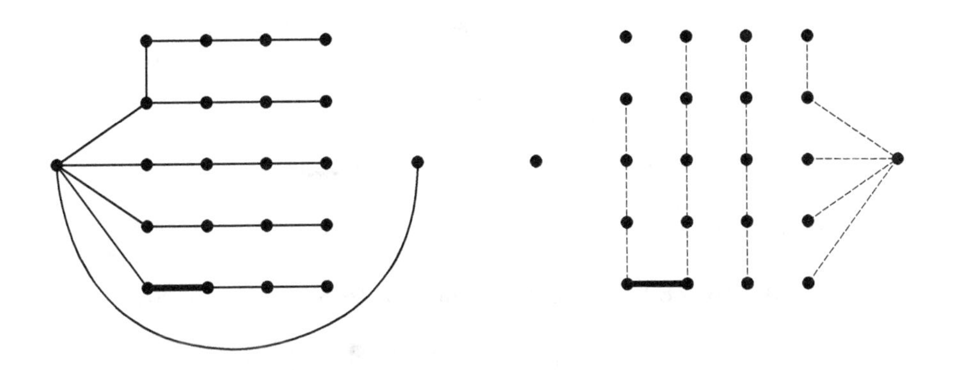

FIGURE 8.7: Pick a subset S_0 of edges that forms a tree (left). From remaining edges, pick a maximal subset S_1 that forms no cycle (right).

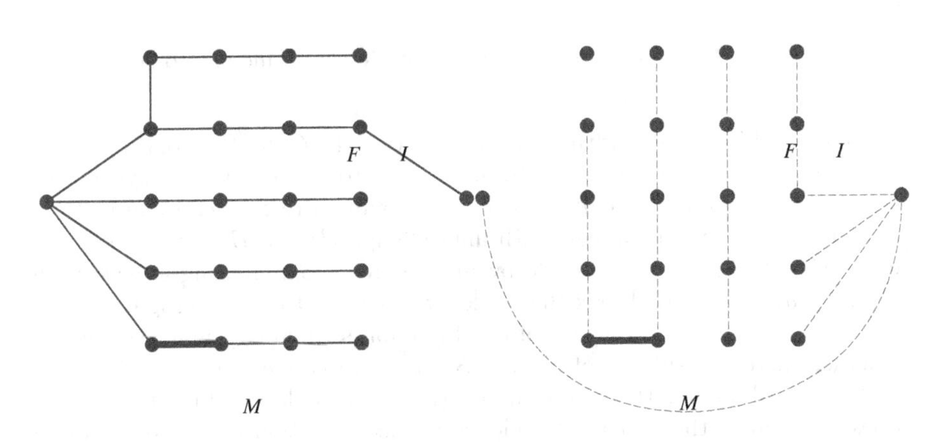

FIGURE 8.8: Add F to S_1, move I to S_0, move M to S_1.

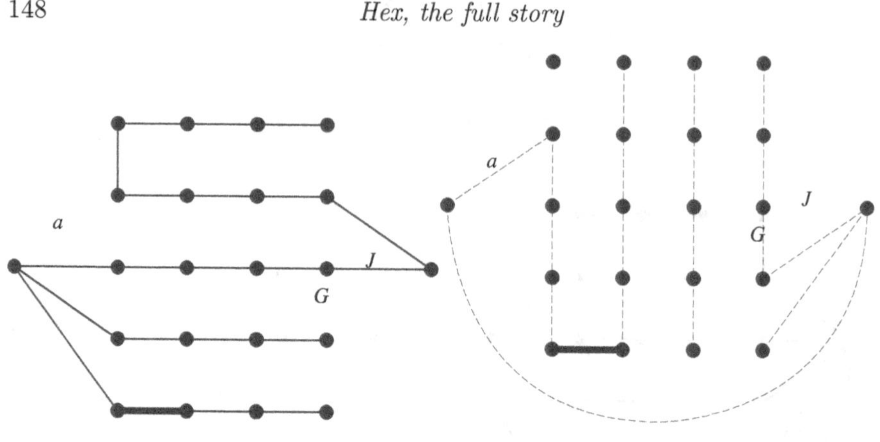

FIGURE 8.9: Add G to S_1, move J to S_0, move a to S_1.

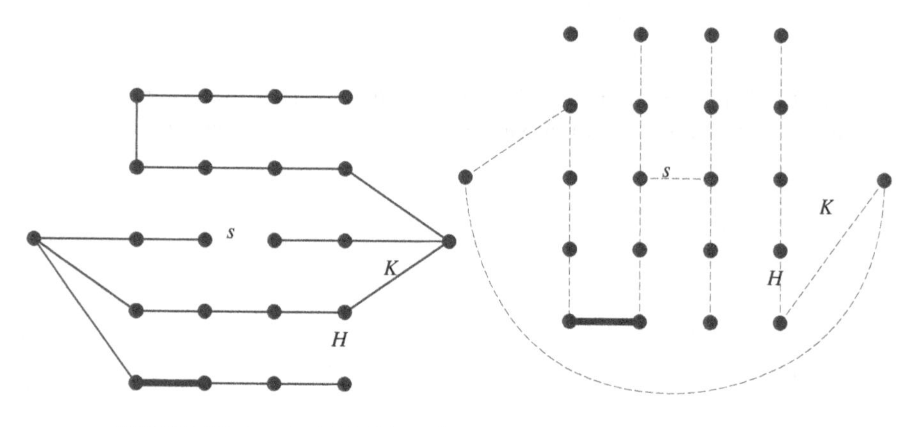

FIGURE 8.10: Add H to S_1, move K to S_0, move s to S_1.

Repeat this process with unused edge G: add G to S_1, creating a cycle with J, K; break this cycle by moving J to S_0, creating a cycle with $a, b, j, k, r, s, A, B, I, J$; break this cycle by moving a to S_1. See Figure 8.9.

Finally, repeat this process with unused edge H: add H to S_1, creating a cycle with K, L; break this cycle by moving K to S_0, creating a cycle with $B, s, k, b, a, j, r, A, l, K$; break this cycle by moving s to S_1. See Figure 8.10.

We have used all the edges of our Shannon switching graph. Now check whether there are subsets of S_0 and S_1 that are connected, span the same node set, and include the two terminals (here, on the left and right sides). S_1 does not include the upper left node, so we need to remove this node and all incident edges from our two trees. This disconnects the remaining top row of nodes in S_0, so these are also removed.

We have found two trees satisfying Lehman's condition, so White wins. Moreover, these trees give a winning strategy. We assume that Black cuts the

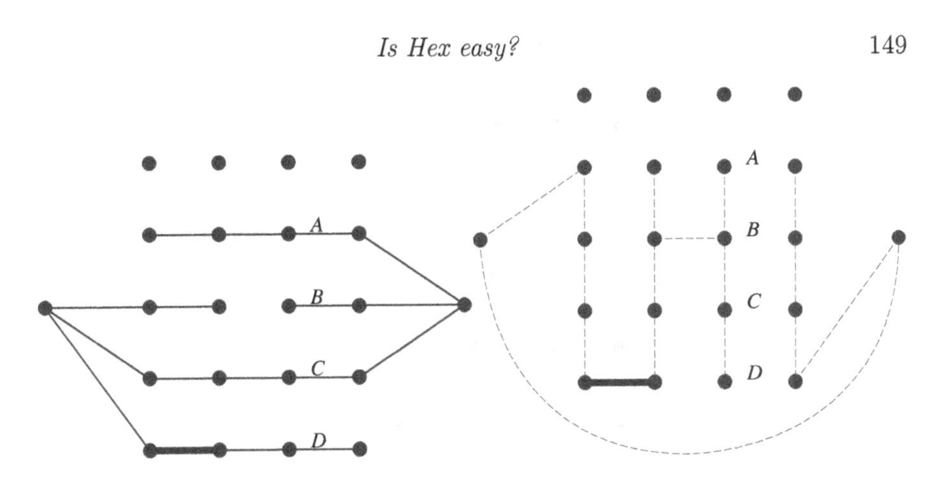

FIGURE 8.11: The final two trees. After removal of the extra edge, White wins by shorting any of A,B,C,D.

extra edge that we added at the start of this process. This edge is in our S_1, and so leaves S_1 in two pieces. So now we pick any edge of S_0 that reconnects S_1: here, notice that any of A, B, C, D suffice. (These are not necessarily the only winning moves: different tree pairs will give different move choices.) So White can make any of these four moves, and continue to use S_0 and S_1 to play the rest of the game. See Figure 8.11.

Once news got out that Bridg-it was "solved", the game became less popular. But we still find Bridg-it fun to play! The solution method that we have described is efficient but far from trivial. Try it yourself: see Figure 8.12.

8.3 Will Hex fall?

Around 1960, the College of Letters and Science at the University of Wisconsin brought in a requirement that every student should take either a foreign language or a year of calculus level math. The math department felt that calculus was not appropriate for everyone, so three professors in the department — Anatole Beck,[7] Michael N. Bleicher and Donald Crowe — wrote the 6-chapter calculus-level non-calculus text *Excursions into Math*, first published in 1969. The 2000 edition includes a foreword by Gardner. Beck included a chapter on games, with a section on Hex. Beck, who received his PhD from Yale in 1956, writes of $n \times n$ Hex:[8]

> *If* n *is odd, then the first move seems always to be in the center. Thus players prefer* n *even, as it was in the Yale Mathematics Common Room in 1952.* ...

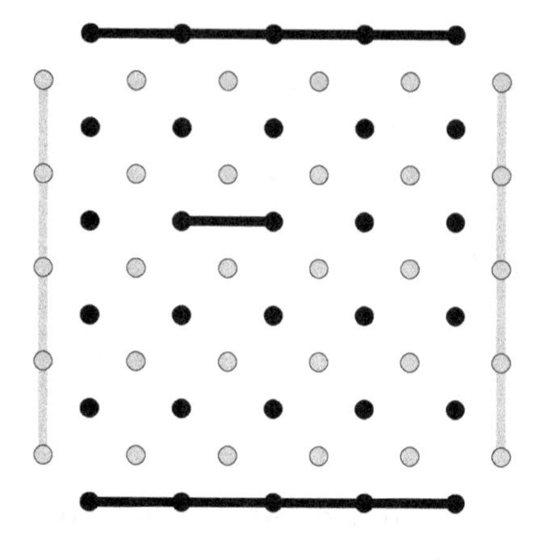

FIGURE 8.12: Does this Bridg-it Black opening move win? Answer this question by finding a two-trees diagram. Solution in Figure 8.14.

> *Until recently, a commercially marketed game called Bridge-it [sic] shared [the property of having a first-player winning strategy, but with no such strategy known]. Now a simple strategy is known whereby one can force a win there.*
>
> *A number of mathematicians have attempted to devise an explicit [winning] strategy for Hex. The question commands some interest, and [whoever] solves it will achieve thereby a worldwide, if fleeting, renown.*

8.4 Losing Hex openings

To offset the advantage of playing first, Beck suggested this Hex variant: the second player is allowed to tell the first player where she must make her first move. When playing this variant — called Beck's Hex — who wins, the first player or the second? This is equivalent to asking whether Hex has any losing opening moves.

From the beginning, Hex players have guessed that, for all $n{\times}n$ boards, opening in the center — or more generally, anywhere on the short diagonal — always wins. No one has been able to prove this. However, Beck did prove that a certain opening always loses.

In *Excursions into Mathematics*, Beck proved that, on each $n{\times}n$ board with n at least 2, the acute corner is a losing opening move. In the millennial

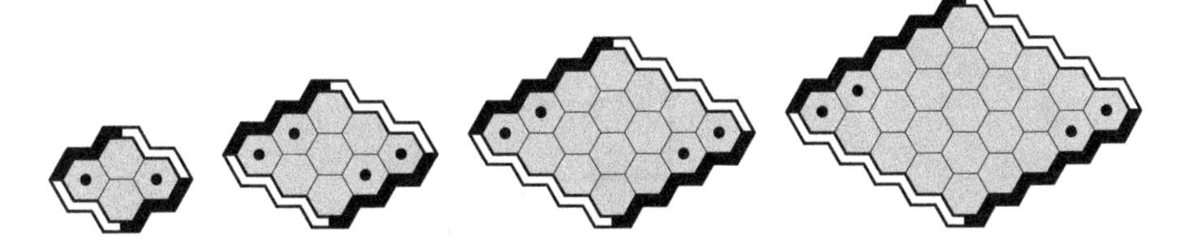

FIGURE 8.13: For all $n \times n$ boards with $n \geq 2$, opening in the acute corner loses. For all $n \times n$ boards with $n \geq 3$, the player-side 2-stone acute-corner opening also loses.

edition of this book, he and W. Charles Holland further proved that, on each $n \times n$ board with n at least 3, the cell beside the acute corner and on the player's border is also a losing opening move. In fact, if the first player is allowed to occupy two cells on the first move, then playing at the acute corner and the neighbouring cell on the player's side is a losing two-cell Hex opening. See Figure 8.13.

As with Nash's proof that the first player wins on $n \times n$ boards, the proofs of these results are existence proofs: using contradiction and strategy stealing, they imply that, for any such opening, there exists a second-player winning strategy. Except for small boards, no such explicit strategies are known.

The proofs are based on curious local properties of Hex play: we explain more in the next chapter.

In Bridg-it, each opening move wins, so the first player wins Beck's Bridg-it. By contrast, except on the trivial 1×1 board, the second player wins Beck's Hex. Is this a sign that Hex is harder than Bridg-it?

8.5 Hex is probably hard

Around this time, scientists were starting to study the computational complexity of so-called hard problems, namely problems for which no efficient (polynomial runtime) algorithm was known. One class of such problems is called NP (non-deterministic polytime): roughly, these are yes-no questions such that, for each question whose answer is 'yes', there exists a proof of the answer that — although it might take a long time to find — can be verified quickly (in time polynomial in the question size). Stephen Cook in Canada and independently Leonid Levin in the Soviet Union proved that one such problem — known as satisfiability — is in a sense as hard as any problem in all of NP: if someone finds a polytime algorithm for satisfiability, it immediately gives a polytime algorithm for every problem in NP.

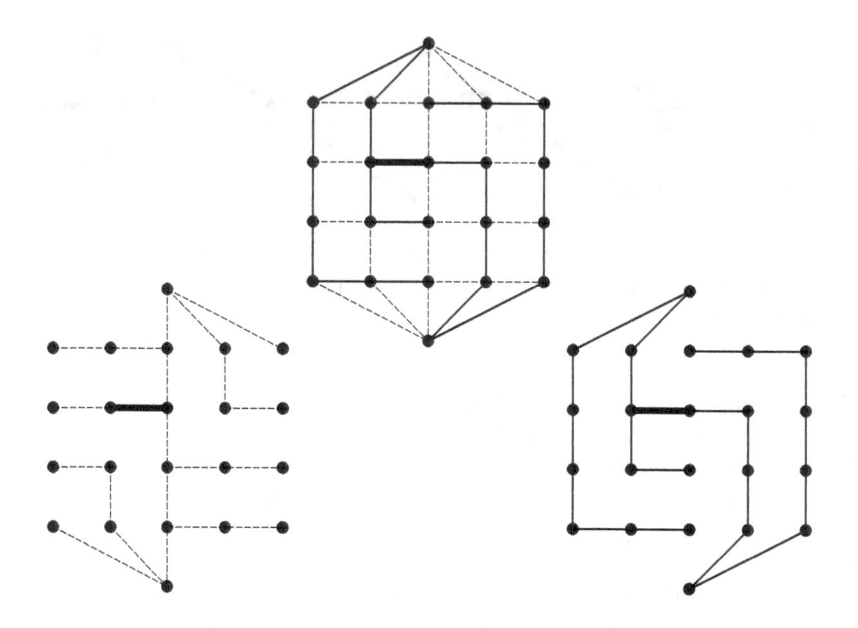

FIGURE 8.14: A solution to the puzzle in Figure 8.12. Black wins.

People immediately started looking for other such problems, which are now called NP-complete. In 1975, Shimon Even and Robert Tarjan announced one such result: the Shannon node-switching game is P-space complete (so as hard as any problem in the class called P-space, a class of problems that includes all problems in NP).[9]

Recall that Hex is a restricted form of Shannon node-switching. So this result does not preclude that solving Hex could be easier than solving Shannon node-switching. But six years later, the student Stefan Reisch of Germany showed that Hex itself is also P-space-complete.[10]

What does this mean for Hex players? Many mathematicians suspect that NP-complete problems are harder than polytime problems. In fact, resolving this is one of the great open problems in mathematics today. Beck remarked that, if you found an explicit strategy for winning Hex on arbitary boards, you would probably receive fleeting attention. But if you found a *polytime algorithm* that solves arbitrary Hex problems, you would be famous much longer: by Reisch's result, this would imply that each problem in P-space — so also each problem in NP — could be solved in polytime. You would have answered the question "does the class NP equal the class P?" with a 'yes', so — once your algorithm had been checked by a panel of professional scientists — you would win a Clay Millenium Prize and receive $1 000 000.

Many who study computational complexity think it unlikely that P equals NP. Rather, they think that the hardest problems in NP are too difficult to solve with a polytime algorithm. So, is Hex easy? Probably not.[11]

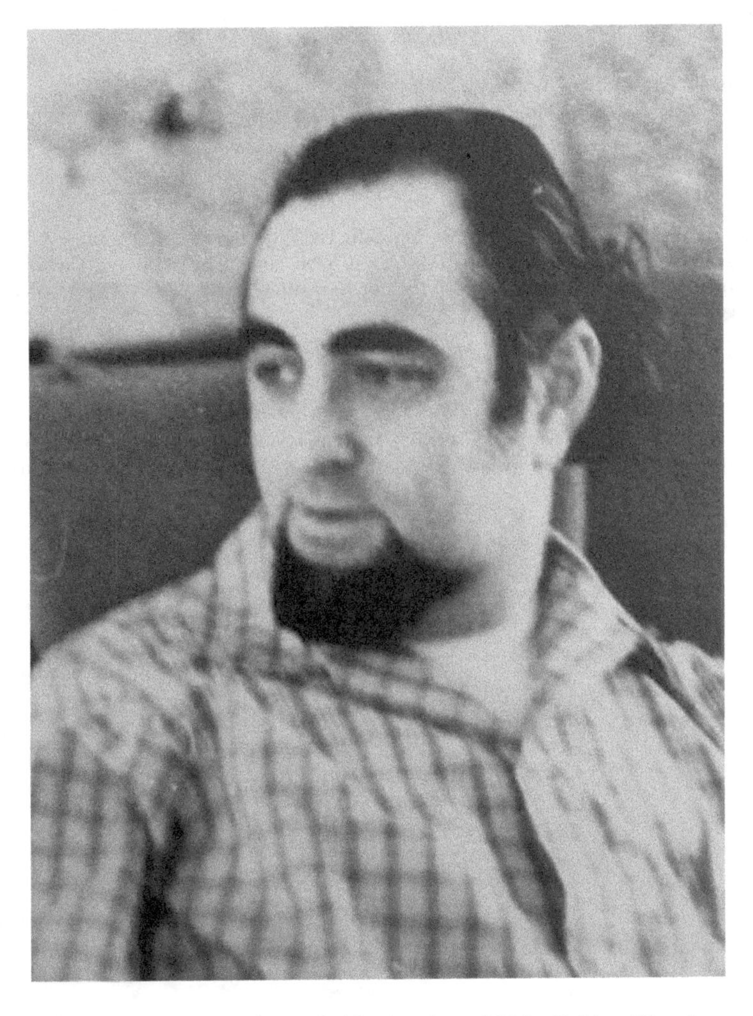

FIGURE 8.15: Anatole Beck, circa 1968. © Eve Siegel.

Notes

[1][24]
[2]1960.02.04 letter Gale-Gardner [23].
[3]Gale later dated his learning of Hex as late winter 1949, so here he presumably means 1949.
[4]1961.03.13 letter Gross-Gardner [23].
[5][47]
[6][15]
[7]Beck was interested in all kinds of games. In 1969 Manifold — a Warwick University game magazine — published *A Pandora's Box of Non-Games* by Beck and David Fowler, which includes the curious naming game Finchley Central. Curiously, that issue also included an article on Hex by Michael Jackson.
[8][5]
[9][18]
[10][62]
[11]Unless P equals NP, there is no polytime algorithm to solve *arbitrary* Hex problems. But some subset of Hex problems — such as finding a winning strategy starting from the empty board — might be easier than the problem of solving arbitrary Hex problems.

Chapter 9

Hex theory

In mathematics, you don't understand things. You just get used to them.

John von Neumann

As Hein's game-playing friends first reported to him, Hex is fun to play. And — as perhaps Lindhard was first to realize — part of the fun comes from mastering logically intricate mathematical properties — Hex theory — that can increase a player's strength.

Of course, to play Hex well, one also needs to know the game's strategic princples, which over time were shared by 'Hexperts' such as Lindhard and the Fine Hall players. Shannon based his machines on one such principle: interrupt your opponent's 'flow' as much as possible. Schensted and Titus proffered principles as proverbs, for example *double trouble* — a move should create at least two threats — and *the best offence is a good defence*:[1]

> *there are often moves which you obviously must make if you are to prevent your opponent from [winning], while there are seldom moves which clearly enable you to [win] yourself.*

In the rest of this chapter we give more strategic observations and introduce the Hex theory of virtual connections and inferior cells.

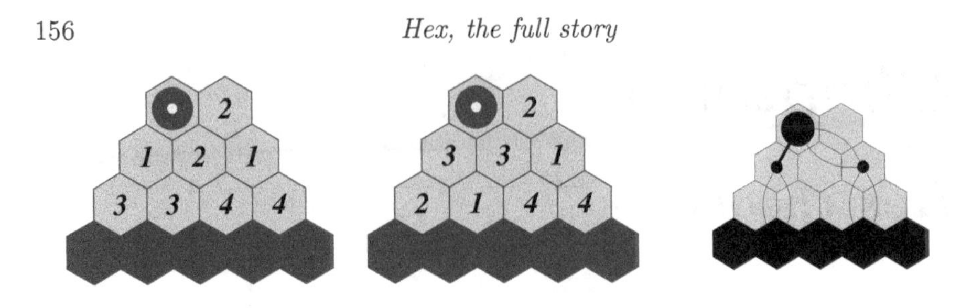

FIGURE 9.1: (left) 4.3.2 connection pairing strategies. (right) 4.3.2 connection formed by two semiconnections.

9.1 Side connections

As Hein and Lindhard remarked in the Polygon salons, and Nash mentioned to Gardner in their evening discussion, Hex play relies on side connections. As Milnor observed, some such connections have a pairing strategy:[2]

> *One important tactical feature of the game is the strategy of "doubles": For example consider the position [Figure 9.1: left] with white to move. (The numbered "squares" are empty.) Here black can connect his stones to the black edge by the following strategy: If white plays in "square" number i, then black should play in the other "square" number i.*

Figure 9.2 shows more side connections: all safely connect, even if the opponent plays next, but not all have a pairing strategy. Those in the top row are called *ladders*: in Hein-Lindhard terminology, the players 'rub shoulders' when contesting such connections. David King maintains an extensive library of side connections, also called templates.[3]

Familiarity with side connections allows a player to join a side from as far away as possible. For example, assume that White wants to attach to the bottom left side of an 11×11 board. Cells in the first row already touch the side: which other cells attach? To answer the question, following Hein and Lindhard at the Polygon salons, use the 2.1, 4.3.2, 7.6.5.2, and 10.9.8.5.3 side connections. See Figures 9.2 and 9.3.

9.2 Art of Hex

After Gardner's 1957 column, interest in Hex soared. Presumably to avoid copyright conflicts with Politiken, Hein renamed Polygon as Con-Tac-Tix — a name he had originally wanted to use — and teamed up with Skjøde Knudsen,

FIGURE 9.2: More side connections, including some ladders (top). Connections in the top two rows have pairing strategies.

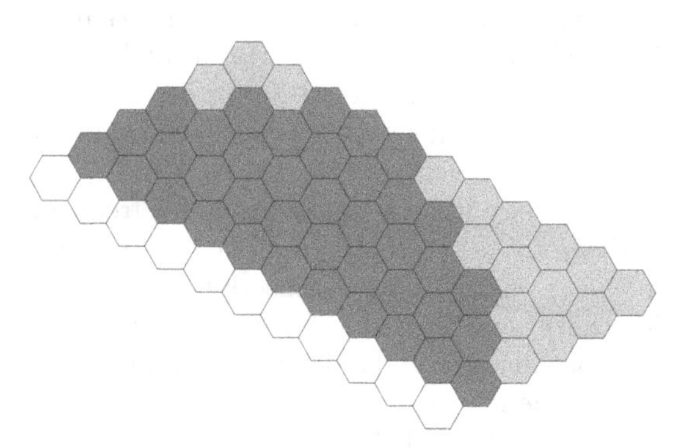

FIGURE 9.3: Each dark cell attaches safely to the bottom left side.

FIGURE 9.4: Con-Tac-Tix board produced by Skjøde Skjern circa 1970.

a designer and craftsman from Skjern in West Jutland.[4] In a 10-year partnership, their company Skjøde Skjern produced many Hein products, including the super egg — a 3-dimensional version of Hein's super-ellipse, Soma game sets, and 12×12 teak Con-Tac-Tix boards. See Figure 9.4. Similar boards are available from `piethein.com`, a company run by Piet Hein's son Hugo.

Around 1970, in *Maison du Denmark* (House of Denmark), the French mathematician Claude Berge (1926-2002) purchased a Con-Tac-Tix board.[5] After a time he found the board small, and so converted it to size 14×14 by drilling extra holes.

Berge was a keen player and took the board on his frequent travels. Thinking that a lighter plastic board would sell well, he had protypes made and also wrote an accompanying instructional manuscript. Unfortunately he failed to find a supplier for the beads he used as markers, so this Hex travel set was never commercially produced. But he did circulate his manuscript — *L'Art Subtil du Hex (The subtle art of Hex)*, 1977 — to friends and colleagues, including the authors. See Figures 9.12 and 9.14.

The manuscript includes strategic advice, including this:

> *Beginning players will to their detriment often play in a short-sighted manner, attempting only to lengthen their most promising chain: they feel as though they are "attacking". Experienced play-*

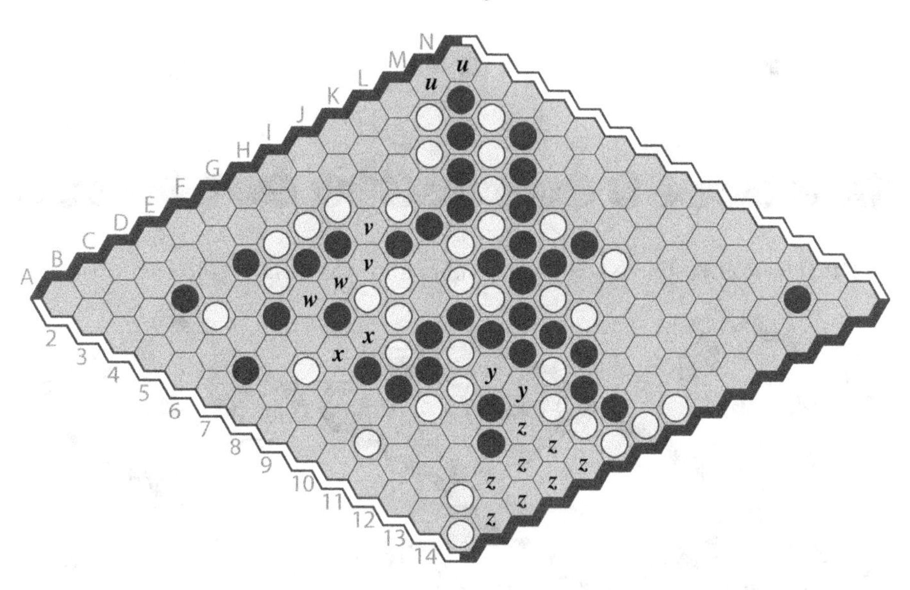

FIGURE 9.5: Berge's illustration of a winning black virtual connection.

> *ers, on the other hand, will play their stones in a more dispersed fashion, and by combinations of "double threat" moves eventually form a connection.*

The manuscript also introduces *virtual connection*, a local subgame between two locations in which a player can force a connection even if the opponent plays first. For example, side connections are virtual connections.

To illustrate, Berge used Figure 9.5. The empty cells u virtually connect the upper left side to the stone at M2: if White ever plays at one such cell, Black can play at the other. Similarly, the respective empty cell sets v, w, x, y virtually connect the stone at M2 to the stone group at E11,D12, and this group is virtually connected to the lower right side via the 4.3.2 connection marked z. So this game is effectively over: combining these virtual connections gives a winning black virtual connection.

Analogous to a virtual connection, a *semiconnection*[6] is a local subgame in which a player can force a connection when moving first.

Sometimes semiconnections can be combined into virtual connections. For example, the 4.3.2 side connection is composed of two semiconnections whose respective cell sets (one of size three, the other size five) do not intersect. So if the opponent plays into the cell set of one semiconnection, the player can play into the other semiconnection.

Notice that a virtual connection's cell set is the union of the cell sets of its semiconnections. For example, the 7.6.5.2 side connection is composed of three semiconnections that overlap, but the combined intersection of their

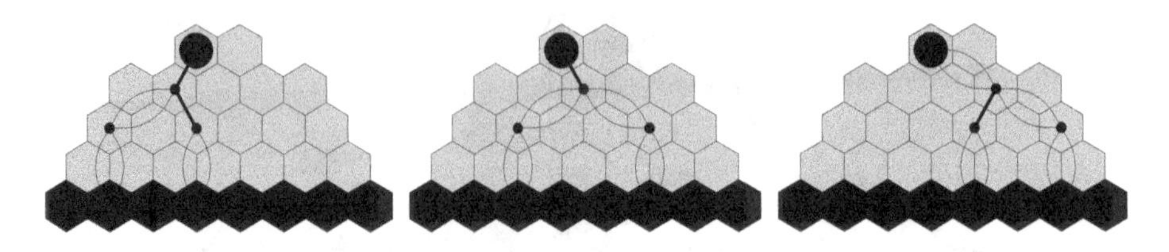

FIGURE 9.6: Three semiconnections form the 7.6.5.2 side connection.

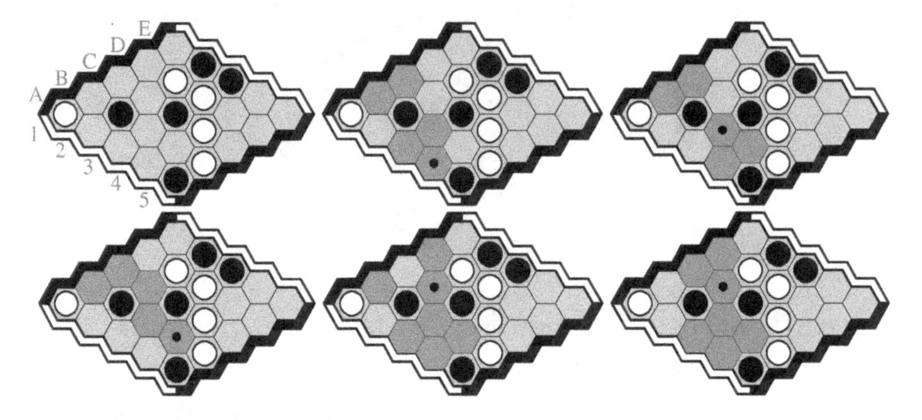

FIGURE 9.7: Berge Puzzle 1: White to play. These five black semiconnections yield a mustplay with just one cell.

respective cell sets is empty. See Figure 9.6. So wherever the opponent plays, the player can reply in a semiconnection whose cell set is untouched.

Assume that it is your turn to play and that you see some opponent win threats, in the form of side-to-side opponent semiconnections. Your next move must interfere with each such semiconnection: otherwise, the opponent's next move can turn a side-to-side semiconnection into a side-to-side virtual connection, and so win. So your next move must intersect the cell set of each such semiconnection: we call this intersection the *mustplay*.[7]

The Berge manuscript includes five puzzles, four of them new:[8] see Figures 9.7 through 9.11. Let us show how to find a mustplay region in Puzzle 1. Each Black move in {A4,B3,B4} yields a Black top-to-bottom semiconnection. See Figure 9.7. The intersection of the cell sets of these three semiconnection is {B1,C1,B3}. So every White move not in this mustplay {B1,C1,B3} loses. As shown in Figure 9.7, after finding two more Black semiconnections we reduce the mustplay to {B3}. So, unless White plays there, Black can win.

And does B3 win? Yes. Figure 9.8 shows a winning virtual connection: the left side connects by {A3,A4} to the marked white stone, which connects by {B4,C2} to the 4-stone white group, which connects by either {E1} or {C5,D4,D5,E4,E5} to the right side. We have solved Puzzle 1: White B3 wins and all other White moves lose.

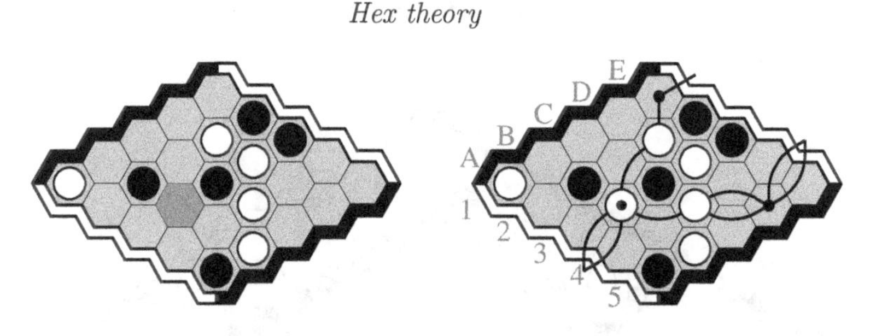

FIGURE 9.8: Solved Berge Puzzle 1: white mustplay (left) and strategy.

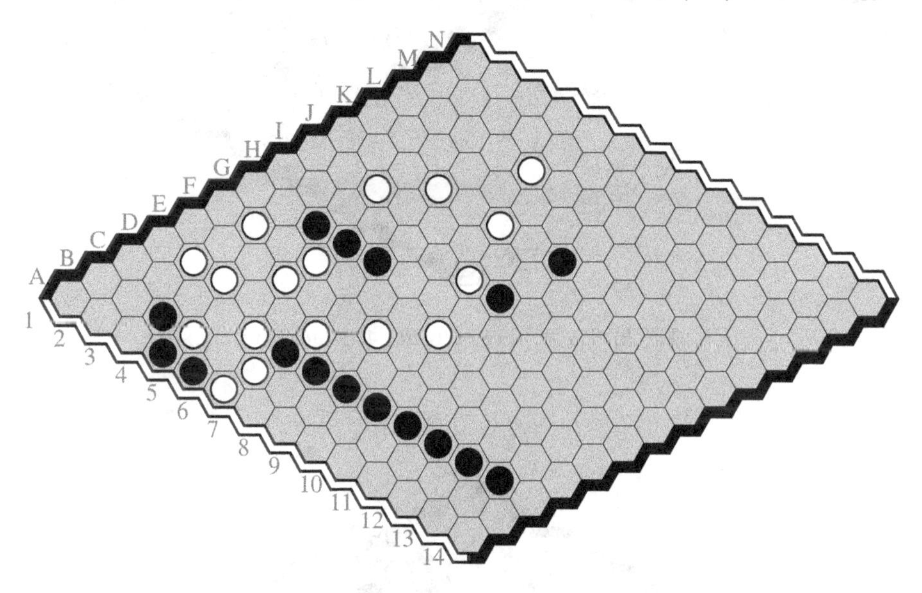

FIGURE 9.9: Berge Puzzle 3. Black to play.

In his manuscript, Berge quotes Sam Loyd, whom Gardner called 'America's greatest puzzlist':[9]

> *My theory of a key move was always to make it just the reverse of what a player in 999 cases out of 1000 would look for.*

In Berge's Puzzle 3 — Figure 9.9 — White seems to have a side-to-side virtual connection. But appearances deceive. Puzzle 4 is also deceptive: the black groups seem isolated by white walls from K1 to N14 and from H2 to A14, so how can Black win? Of Puzzle 5 — Figure 9.11 — Berge wrote:

> *[This] is a study rather than a puzzle. The interesting question is whether White can connect [K6] to the [upper-right].*

In 1980, Berge's Puzzle 4 appeared in *Some Remarks about a Hex Problem*, a chapter of *The Mathematical Gardner*, a collection of articles secretly

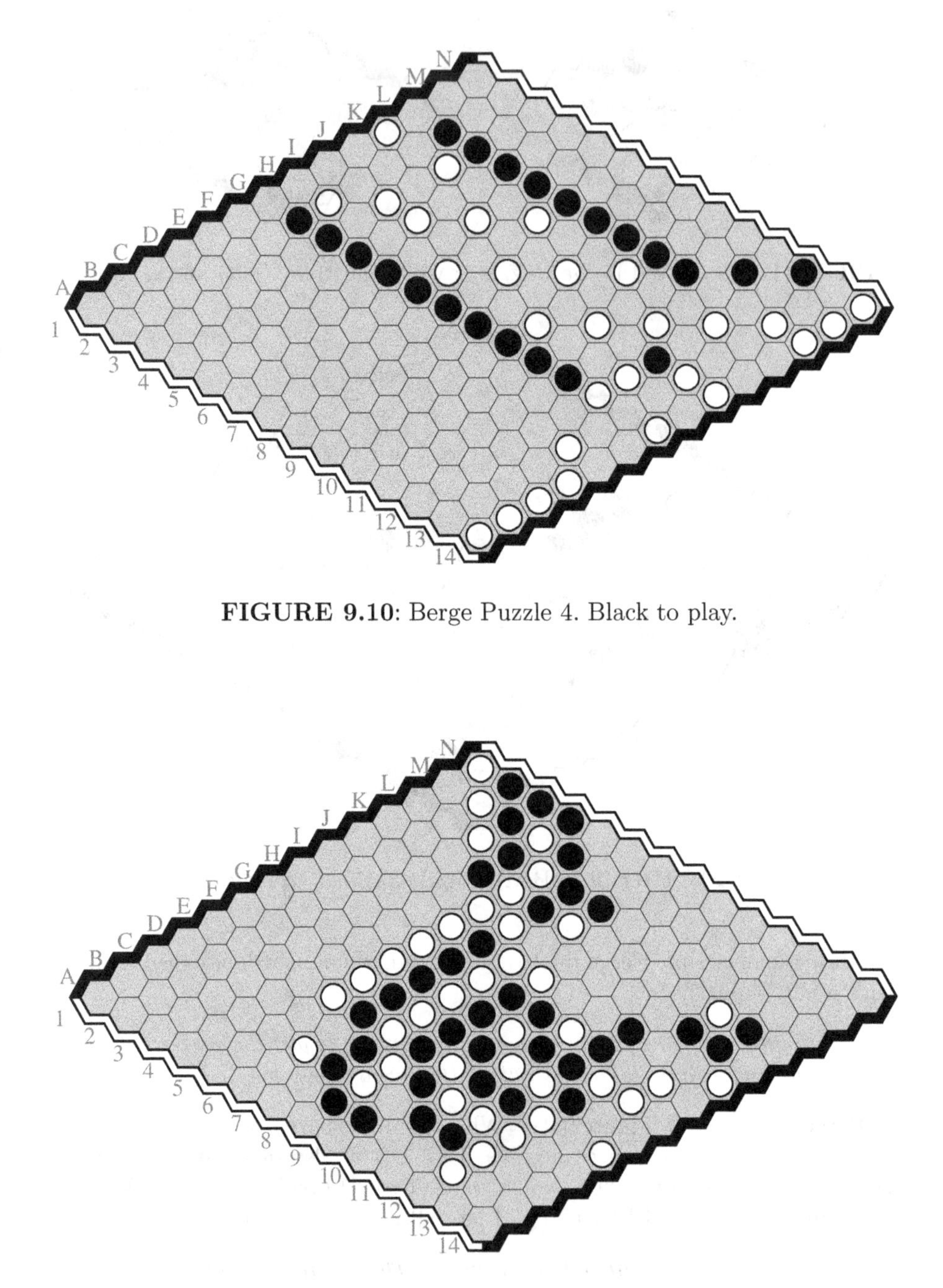

FIGURE 9.10: Berge Puzzle 4. Black to play.

FIGURE 9.11: Berge Puzzle 5. White to play. A study rather than a puzzle. Can White join K6 to the upper right?

FIGURE 9.12: Claude Berge and Ryan Hayward playing Hex on Berge's travel board, perhaps 1995 or 1996. © Chính Hoàng.

compiled by David Klarner, Don Knuth and Ron Graham to mark the occasion of the announced end of Gardner's *Mathematical Games* column.[10] Unfortunately, the black stone at G11 was inadvertently omitted. In his book *Hex Strategy: Making the Right Connections*[11] Cameron Browne shows that, without this stone, Black loses.

In his chapter, Berge commented:[12]

> *It would be nice to solve some Hex problem by using nontrivial theorems about combinatorial properties of sets (the sets considered are groups of critical [board cells]). It is not possible to forget that a famous chess problem of Sam Loyd (the 'comet'), involving parity, is easy to solve for a mathematician aware of König's theorem on bipartite graphs; also, in chess, the theory of conjugate squares of Marcel Duchamp and Alberstadt is a beautiful application of the algebraic theory of graph isomorphism (the two graphs are defined by the moves of the kings).*
>
> *The use of mathematical tools may be unexpected and therefore add some new interest to a game; but Hex exists as a most enjoyable game in its own right, for mathematician and layman alike.*

We close this section with a puzzle from a Berge-Hayward game[13] shown in Figure 9.13: see if you can reconstruct the move order.

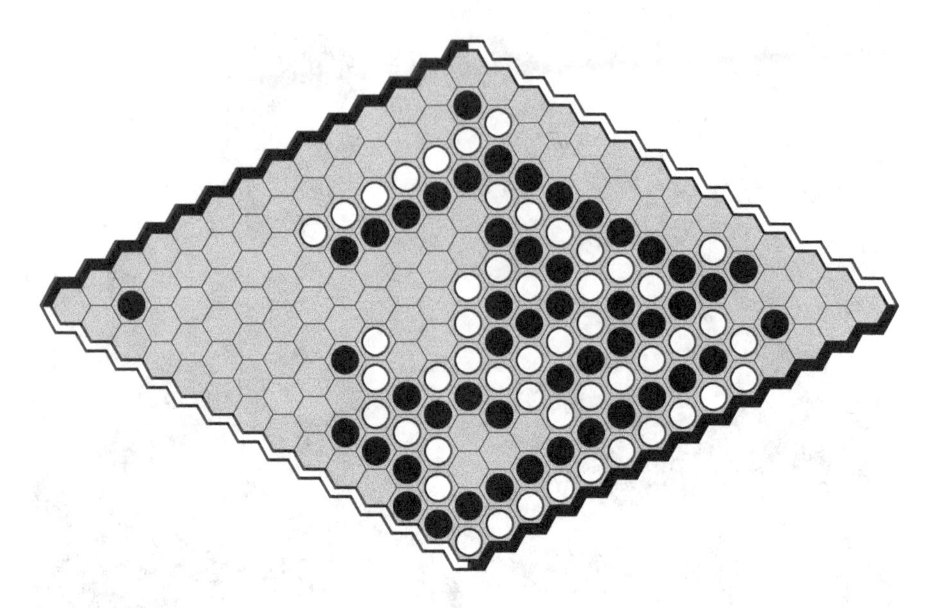

FIGURE 9.13: A reconstruction puzzle: find the move order for this Berge-Hayward game. Solution in Figure 9.24. To offset the first player advantage, Berge often opened near the acute corner.

FIGURE 9.14: Bjarne Toft, Birgit Bock and Claude Berge in 2000 at a graph theory conference in Marseille-Luminy, France. © Kathie Cameron

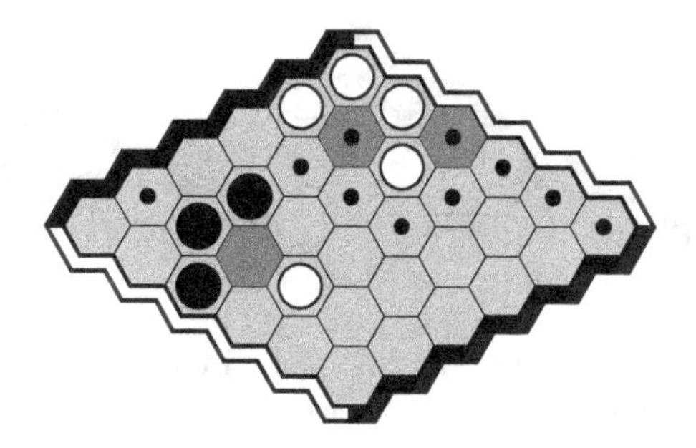

FIGURE 9.15: A black joinset (dots) and a minimal black joinset (non-dark dots). The three dark cells are dead, all other empty cells are live.

9.3 Inferior cell analysis

A challenge in playing Hex is that there can be many move options. For a position, the number of legal moves is the number of empty cells. For example, on the 19×19 board, there are $\lceil 19 \times 19/2 \rceil = 181$ asymmetric first moves and then $19{\times}19 - 1 = 360$ second moves: being able to safely prune this move set would be useful. Happily, some moves — e.g., the acute corners on boards 3×3 or larger — are provably useless, while others are provably inferior to a neighbouring move. We now explain *inferior cell analysis.*

The dots in Figure 9.15 show a Black *joinset*, i.e. a set of empty cells that join Black's sides. This joinset is not *minimal*, as a strict subset — found by removing the dark cells — is still a joinset. We leave it as an exercise to show that a move is in a minimal joinset for one player if and only if it is in a minimal joinset for the other, so being in a minimal joinset is independent of colour. An empty cell is *live* if it is in a minimal joinset, otherwise it is *dead.*

Winning strategies never need dead cells. Here's why. Assume that a player has a winning strategy, with some move to a dead cell. Modify that strategy by instead moving to any live cell (if no cells are live, then the game is over). The result is a new winning strategy in which the number of dead moves has been reduced. Repeating this process eventually yields a strategy with each move to a live cell.

Vulnerable and captured cells can also be pruned from consideration. A cell set is *captured* by a player if — restricted to the subgame played on that set — the player has a replying strategy that leaves each opponent stone dead. A cell set is *vulnerable* for a player if the opponent has a move that captures all cells in the set. For example, in Figure 9.16, C1 is white-vulnerable to D1, and vice versa, so {C1, D1} is black-captured. Similarly, {D3,D4} is white-captured, as is {B1,B2,A2,A3} using the pairing strategy {{B1,A2},{B2,A3}}. So, for this position, White can prune these eight captured cells from the list of move options. Figure 9.17 shows more inferior patterns.

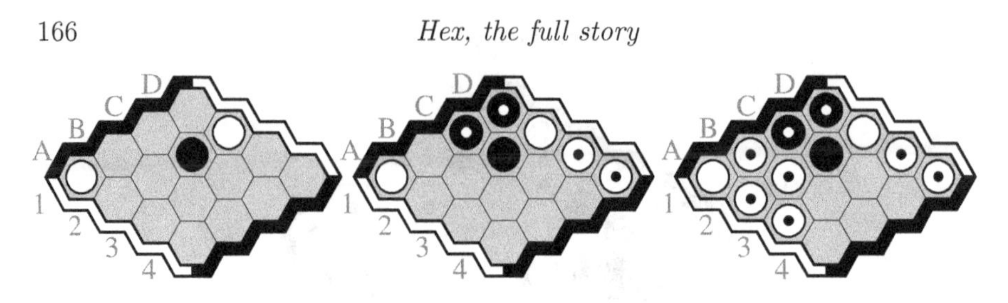

FIGURE 9.16: Inferior cell analysis of a Lindhard puzzle. White to play. Filling captured cells makes it easy to see that B4 wins.

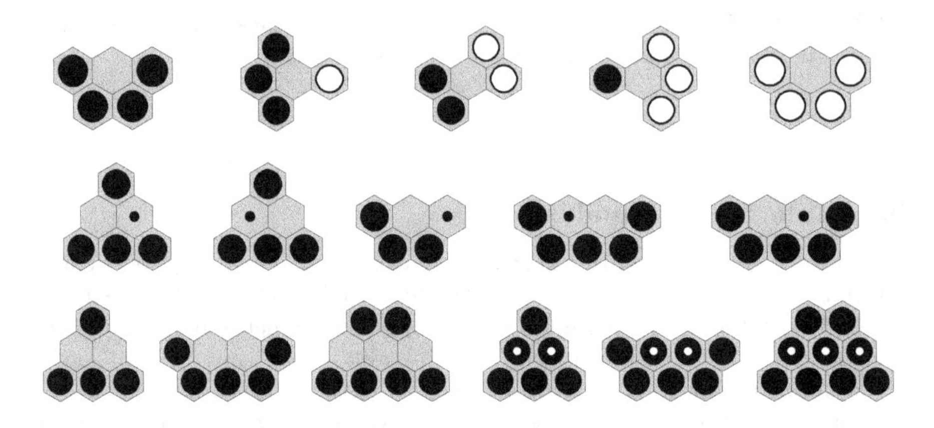

FIGURE 9.17: Some inferior cell patterns: (top) dead, (middle) white-vulnerable: killed by the dot, (bottom) black-captured: filling does not change the winner.

Dead and captured cells simplify puzzle analysis:[14]

- colouring a dead cell does not change the outcome of a game,
- if uncolouring a cell leaves it dead then uncolouring it (or changing its colour) does not change the outcome of a game,
- if a player can win, the player has a winning move to a cell that is live and not opponent-captured and (unless there are no other cells) not self-captured,
- colouring a captured set for the capturing player does not change the outcome of a game.

Our terminology to this point follows Hayward et al.[15] and is implicit in the proofs of Beck[16] and Beck and Holland.[17] Here is Beck and Holland's proof that the two-cell player-side acute corner opening loses.

Figure 9.18 shows the acute corner of an $n \times n$ Hex board with n at least 3: P_1 is the position after the two-cell player-side acute-corner opening, with

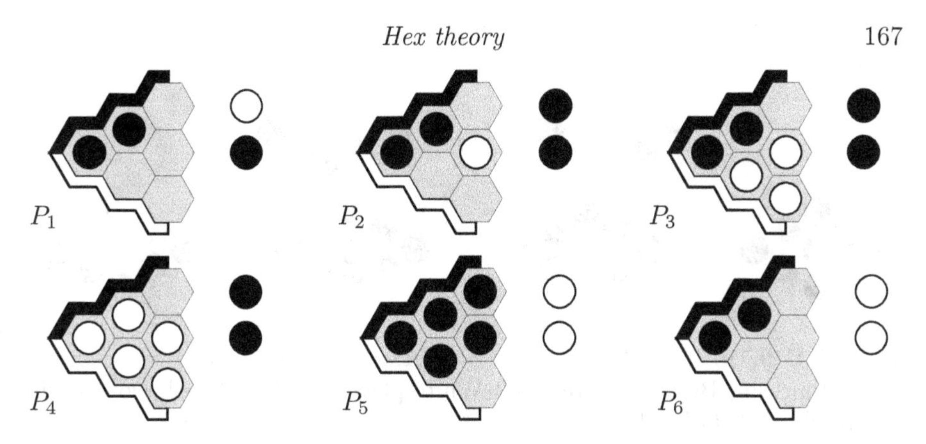

FIGURE 9.18: Beck-Holland proof: two-stone player-side acute-corner opening loses. Off-board circles above/below show player to move/win. P_1: B opens, W to move, assume B wins. P_2: W replies, B to move, B wins. P_3: colour captured cells. P_4: recolour dead B cells W. P_5: exchange colours, flip board. P_6: remove B cells, contradicts P_1.

White to play. Argue by contradiction: assume Black wins. White replies, leaving P_2: Black to play and win. Fill the white-captured set: P_3, still Black to play and win. Recolour the dead black stones: P_4, Black to play and win. Exchange black and white colours and flip the position along the horizontal axis so that the upper-left side is, as usual, black: P_5, White (was Black) to play and win. Remove some black stones: still White to play and win. But now P_6 with White to play and win identical to P_1 with White to play but Black to win, contradicting our assumption that Black wins from P_1. So we have proved that from P_1, with White to play, Black loses.

As a way to offset the advantage of playing first, Beck proposed a variant now called *Beck's Hex*: the second player picks the first player's first move. As Michael N. Bleicher observed, knowing that the acute corner is a losing opening move "*wrecks Beck's Hex*".[18]

9.4 Handicap strategy

In his 2010 Ph.D. thesis *Playing and Solving Hex*[19], Philip Henderson found local patterns with other kinds of inferior moves. Figure 9.19 shows one such pattern, where the dot is a *permanently inferior cell* that can be coloured black. Here's why. If Black plays first in this region, we can assume that it is at the dark cell, since this captures the other three: now the dot is black. If White plays first in this region, again we can assume it is at the dark cell, since the other three cells are vulnerable to a black reply there: now this white move kills the dot, which can be coloured black.

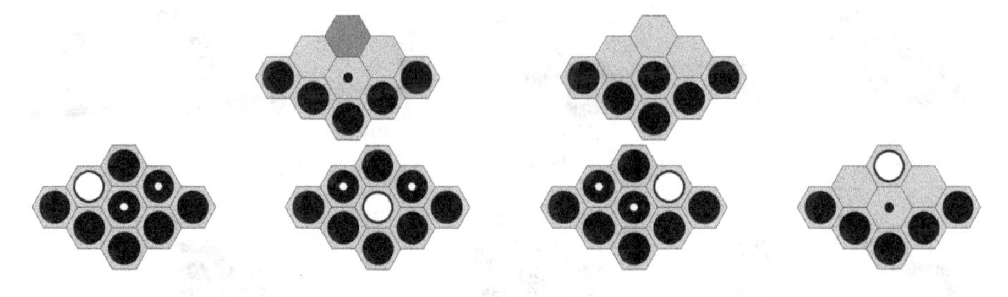

FIGURE 9.19: (top) A permanently inferior pattern: dot can be coloured black. (bottom) Black captures all if White plays not at dark cell. Dot dies if White plays at dark cell (right).

Using this permanently inferior pattern, Henderson found an explicit handicap strategy for $n \times n$ Hex: for $n \geq 1$, the first player wins if allowed to place $\lceil (n+1)/6 \rceil$ stones on their first move. Roughly, the strategy is to use captured and permanently inferior patterns on the last row, and use the usual $n \times (n+1)$ mirror strategy as shown in Figure 5.8 on the rest of the board.

Here is the strategy for the 11×11 board. Play the first two stones as shown in Figure 9.20; if White plays at a dead cell, play anywhere; if White plays a vulnerable move (in row 11 and/or a permanently inferior region), follow a local strategy that eventually kills it; if White plays any other move, reply as in the usual 10×11 mirroring strategy.

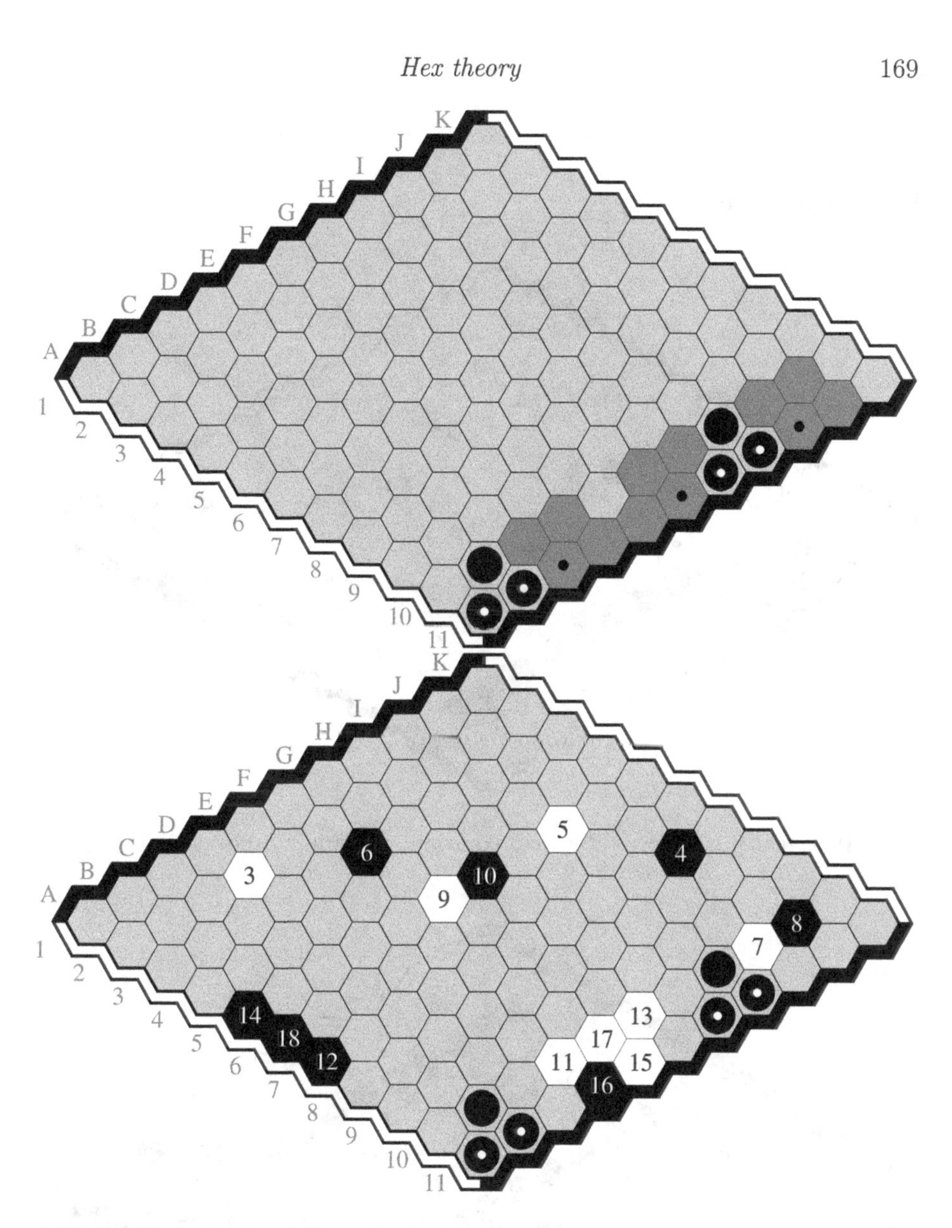

FIGURE 9.20: (top) The 11×11 handicap strategy uses three permanently inferior patterns. (bottom) The start of a game. Moves 4, 6, 10, 12, 14 mirror moves 3, 5, 9, 11, 13 respectively and 8 kill-captures 7. Now C11, F11 are dead — killed by moves 11, 13 — so {D11, E11} is black-captured and move 16 kills move 15. {I11, J11} is also black-captured, so Black now owns row 11.

9.5 Solutions

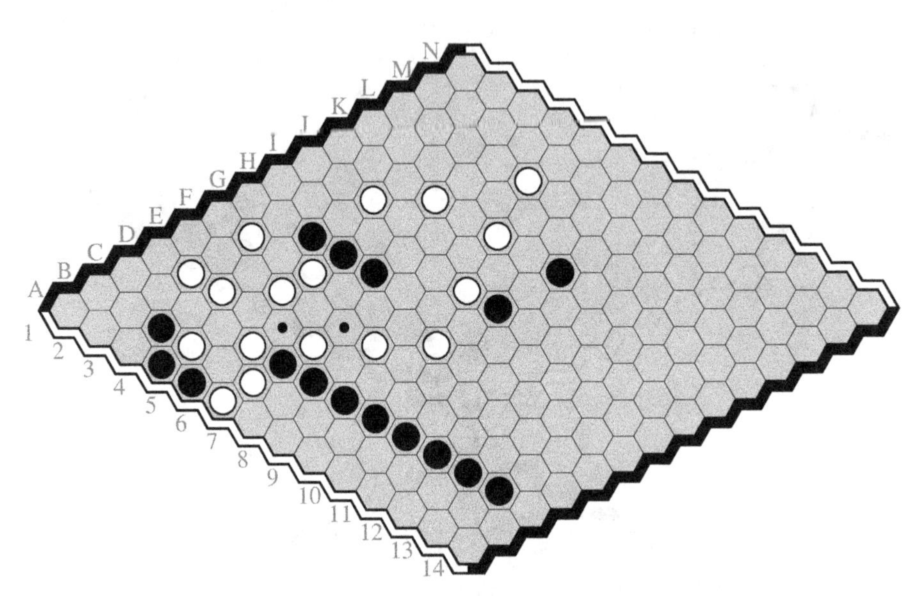

FIGURE 9.21: Berge Puzzle 3 solution: all winning moves (dots).

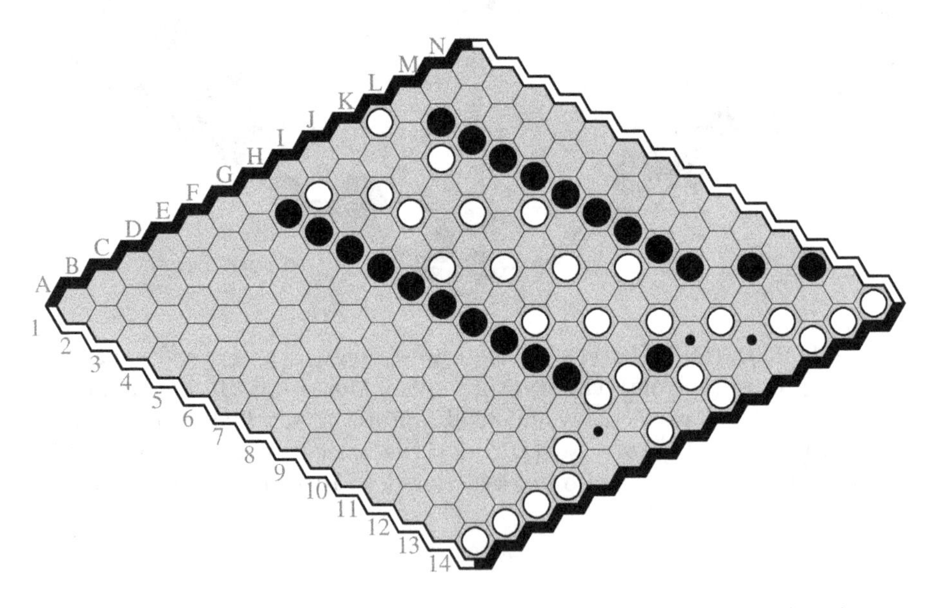

FIGURE 9.22: Berge Puzzle 4 solution: all winning moves (dots).

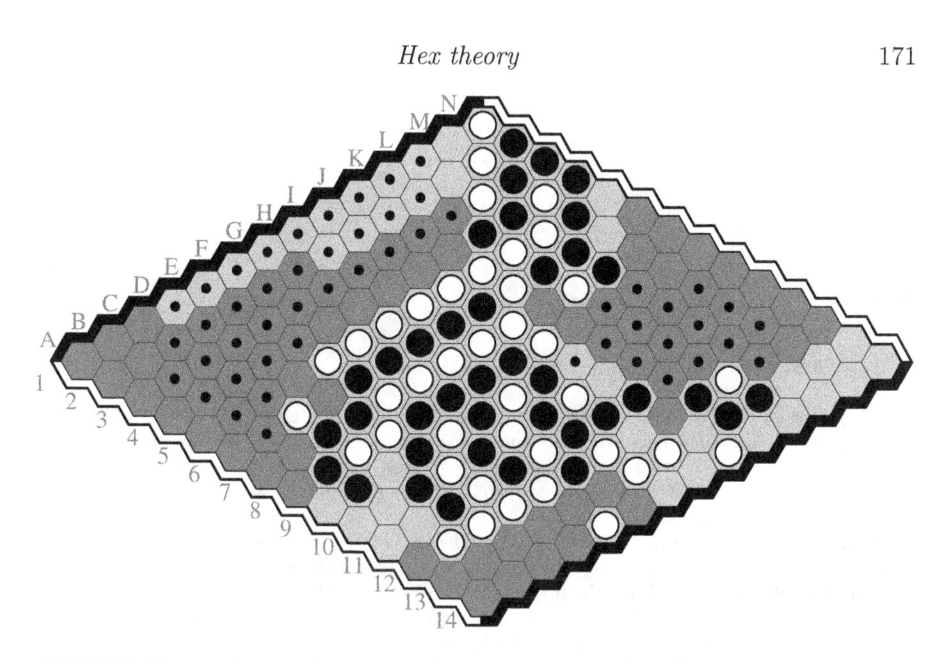

FIGURE 9.23: Berge Puzzle 5: all winning moves (dots). K3 wins using dark cells at left, K8 wins using dark cells at right and bottom.

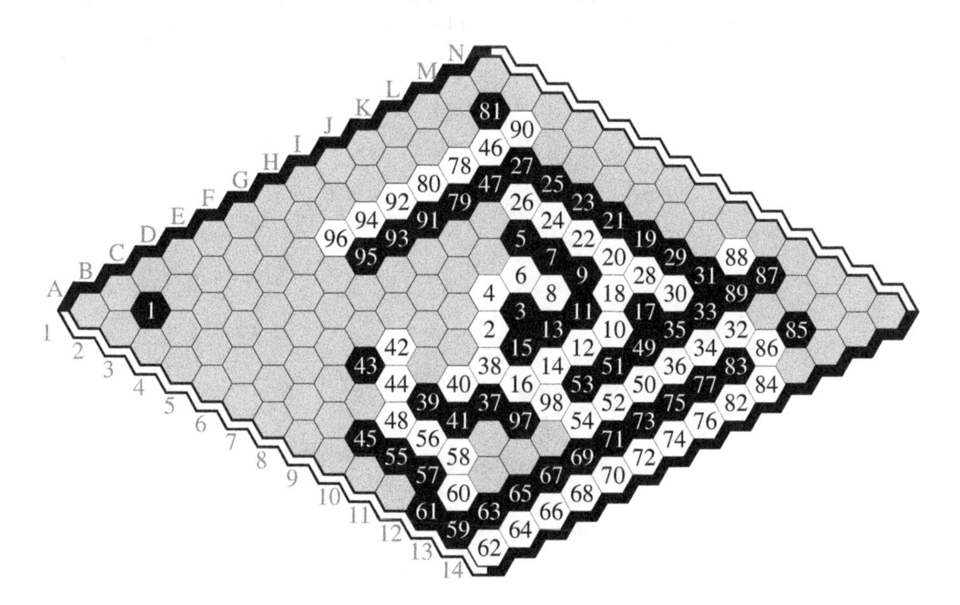

FIGURE 9.24: A reconstruction of the Berge-Hayward game.

Notes

[1] [64] p 154.

[2] 1957.03.26 letter Milnor-Gardner [23].

[3] [44]

[4] They met at the Fredericia Messe design fair in the late 1960s.

[5] Berge told Toft that he purchased the board there. With its shop, meeting rooms, and restaurants *Flora Danica* and *Copenhague*, *Maison du Danemark — 142 Avenue du Champs-Élysés*, near *Place de l'Étoile* — brings a taste of Denmark to Paris. `www.maisondudanemark.dk`

[6] In presenting an algebra to compute virtual connections, Anshelevich introduced the term *semi virtual connection*, which we shorten to semiconnection.

[7] The mustplay notion is implicit in the puzzle solutions of Berge's manuscript.

[8] Puzzle 2 is Politiken Puzzle 13, which also appeared in Gardner's column.

[9] *Lasker's Chess Magazine*, December 1904, p 84.

[10] Due to popular demand, the column continued intermittently until 1986.

[11] [9]

[12] [7]

[13] This was perhaps at the 1995 Conference on Graphs and Combinatorics in Marseille-Luminy, France or the 1996 Colloquium on Graphs and Optimization in Loèche-les-Bain, Switzerland in honour of Claude Berge on the occasion of his 70th birthday.

[14] [30]

[15] [31, 32]

[16] [5] pp 332-333.

[17] [6] pp 495-497.

[18] [5] p 334.

[19] [37]

Chapter 10

Rex theory

Reverse every natural instinct and do the opposite of what you are inclined to do, and you will probably come close to having a perfect golf swing.

Ben Hogan

'That's the effect of living backwards', the Queen said kindly: 'it always makes one a little giddy at first –'

from *Through the Looking Glass* by Lewis Carroll

As we mentioned earlier, Reverse Hex or Rex is the misère version of Hex: whoever joins their two sides loses. Gardner mentioned Rex in his 1957 Hex column and again in a column in 1975.

Rex is a great game to play if you are travelling, because it is hard even on small boards. Try a game yourself. What do you think of the game in Figure 10.1? Did either player miss any good moves? We will analyze the game later: first, some hints.

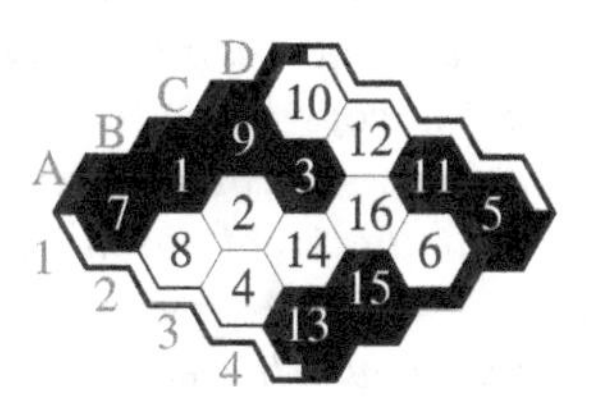

FIGURE 10.1: A Rex game. White joins her sides, so Black wins.

10.1 Winning openings

As you might have guessed, some Hex properties from the previous chapter have Rex analogues:

- Rex never ends in a draw (the proof is the same as for Hex),
- for a Rex position and the player to move, adding an even number of opponent stones — or removing an even number of the player's stones — does not change a player-win to a player-loss (the proof is similar to that for Hex).

And for $n \times m$ Rex, the winning player is known:

- for $n < m$, the player whose sides are further apart — regardless of whether they play first or second — has a winning strategy,
- for $n = m$ with n even (respectively odd), the first-player (second-player) has a winning strategy.

Robert O. Winder proved the last property above while a student at Princeton. Gardner learned of Winder's proof — which he called 'involved'[1] in his July 1957 Hex column — during his evening conversation with Nash.[2] Mathematicians can be sticklers for precision, and Winder wrote Gardner in August 1957:[3]

> *Dear Sir:*
>
> *I've belatedly had called to my attention your July article on the game of Hex, and would appreciate a copy of the issue if it could be sent.*
>
> *I feel I should take mild issue with the characterization of my reverse-Hex proofs as "involved", since indeed they're strikingly similar to the usual forward Hex proof. Also, your report that the first player of the odd-board game "can lose" seems an understatement, since he can in fact always be beaten.*
>
> *Just in case you didn't have at hand this latter proof, and to support my contention that the proofs aren't much more involved, I'll write them down:*
>
> *For the reverse game, in just the same manner as in the forward, we show the impossibility of a draw — and immediately have the corollary that one player must have a forced win.*
>
> *In the even-board case we assume B (the second player) has a win, and describe a strategy for A (the first player) that must win: he initially plays arbitrarily, and regards himself thereafter as the* <u>second</u> *player, following the hypothesized winning pattern. (He*

plays again arbitrarily if called upon to play where his previous arbitrary move was made.) At the end of the game (we may assume, of course, that the game is over only when the board is completely filled), even though he can't actually play the required final stone, he must have won, since his last arbitrary move must occupy the final hex called for by the pattern (this inasmuch as it's the only hex remaining!).

This proof clearly applies to a large system of possible games, including of course the original even-board forward game.

In the odd-board case we assume A has a win. B now practices a slightly more complicated deception than previous winners found necessary: he plays his first move arbitrarily, pretends A's initial move was his, and that his own second move was only the third move in a game in which he's playing the hypothesized first-player strategy. The game proceeds and ends exactly as in the last case — B is unable actively to play the final move, but can count his last arbitrary move as the required reply. In this case, however, the initial stone played is the 'wrong' colour — B would have the hypothesized win if it were his colour instead of A's. But in the reverse-Hex game this difference can do B no harm, so again we have a contradiction.

Gardner replied to Winder a month later:[4]

Thank you for your long and fascinating letter, which I found in a large batch of mail sent to me yesterday by Scientific American. A notation on the letter indicates that the magazine has sent you a copy of the issue you requested, so I assume you have received it. ... The "can lose" phrase was due to an editor's revision of my copy. It's amazing how simple editorial changes that seem innocent to the person doing the revising can be misleading from a mathematical standpoint. ...

In 1999 Jeffrey Lagarias and Daniel Sleator proved a curious Rex property that immediately implies Winder's property: each player can prolong the game until the bitter end:[5]

(∗) for $n \times n$ Rex starting from the empty board, each player has a strategy that avoids losing until the board is filled.

We will show a proof of this later.

Around 1974 at the University of Wisconsin-Madison, Ronald J. Evans taught a math course based on *Excursions into Mathematics* by Anatole Beck, Michael N. Bleicher and Donald W. Crowe.[6] The text's mention of Rex piqued his interest, and Evans proved an analogue of Beck's acute-corner-loses-Hex theorem:[7]

- for $n \times n$ Rex with n even, the acute corner is a winning first move.

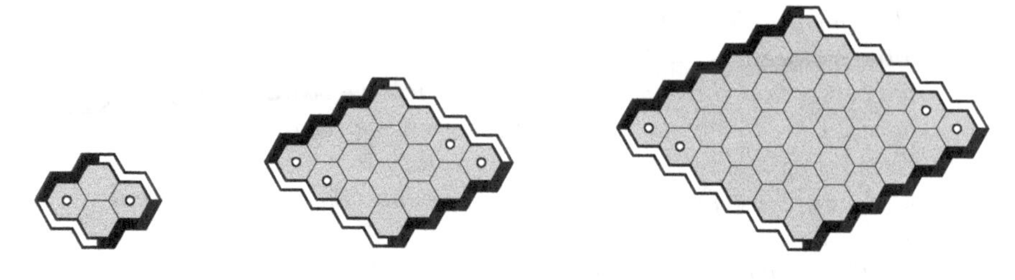

FIGURE 10.2: Some Rex-winning White opening moves.

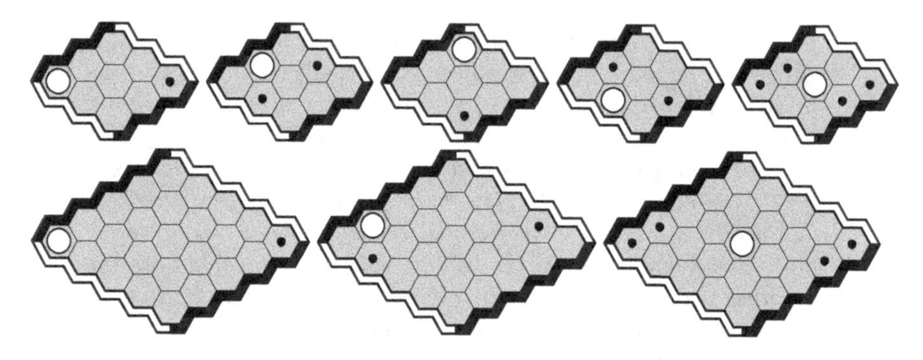

FIGURE 10.3: After White opens, some Rex-winning Black replies.

Almost 40 years later, the authors and Philip Henderson found another even-board winning opening,[8] as shown in Figure 10.2:

- for $n \times n$ Rex with n even and at least four, each player's-side cell adjacent to an acute corner is a winning first move,

They also found some odd-board winning replies, including those shown in Figure 10.3:

- for $n \times n$ Rex with n odd and at least three, each second-player move that reflects the first-player's move — through either the horizontal axis between the acute corners or the vertical axis between the obtuse corners — is a winning reply, and if the first-player's move is to the center then the acute corner or its player-side neighbour is a winning reply.

10.2 Terminated Rex

Property (∗) epitomizes Rex, so we give a proof here. Readers who wish can skip this section without any loss of continuity.

FIGURE 10.4: TRex, Black to play, and a winning strategy.

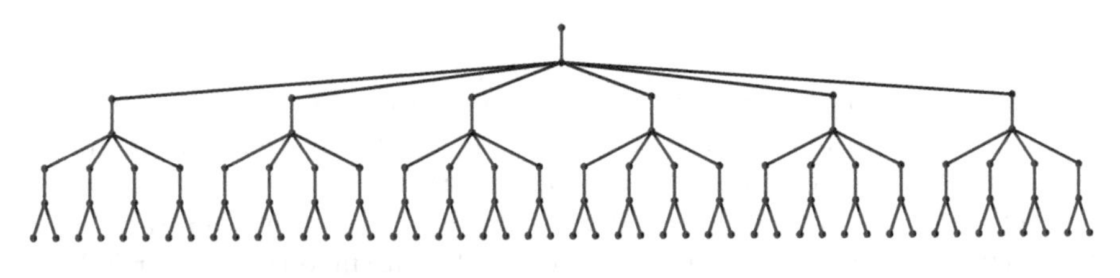

FIGURE 10.5: Black's strategy tree (unlabelled) for Figure 10.4.

Our proof is similar to that of Lagarias and Sleator. To simplify matters, we introduce Terminated Rex (TRex), a Rex variant with this extra rule: during the game, if neither player has won and there remains exactly one empty cell, then the game ends in a draw. Figure 10.4 shows a TRex position with Black to play. After the first move, Black can win with the pairing.

For a particular state — so, a position and the player-to-move, a strategy can be described by giving the set of possible game continuations. Such a set is often depicted as a tree diagram as in Figure 10.5. Each tree node corresponds to a continuation state, with the root node usually drawn on top. For each state with the player to move, the move is represented by the single down-edge. For each state with the opponent to move, the set of possible moves is represented by the set of down-edges. Figure 10.5 shows a first-player strategy, so the root node has one down-edge. Then the opponent has six possible moves, and then four, and finally two.

Our proof of $(*)$ uses these TRex properties:

- for a position and player with a winning TRex strategy, removing or adding a stone leaves a non-losing strategy,
- the empty $n \times n$ board is non-losing for each TRex player.

The first TRex property is illustrated in Figure 10.6, where a black stone was added to the position in Figure 10.4. So the Black TRex winning strategy for the original position yields a TRex non-losing strategy for the new position: make the same first move as before; then avoid cell a and force White to match the other pairs.

The second TRex property implies $(*)$: each player can follow a non-losing TRex strategy, so the TRex game ends in a draw, namely with one empty cell and with neither player having joined their sides. So in the Rex game,

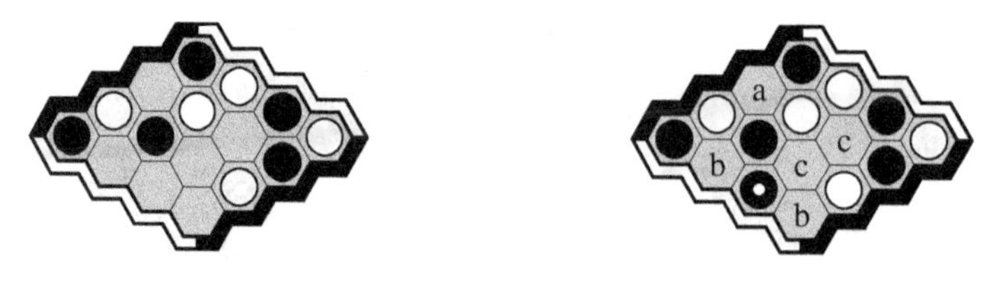

FIGURE 10.6: TRex, Black to play. Adding a stone changes a winning TRex strategy to a non-losing TRex strategy.

whoever has to move last — the second player if n is even, the first player if n is odd — will have to fill the last cell. And since there are no draws in Rex, this last move just join that player's sides, so they lose.

Now let us prove the first TRex property in the case where a stone is removed. For a position P with a stone at cell c, assume that player X has a winning strategy S. Now remove the stone at c from P. When it is X's turn, X pretends there is still a stone at c and follows S. So what happens if at some point Y plays at c? X then pretends that Y has instead played at some empty cell d, and follows S (where in reality there is a stone at c but none at d). X continues in this way: at each point, the difference between the real game and X's pretend game is always that one stone is missing. In the pretend TRex game, X wins by following S, so Y eventually joins her two sides. If this also happens in the real game, X wins. If it does not happen, it is because the missing stone q is part of a path of stones joining Y's two sides. Then, from this point, X changes to this strategy: play anywhere except q (so, take advantage of the TRex property of always having at least two cells to choose from). This guarantees that X does not lose (since any X-chain joining X's sides must use q).

The case in which a stone is added follows from the case in which a stone is removed. For a position P with no stone at cell c, assume X can win. Change P to P′ by adding a stone for either player at c. For P′, assume by way of contradiction that X loses. So for P′, Y wins. Now remove the stone at c: for P, Y does not lose, contradicting our assumption that X wins.

Now we prove the second TRex property: the board is empty, the players are X and Y, and P — either X or Y — moves first. Assume by way of contradiction that Q — either P or not-P — can win. After P's first move, it is not-P's turn and Q can still win. Now remove P's stone: the board is empty, it is not-P's turn, and — by the first TRex property — Q can avoid defeat. Now exchange the colours and players: the board is empty, it is P's turn, and not-Q can avoid defeat. This contradicts our assumption that Q can win.

So we have proved (∗). For $n \times n$ Rex, by following a TRex non-losing strategy as long as there remains at least two empty cells, each player can avoid defeat until the board is filled. □

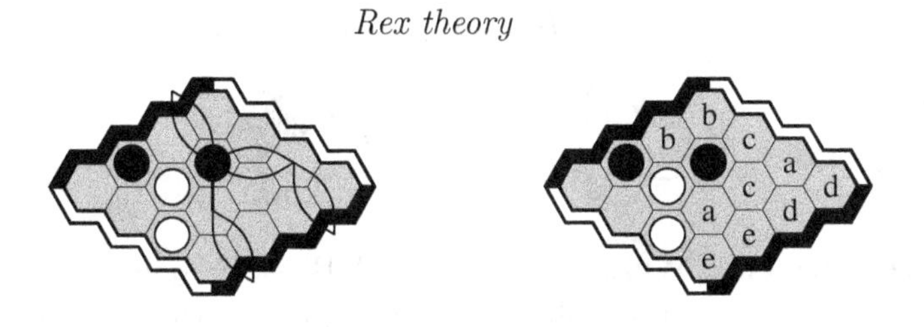

FIGURE 10.7: White can force Black to match pairs, so White wins.

10.3 Pairing strategies

These opening-move results we have discussed so far are based on strategy stealing: they imply that certain moves win, but reveal nothing about the corresponding winning strategy. So, how does one find such a strategy? More generally, what Rex properties are helpful when playing?

To play Hex, it helps to understand virtual connections and inferior cell analyis; to play Rex, it helps to understand the Rex analogues of these notions. In this section we discuss a Rex analogue of virtual connections, namely pairing strategies.

A *join-pair* for a player is a pairing strategy that joins the player's two sides. If you learn only one thing about Rex, learn this:

- you can force an opponent to follow any pairing strategy, so if you find an opponent join-pair, the opponent will lose.

Consider for example Figure 10.7, where Black has a join-pair. White plays like this: if Black takes a cell of an empty pair, White takes the other; otherwise, if some cell is in no pair, play there; otherwise — so each empty cell is in an empty pair — play in any such cell. We leave it to the reader to check that this is always possible, and results in Black playing at least one cell of each pair, thus joining the two black sides. So, regardless of who plays next, White wins.

For a Rex position, a *pre-join-pair* is a set of empty cells that, when one is coloured, leaves a join-pair for that colour. The three marked cells in the left diagrams of Figure 10.8 form a black pre-join-pair, since colouring the 0-cell black leaves a black join-pair. We call the player who will make the last move if the game fills all cells *last-player*. Here is why pre-join-pairs are useful:[9]

- in Rex, if last-player has a pre-join-pair, she loses.

Consider for example the empty 2×2 board with White to play. The board has $2 \times 2 = 4$ cells, so Black is last-player. And from Figure 10.8 we see that Black has a pre-join-pair, so White can win: avoid the pre-join-pair's 0-cell (here, the top obtuse corner) and force Black to follow the pairing. The number of empty cells is even, so Black plays last if the board is filled, so Black eventually plays

FIGURE 10.8: Rex, White to play. (left) A Black pre-join-pair that White uses to win: (next) avoid cell 0, force Black to match the pair. (next, right) Possible game sequences.

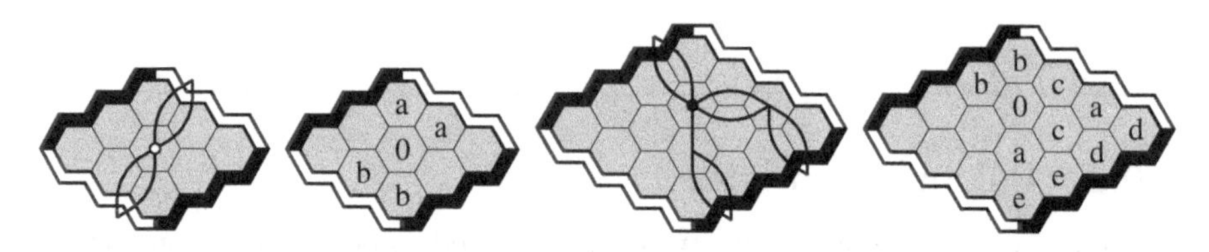

FIGURE 10.9: Rex, White to play. Black wins with these pre-join-pairs.

the 0-cell, and White forces Black to follow the pairing strategy, so Black joins her two sides and loses. Figure 10.8 shows two possible lines of play.

Similarly, consider 3×3 Rex with White to play first. The board has nine cells, so White is also last-player, and Black can use the left pre-join-pair in Figure 10.9 to win. Figures 10.9 through 10.11 give winning strategies for 4×4 and 5×5 Rex, while Figure 10.12 shows the opening moves that work with these — or symmetric — 4×4 strategies.

10.4 Inferior cells

In Rex as in Hex, inferior cell analysis starts with dead cells. Recall that a cell is dead if it is not in a minimal set of empty cells joining a player's two sides. So, in Hex play, you should avoid dead cells, while in Rex play, you should play them.

Similarly, in Hex, if you have a captured set — empty cells with a replying strategy that leaves everything in the set yours or dead — then you should fol-

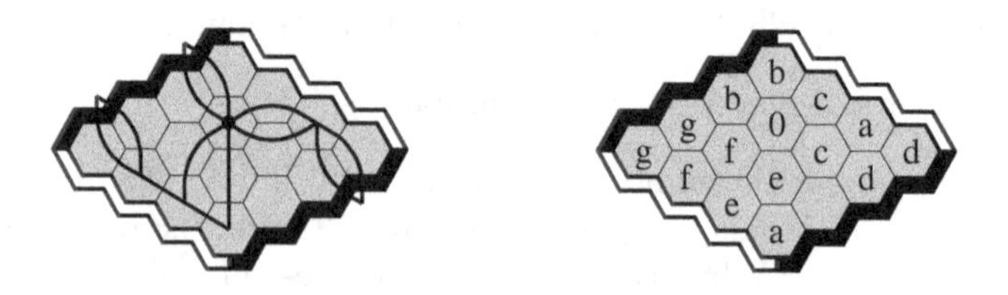

FIGURE 10.10: Rex, White to play. White wins with this pre-join-pair.

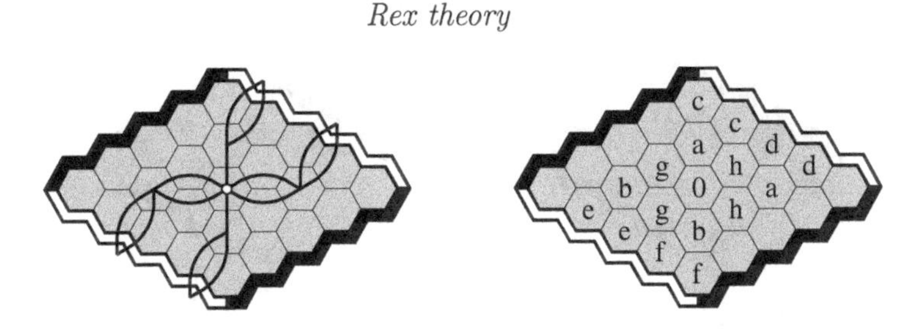

FIGURE 10.11: Rex, White to play. Black wins with this pre-join-pair.

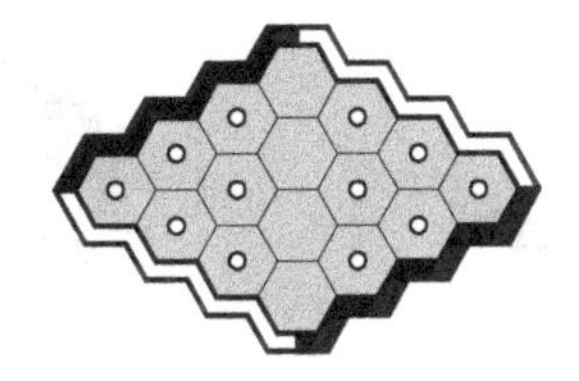

FIGURE 10.12: Rex, White to play. Some winning opening moves.

low the capturing strategy. In Rex, if your opponent has a captured set whose replying strategy is a pairing strategy, then you should force your opponent to follow that strategy.

Follow these Rex proverbs as much as possible:[10]

- bury the dead (if a cell is dead, then play there),
- do not kill or capture,
- force your opponent to capture.

The first proverb has highest priority: it is always safe to play at a dead cell — if such a move loses, then all moves lose. For example, this proverb suffices to solve the Rex puzzle of Ronald J. Evans we saw in an earlier chapter. See Figure 10.13.

To illustrate the Rex proverbs, consider the game in Figure 10.1. Which moves are weak? In a position with at least one winning move, a *blunder* is a losing move. Can you find any blunders?

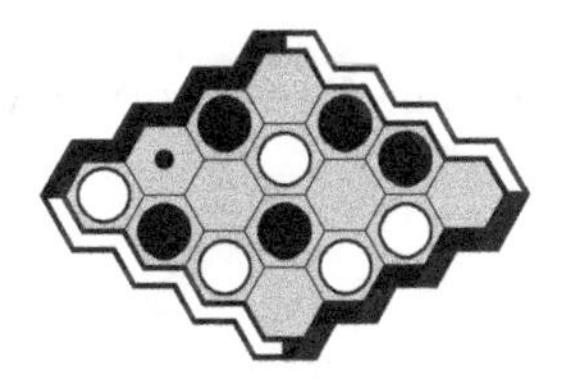

FIGURE 10.13: Evans's Rex puzzle: White to play. The dead cell is the only winning move.

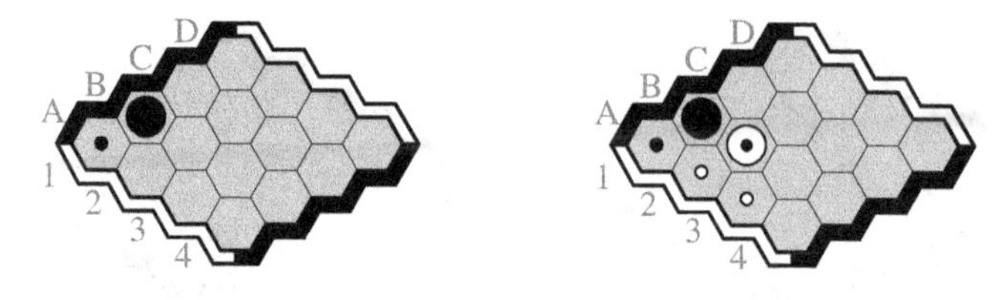

FIGURE 10.14: Rex proverbs advise against these moves, which kill (left), do not bury the dead (right), or capture (right).

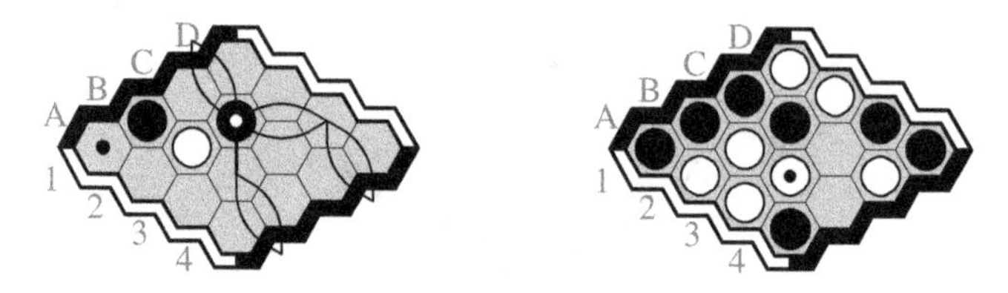

FIGURE 10.15: Rex blunders. (left) Instead of burying the dead, Black makes a join-pair. (right) b4 was dead.

The game starts with Black 1.b1. From Figure 10.12 after colour exchange, we see that b1 — on the first player's side — is a winning opening. But this move kills cell a1, violating the second Rex proverb. In Rex, playing at the cell that is killed is at least as good as playing at its killer,[11] so 1.a1 is at least as good a move as 1.b1. Figure 10.12 confirms that both moves win. See Figure 10.14.

White 2.b2 also violates the Rex proverbs: it ignores the dead cell at a1, and captures {a2,a3}. 1.b1 wins, so all White replies lose, but 2.b2 makes life easy for Black: of Black's 14 possible replies, 13 win. By contrast, after 2.a1, only 7 Black replies win.

Similarly, Black 3.c2 violates the Rex proverbs, still ignoring the dead cell at a1, and capturing {c1,d1}. In fact, 3.c2 is a blunder: it creates a black pair-join, so now White can win easily. See Figure 10.15.

White forces Black to follow this pairing strategy until 12.d2, but this wins, so is not a blunder. But after 13.a4, White's only winning reply is b4: instead, White blunders with 14.b3, and Black wins by avoiding c3. So this Rex game had many weak moves, but only two blunders: moves 3 and 14.

Now that you have seen some Rex theory, try the new Rex puzzles we created in Appendix D.

Notes

[1] In the outline for his Hex column, Gardner writes of the Rex first-player win/loss theorem: *Proof is complex, since it differs for even & odd, etc.*

[2] 1957.03.23 letter Gardner-Shannon [23].

[3] 1957.08.19 letter Winder-Gardner [23].

[4] 1957.09.11 letter Gardner-Winder [23].

[5] [46]

[6] [5]

[7] [17]

[8] [29]

[9] [29]

[10] Kenny Young and Ryan B. Hayward found more Rex inferior cell proverbs [77].

[11] [77]

Chapter 11

Quest for strategies

What is your quest?

Bridge-keeper in *Monty Python and the Holy Grail.*

To do something, say something, see something, before anybody *else — these are things that confer a pleasure compared with which other pleasures are tame and commonplace, other ecstasies cheap and trivial.*

Mark Twain

If there is a problem you can't solve, then there is an easier problem you can solve: find it.

George Pólya

From the beginning, as Hex players realized that first-player winning strategies must exist, they sought to find them. As Hein told Politiken readers in 1942[1] and Nash explained to Gardner in 1957,[2] finding such strategies on small boards is easy. But — as Hein remarked — things get interesting as board size increases:[3]

> *With the 5×5 board it is already possible to have interesting games and difficult puzzles.*

Try it yourself: on what boards can you find a winning first-player strategy? And on what boards can you find *all* opening winning moves? Hint: use virtual connections and mustplay reasoning. Check your answers in Figure 11.1.

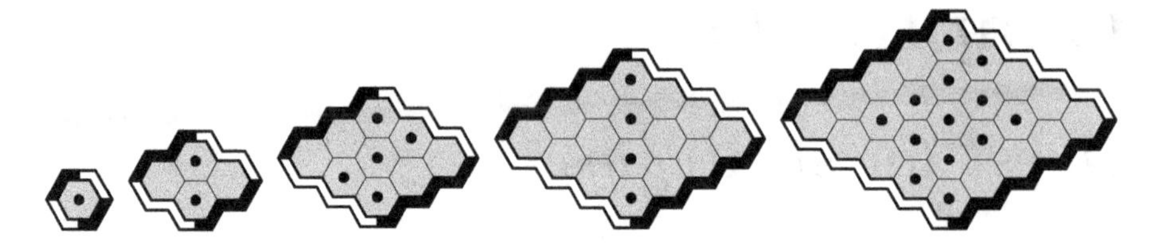

FIGURE 11.1: Black to play. On these boards, all winning first moves.

11.1 Lindhard's 6×6 strategy

As Gardner reported in 1957,[4]

> *On larger boards the analysis becomes enormously difficult. So far as I know no one has fully analyzed the possibilities in a game of Hex played on a six-by-six board. Of course the standard 11-by-11 board introduces such an astronomical number of complications that a complete analysis seems out of the question.*

Unbeknownst to Gardner, Lindhard had already analyzed 6×6 Hex by 1943! We only realized this ourselves in October 2017, when we deduced the meaning of the six diagrams from the Lindhard archive shown in Figures 11.2 through 11.4.

These beautiful diagrams — perhaps intended for the unfinished Hein-Lindhard booklet — outline a winning 6×6 first-player strategy using mustplay analysis! In Diagram A, White — the first player — has played the strategy's opening move at cell 1: if Black now plays at any of the twenty-two cells marked 2, then White can win by playing at cell 3. So, Diagram A reduces Black's mustplay to size $6 \times 6 - 1 - 22 = 13$: after White's opening move, Black must play at one of the thirteen cells not marked 1 or 2, or lose.

In similar fashion, Diagrams B through F further reduce Black's mustplay to size ten, nine, eight, three and finally zero, confirming that Lindhard's strategy is a winning strategy.

Diagram D shows the most complicated substrategy. After moves 1-9, there are six possibilities for move 10. The diagram shows the line of play for only one case: the other five cases are described in the diagram's writing. In some of these cases White needs to use cells f4-f6.

While preparing this book, before discovering the Lindhard archive we found our own 6×6 strategy. It is similar to Lindhard's: see Figure 11.5. The bottom left diagram there corresponds to Lindhard's Diagram D. There, if White next blocks near E4 (respectively C4) Black replies B5 (respectively E4). So Black wins.

Figure 11.6 shows a position from a first-player 6×6 strategy that uses only 27 of the board's 36 cells. Can you find the next move?

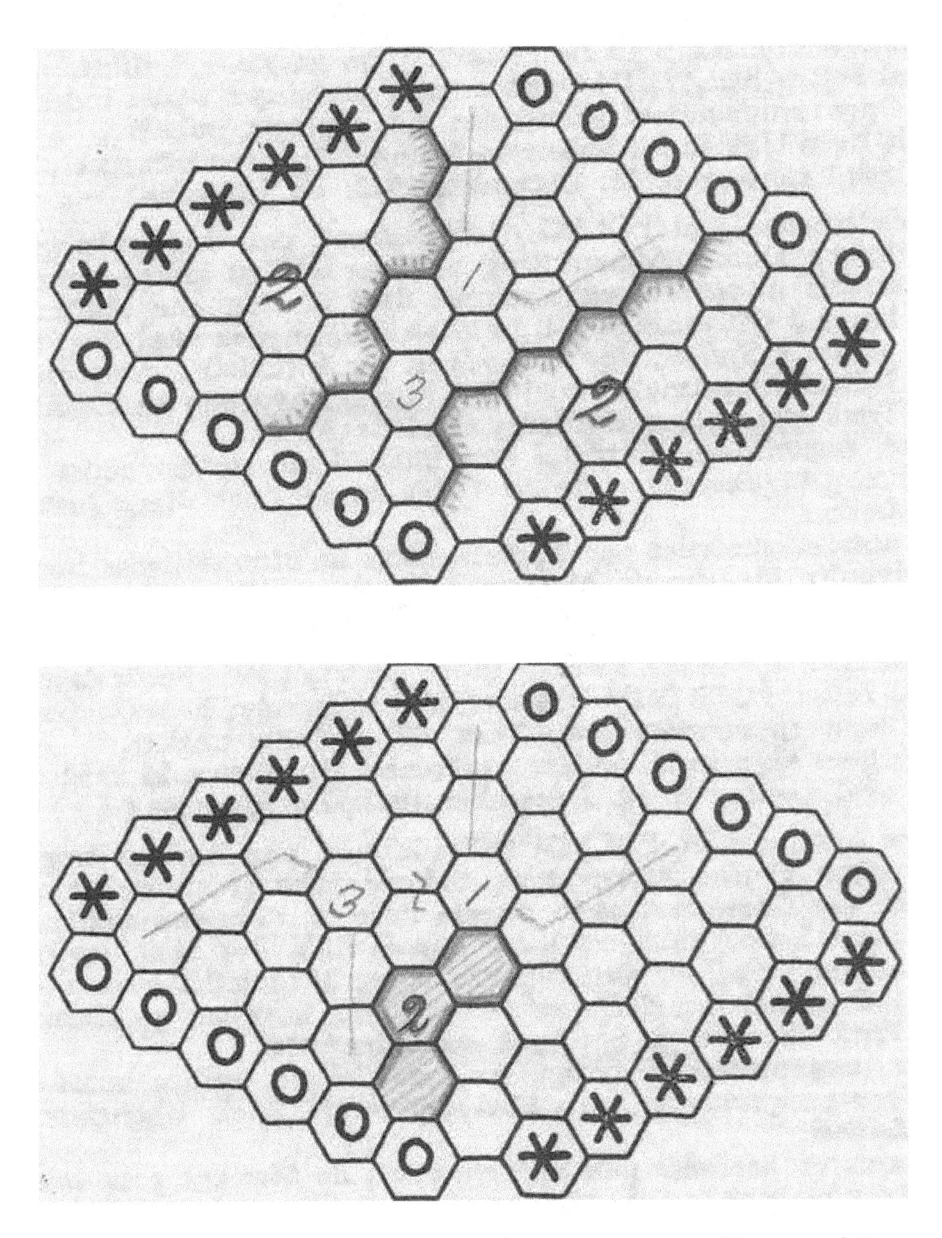

FIGURE 11.2: Lindhard's 6×6 1st-player-win strategy. Diagrams A,B.

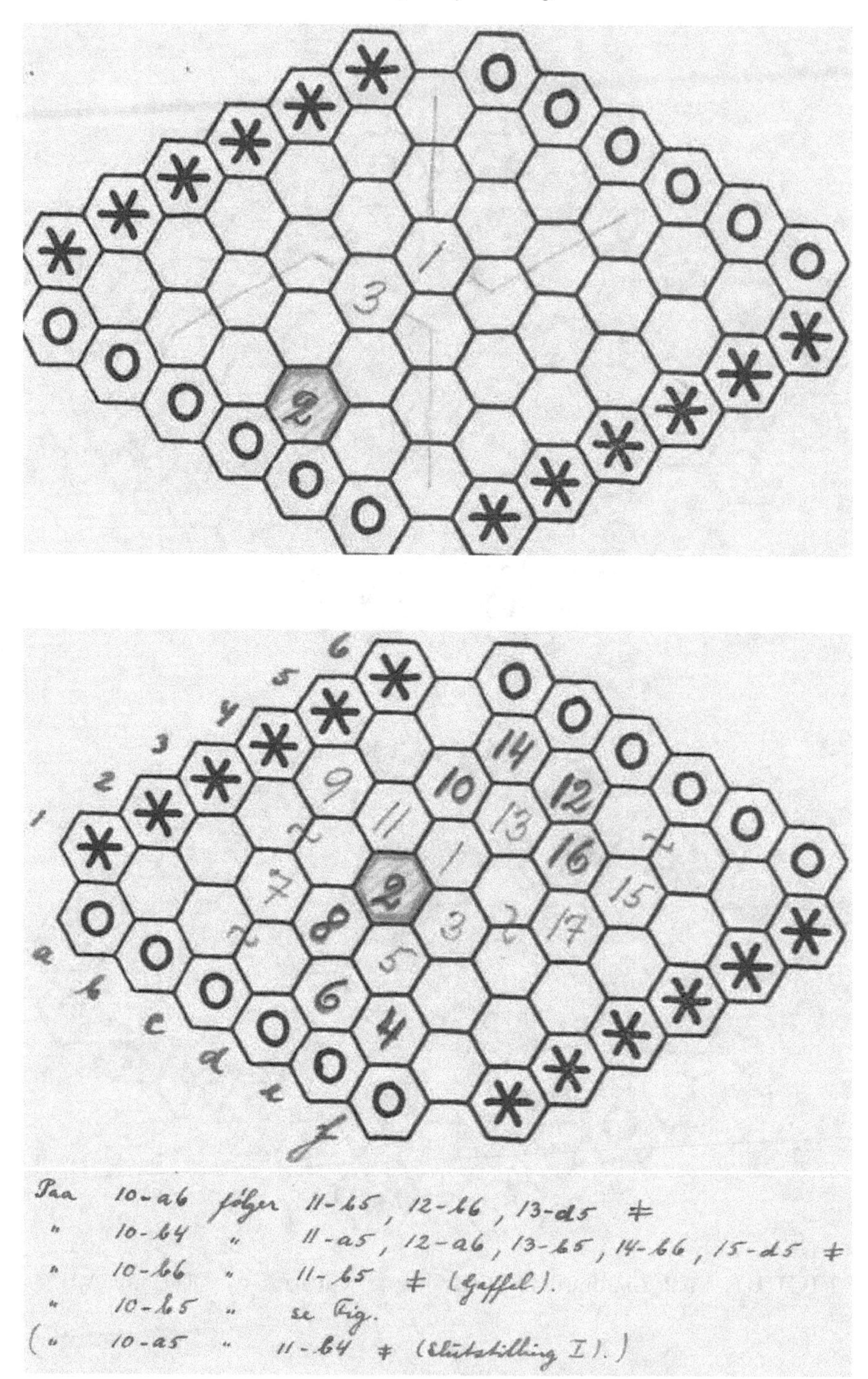

FIGURE 11.3: Lindhard's 6×6 strategy. Diagrams C,D.

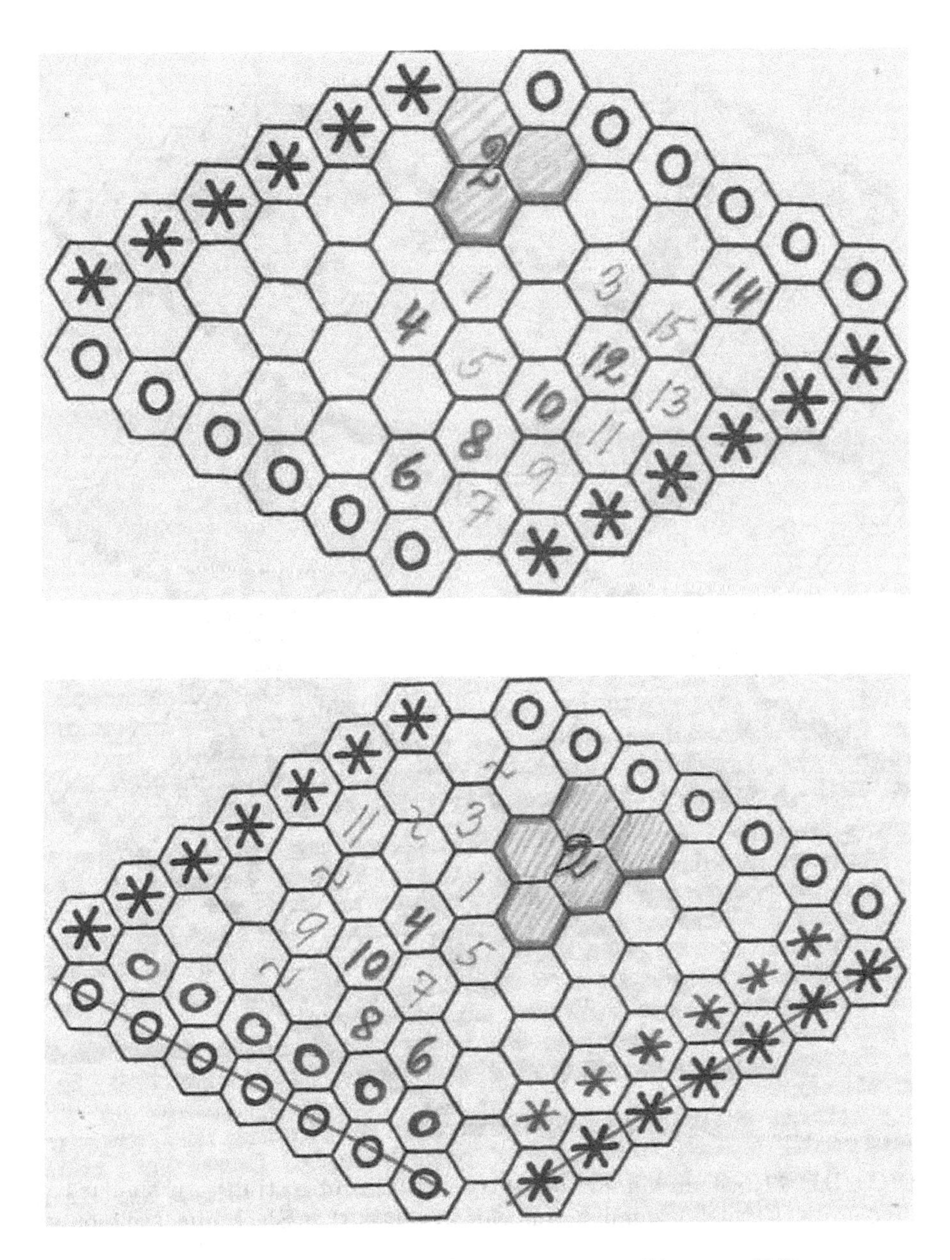

FIGURE 11.4: Lindhard's 6×6 strategy. Diagrams E,F.

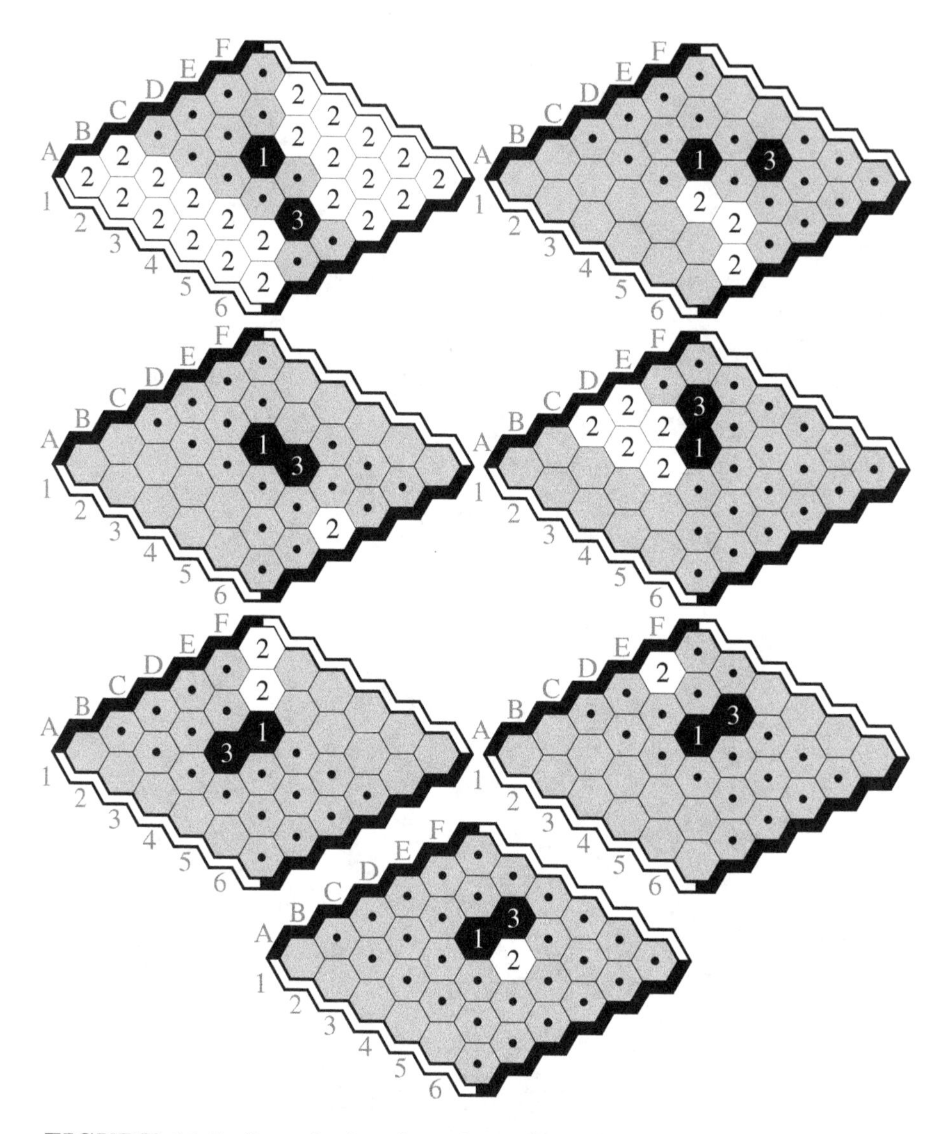

FIGURE 11.5: Our winning first-player (Black) 6×6 strategy is similar to Lindhard's. In the bottom diagram, can you show that Black wins using only the dotted cells?

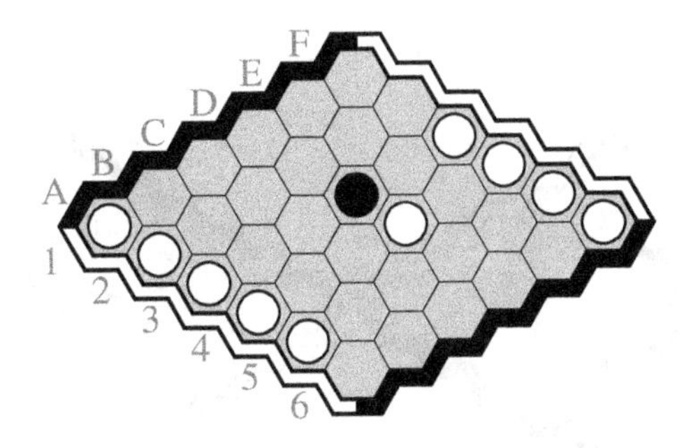

FIGURE 11.6: Black to play. Find all winning moves. Solution later.

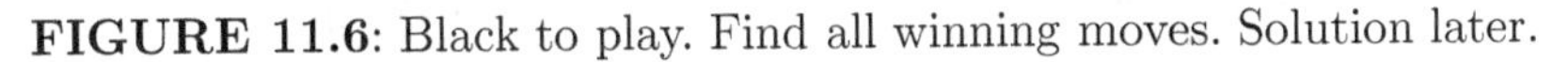

11.2 Letters to Gardner

Gardner's unstated challenge — find a winning Hex strategy on boards 6×6 or larger — was met by at least three readers: Robert O. Winder in August 1957,[5] Jerry Weaver in August 1966,[6] and Edward E. Max in February 1969.[7] Weaver, at the time a 21-year old math student at Bluffton College in Ohio, found quick centermost-opening strategies on both the 6×6 and 7×7 boards:

> *... after I came across your article two summers ago, I started playing the game on a board of [six-by-six] size. It soon became obvious that the first player could always win if he played correctly, and I believe that I have now made a sufficiently exhaustive (and exhausting!) analysis of the possibilities to be able to say that the first player can always win in no more than eleven moves. ...*
>
> *The seven-by-seven board also seems to allow the first player a sure win, this time in no more than thirteen moves. ...*

Weaver limited his description of each strategy to its main line of play, plus one variation. Figure 11.7 shows the initial moves of each main line: can you guess the remaining moves? Figures 11.8 and 11.9 have the answers.

In subsequent updates of the Hex column,[8] Gardner omitted any reference to 6×6 analysis. As far as we know, he did not publicly report on any of these readers' strategies, although he did put Max — who also found both a 6×6 and a 7×7 strategy — in contact with Weaver.

11.3 Back to Pittsburgh

In 1948, before starting his doctoral studies in Princeton, John Nash was a student at Carnegie Institute of Technology in Pittsburgh. Around forty

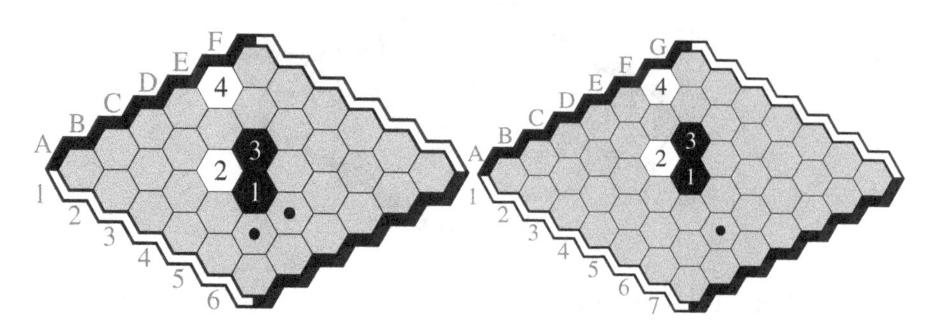

FIGURE 11.7: Start of Weaver's strategies. Black to win by move 21 (left), move 25 (right). Dots show possible next moves.

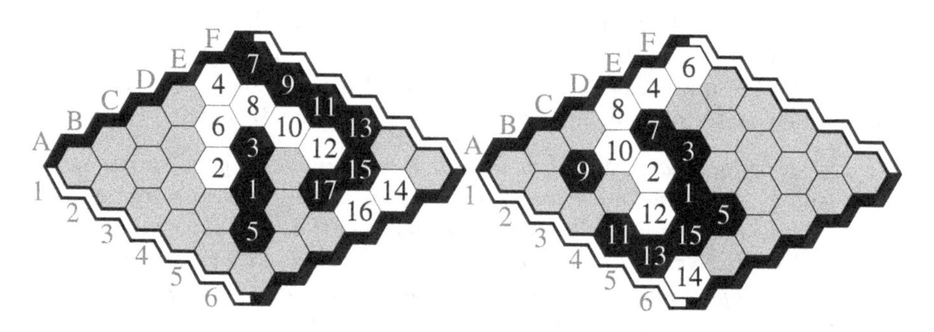

FIGURE 11.8: Weaver's two 6×6 variations. Black wins by move 21.

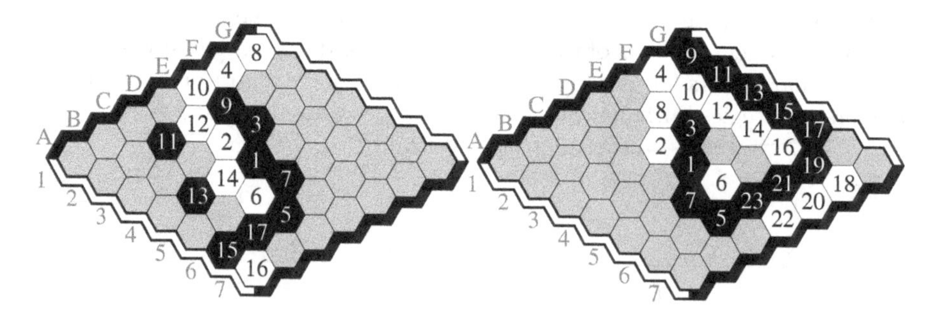

FIGURE 11.9: Weaver's two 7×7 variations. Black wins by move 25.

years later the institute had merged with another and formed Carnegie Mellon University, where student Robert Enderton was automating the process of solving Hex positions:[9]

> *[My] algorithm that solved 6x6 Hex openings was kind of complicated. The main idea was this: say you've got a position like in the diagram below with [Black] to play.*
>
> 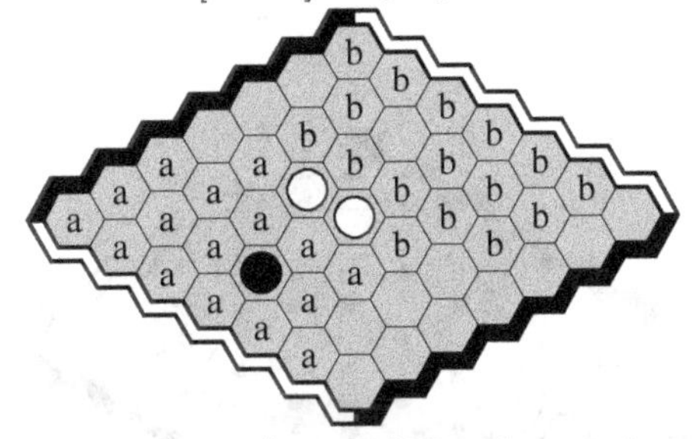
>
> *If it can prove (via search) that [White] can always connect his stones in the middle to the bottom edge, using some set of [cells] (marked a), and separately prove that he can connect his stones to the top edge, using a different set of [cells] (b), and the two sets are disjoint, then it can combine those lemmas into a proof that [White] has a winning position. ... [My algorithm] had a few other tricks too, such as places where it could reason that a particular exchange of moves could never be helpful. But my program didn't have good heuristics for where to play when it couldn't find a provable win. So if you played it on a big board, typically it would make a few bad moves, and then it was able to resign because it could prove to itself that it was in a losing position.*

By 1995, Enderton's program had found all 6×6 winning opening moves.[10] The program also found three 7×7 winning openings. Enderton recorded his results on his CMU webpage, *Answers to infrequently asked questions about the game of Hex*, which also includes the challenging puzzles shown in Figure 11.10.[11]

To understand the search process performed by Enderton's solver — find the win/loss value of each opening 6×6 move — let us return to the winning 6×6 strategy in Figure 11.5. We showed that d3, the strategy's first move, is a winning opening. By symmetry, so is c4, its reflection through the board's center. Also, the strategy does not use any cell in {a1, ..., a5}, so these are losing White openings, since Black can use the strategy in response. So, by colour exchange and board relabelling, {a1, ..., e1} are Black losing openings. By board symmetry, {f6, ..., b6} are also Black losing openings.

To this point, we know the win/loss value for 12 of the 36 opening moves. Now we want to solve the $24/2 = 12$ asymmetric remaining openings, either by hand or with a computer. When our search tries the opening Black 1.a2, it finds a winning White reply. Can you find it?[12] By symmetry, Black 1.f5 also loses. We continue with the 11 asymmetric remaining cases. Eventually

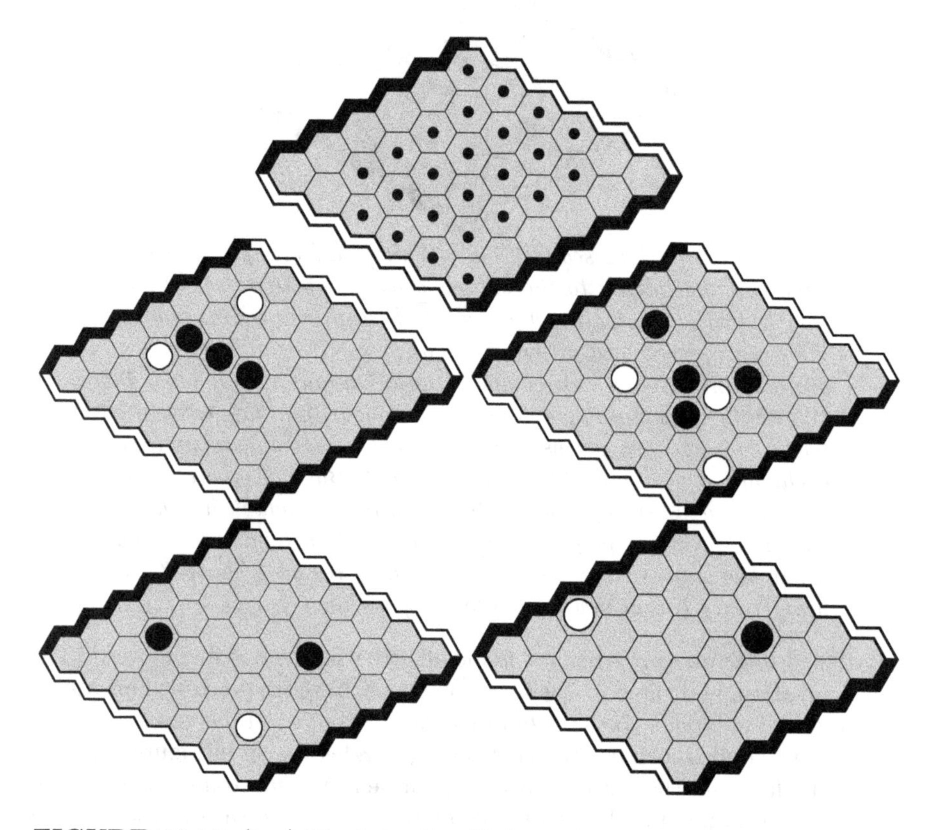

FIGURE 11.10: (top) Black to play. Enderton's program found all winning 6×6 openings. (bottom) White to play. Four Enderton puzzles.

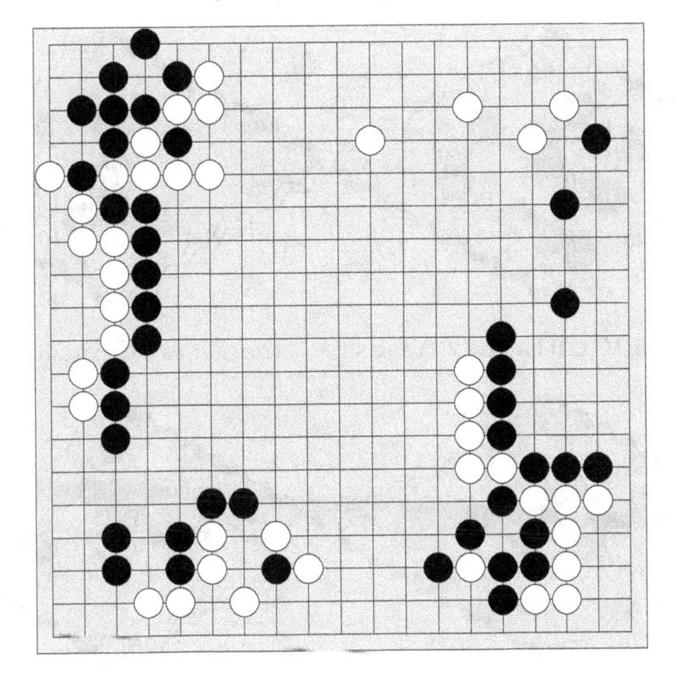

FIGURE 11.11: A Go game in progress.

we find a winning strategy for each of these openings, and we are done. See Figure 11.10.

11.4 Go lessons

As a board game, Hex — less than 100 years old — is an infant. By contrast, Go is an oldtimer, having been played continuously for over 3000 years. Go is a game of territory: whoever surrounds more at the end is the winner. In Go as in Hex, ladders — forcing sequences in which each move 'rubs shoulders' with the opponent's last move — arise often. And like Hex, Go can be played on any $m \times n$ board: 9×9, 13×13, and especially 19×19 are most common, with smaller boards sometimes used by beginners.

Figure 11.11 shows a Go game in progress. Play has started near the board's edges, where each player can use the corners and sides to help surround territory.

For a few months around 1990, Enderton took weekly Go lessons from Jing Yang, who later won the Canadian Open Go championship three times. In 1991, Yang became interested in Hex after hearing a talk by Enderton on

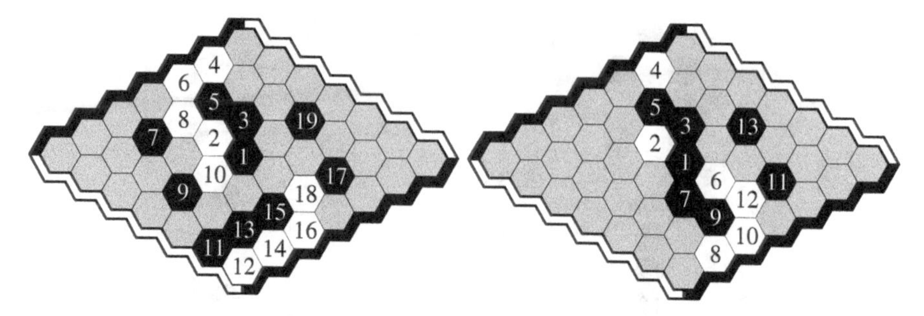

FIGURE 11.12: Yang's 7×7 strategy: two variations.

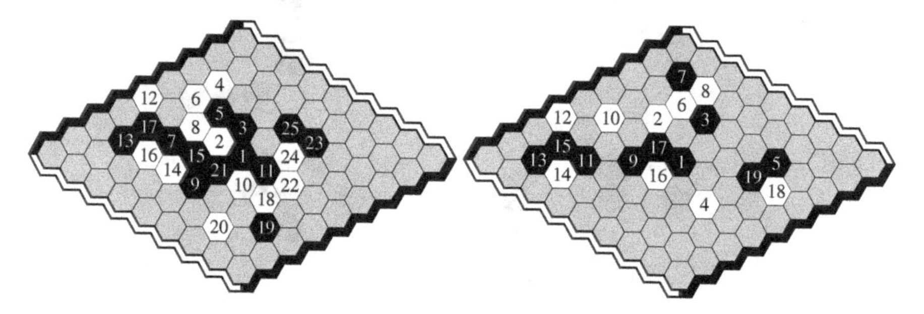

FIGURE 11.13: Yang's 9×9 strategy: two variations.

his Hex solver. A month later Yang had found a winning 7×7 strategy. See Figure 11.12.

Around 2000, Yang — then a graduate student at the University of Manitoba in Winnipeg, Canada — was seeking strategies that could be easily verified — either by a human or a computer — as correct (always winning) and complete (at each point in the strategy, each possible opponent reply is accounted for). Yang's method combines mustplay reasoning, case analysis, and repeated use of local strategy patterns. His initial center-opening 7×7 strategy uses 41 patterns.[13] A complete description of the strategy fills 16 2-column pages, so about $32*60 = 1920$ lines. By 2003, Yang had found another 7×7 strategy,[14] an 8×8 strategy, and a 9×9 strategy.[15] Figure 11.13 shows two variations of Yang's 9×9 strategy: he also wrote an applet on which you can play against the strategy.[16]

Others later found other 7×7 and 8×8 strategies with compact descriptions. For example, Kohei Noshita's strategies exploit local join-or-win strategies, in which a player either forces a win or joins two particular groups.[17]

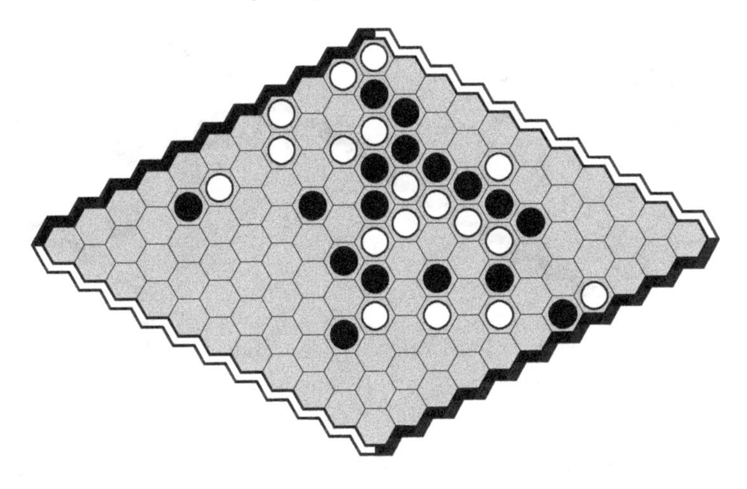

FIGURE 11.14: Silverman puzzle. Black to play.

11.5 Computers and games

In the 1990s, as computers got faster and cheaper, the search for Hex strategies shifted from computer-assisted human efforts to human-assisted computer efforts.

In 1992 in the Netherlands, Jack van Rijswijck was looking for a project for his master's thesis. A professor suggested that he try to get a computer to play Go. But Van Rijswijck did not know Go, so instead he went to the library, where he found David Silverman's games book *Your Move.*[18] Inside was a page on Hex, with the puzzle in Figure 11.14. Van Rijswijck was hooked: for his project, he and a classmate used reinforcement learning to try to teach a neural network to play Hex.

Van Rijswijck later made his way to the University of Alberta in Edmonton, Canada, where a team led by Jonathan Schaeffer had built a stronger-than-human checkers program:[19]

> *After my master's I kept on working on my Hex program as a hobby. I read up on everything to do with game artificial intelligence, including [Schaeffer's] book* One Jump Ahead. *I wanted to go back and get a PhD, and I wanted to do it at the UofA because that was the epicenter of game AI research.*[20]

Starting in 1998, Van Rijswijck wrote a computer program that extended Enderton's 6×6 results by finding win/loss values for multiple-move openings. But 7×7 strategies, especially for weaker opening moves, proved elusive, and Hex solving progress stalled. Consider for example the 7×7 position in Fig-

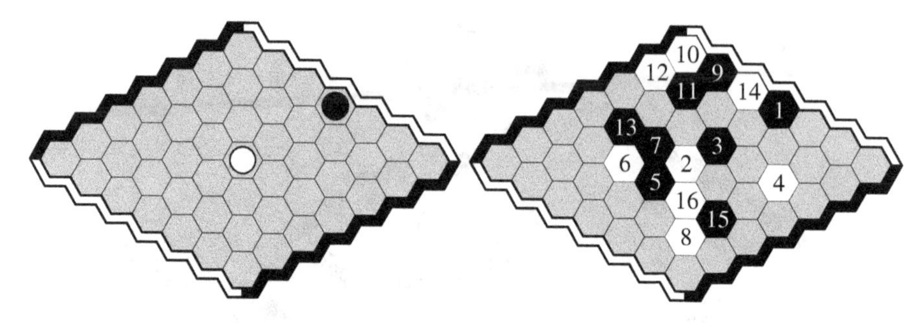

FIGURE 11.15: (left) A hard opening: can Black win? Find good Black replies. (right) Find Black's error in this sequence.

ure 11.15: can Black win? White wins the variation shown, but maybe Black made a mistake?

11.6 Automated solvers

In 2001, two years after joining the University of Alberta, Ryan Hayward started a project to build the eponymous Solver for Hex. Yngvi Björnsson wrote the initial version, helped by Mike Johanson and Nathan Po and Jack van Rijswijck. At first, the program could not solve 7×7 problems, but that changed once Hayward realized that filling captured cells does not change a position's win/loss value.

When Van Rijswijck heard this news, he was astounded, and reminded of a flaw in his program Queenbee that seemed related: Queenbee sometimes attacked an opponent's bridge (angle connection) from the wrong side. He went off and worked out the general theory of these observations, not just for Hex but more generally for the Shannon switching game, computing the first taxonomy of captured and dominated cell patterns.[21]

Consider for example Figure 11.16, which shows a continuation from a black 7×7 opening. The dots are white-captured and can be filled; this in turn captures two more cells, which can also be filled.

These observations — especially filling captured cells — greatly strengthen Solver: the number of available moves drops and the number of discovered virtual connections soars. Solver found the win/loss values for all 7×7 openings by 2003.[22] See Figure 11.16.

Over the next few years, programmer Broderick Arneson and doctoral student Philip Henderson strengthened Solver further.[23] Consider for example Figure 11.17. The game outcome does not change if adjacency is cut between the dotted cells, so the black chain effectively splits the game into left and

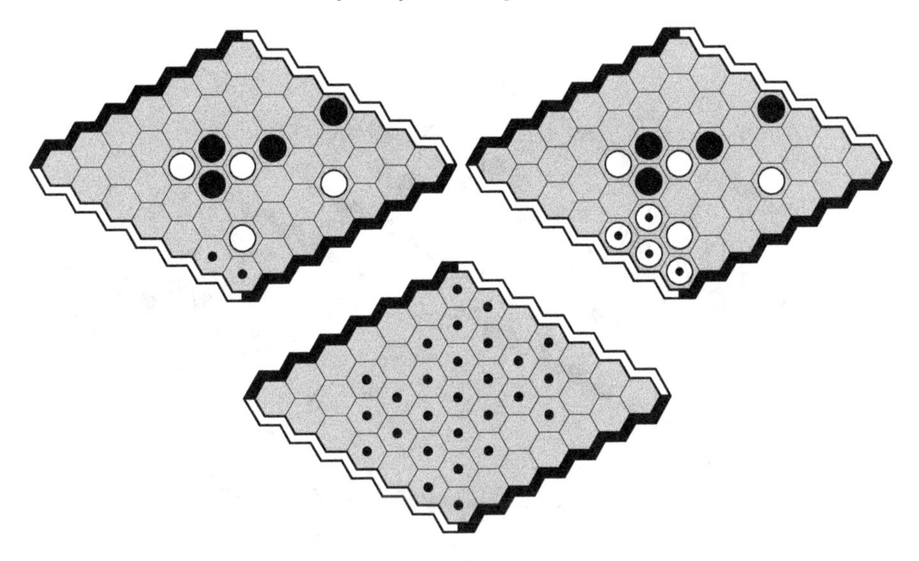

FIGURE 11.16: Filling captured cells (left) captures two more cells which can then be filled (right). This helps Solver, which found all 7x7 winning opening moves by 2003 (bottom, Black to play).

right parts. Black wins only by joining both sides to the black chain, so to solve the game, it suffices to solve the parts independently: this saves time, as continuations with moves from both parts can be ignored. By 2009 Solver had found all winning 8×8 openings. See Figure 11.17

Figure 11.18 illustrates another helpful observation. An empty region of the board is bounded by four alternately coloured chains, and the black chains are virtually connected. So the region is black-captured and can be filled.

Solving 9×9 openings required big improvements, including changing the search to a method based on a position's *proof number*, namely — in the current partly-expanded search tree — the minimum number of leaf nodes whose win/loss value must be a win in order for the root node to be a win. Often, moves with smaller proof numbers require less search to solve.

Another improvement was the use of more inferior cell patterns. Consider for example Figure 11.19. In each top row pattern, the dotted cell is permanently inferior and can be coloured black. In each bottom row pattern, each dotted cell is black-inferior to the empty cell, so when picking a move Black can ignore the dotted cells. For example, consider the leftmost pattern in the bottom row: a white move at the empty cell black-captures — yes, this is counter-intuitive — the dotted cells. Compare two cases: in one, Black plays at a dotted cell and White plays at the empty cell, leaving the other dotted cell dead; in the other, Black passes (skips her move) and White plays at the empty cell, capturing the other two. These two cases are equivalent, from which it follows that Black is better off *not* playing at a dotted cell.

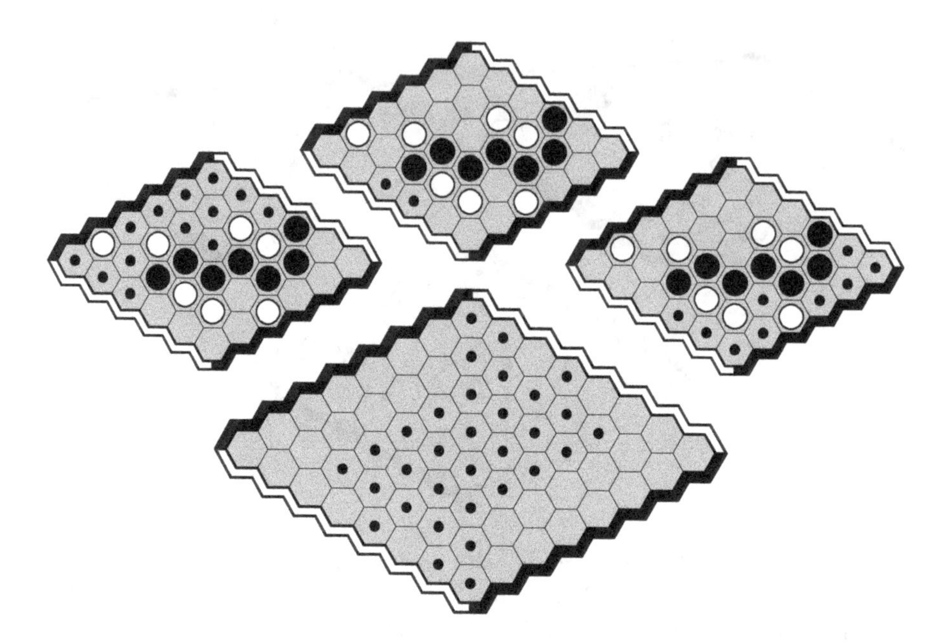

FIGURE 11.17: (top) The black chain splits the board in two. Black wins only by winning both subgames. This helps Solver, which found all winning 8×8 openings by 2009 (bottom, Black to play).

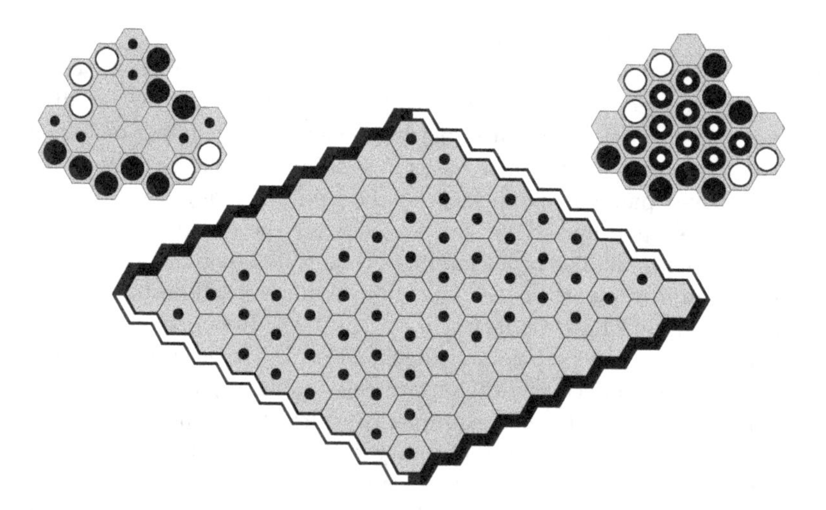

FIGURE 11.18: (left) The four chains isolate the empty region. Black can join her chains and so captures the region (right). This helps Solver, which found all winning 9×9 openings by 2012 (bottom, Black to play).

FIGURE 11.19: (top) Black can add a stone at each dot. (bottom) When moving, Black can ignore each dot.

Another improvement was, within the process for computing virtual connections, to fill-in captured cells when detected. By 2010, Solver had solved about half of the 9×9 openings.[24] More work was needed to solve the rest. Jakub Pawlewicz implemented a parallel version of the proof number search[25] and speeded up the search for virtual connections.[26] By 2012, all 9×9 openings were solved. See Figure 11.18. And on 20 March 2013 Hayward received an email from Pawlewicz with this footnote:

Meanwhile, 10x10 f5 solved, win for black.

The first 10×10 opening was solved! By November 2013 a second opening was solved.[27] As on smaller boards, the two strongest openings — and so easiest to solve — are the centermost, and in the second row of the inter-obtuse-corner diagonal. See Figure 11.20.

Solver has come a long way since 2003: the current version solves the hardest 7×7 opening in under one minute on a single-processor laptop. The 10×10 openings took several months on a 24-processor machine. But finding any winning opening on the original 11×11 board is probably decades away.

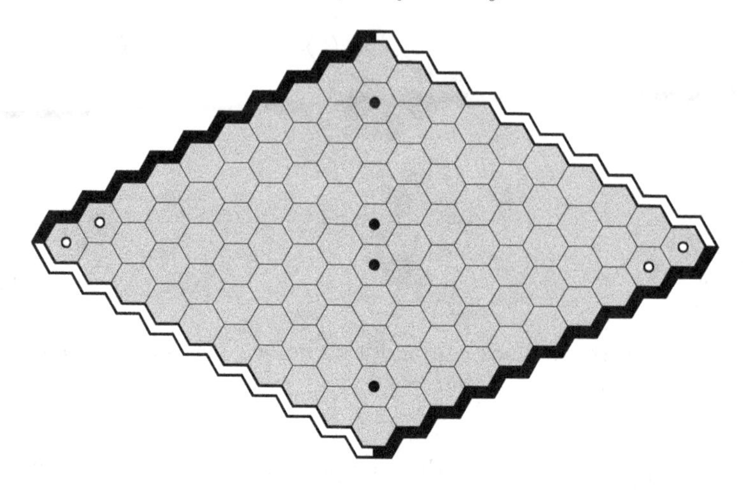

FIGURE 11.20: Solver found these winning 10×10 openings by 2013. By Hex theory, the white dots lose.

11.7 Solutions

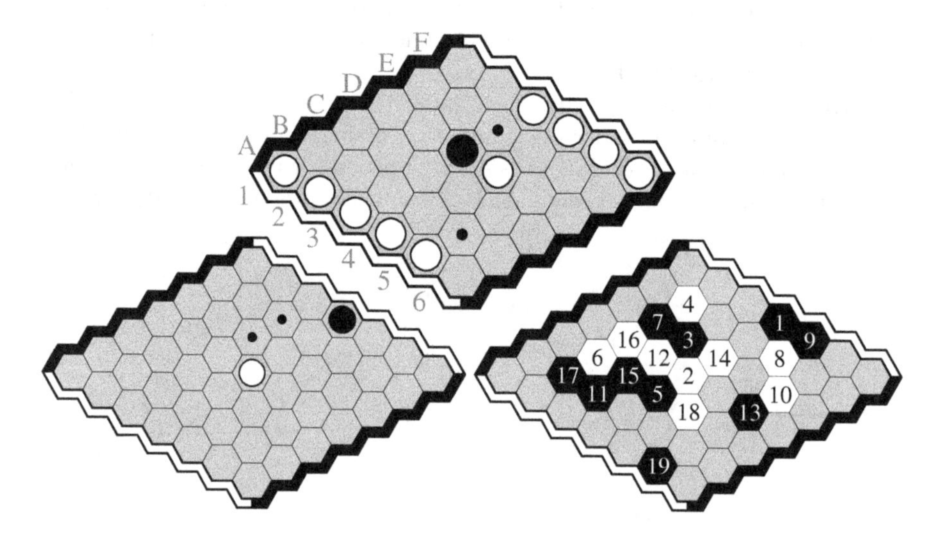

FIGURE 11.21: (top) Solution to Figure 11.6: all winning Black moves. (bottom) Solutions to Figure 11.15: (left) Black's move 3 was not a winner, (right) an error-free continuation.

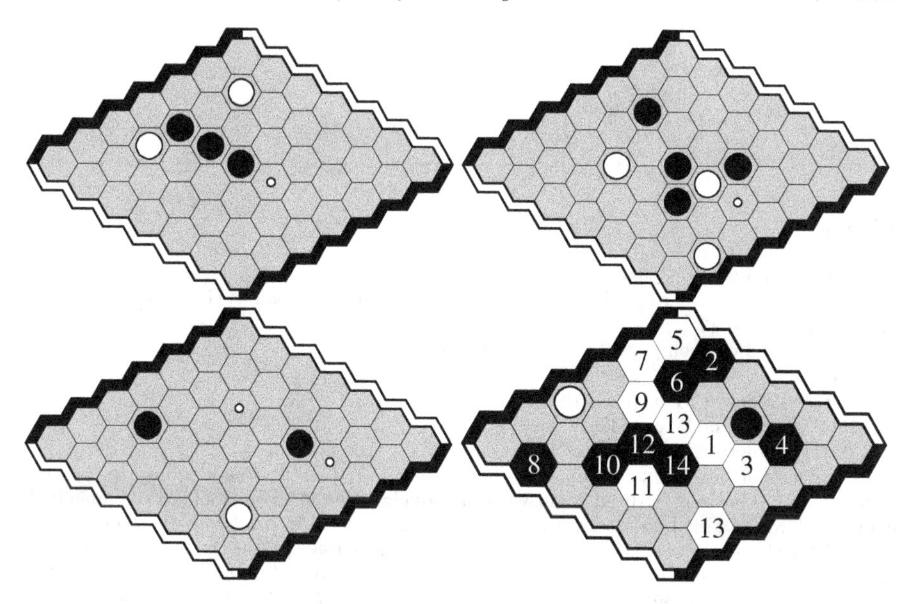

FIGURE 11.22: Enderton puzzle solutions. All winning White moves. The last puzzle has only one winning move, we show a sequence of play.

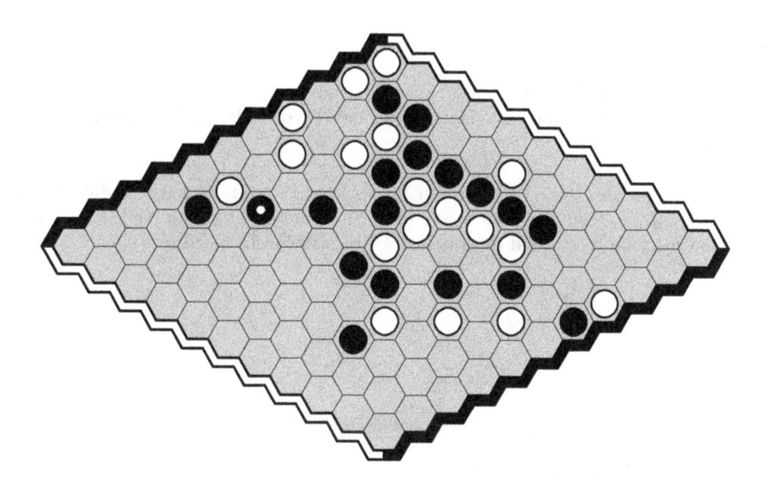

FIGURE 11.23: Silverman puzzle solution: one of many winning moves.

Notes

[1]1942.12.27 Politiken page 4. Second Polygon column.

[2]1957.03.23 letter Gardner-Hein, evening conversation with Nash at Gardner's New York apartment in 1957 [23].

[3]1942.12.26 Politiken page 4. First Polygon column.

[4][21]

[5]1957.08.19 letter Winder-Gardner [23].

[6]1966.08.23 letter Weaver-Gardner [23].

[7]1969.02.08 letter Max-Gardner [23].

[8]Gardner's original Hex column appeared in July 1957 [21]. In an addendum to the October column, Gardner credited Nash with introducing Hex in Princeton. To the amended column, Gardner added an afterword in 1959 (in *The Scientific American Book of Mathematical Puzzles and Diversions*), a further addendum in 1988 (when the book was republished as *Hexaflexagons and Other Mathematical Diversions*) and a final postscript in 2008 (when the book was republished as *Hexaflexagons, Probability Paradoxes, and the Tower of Hanoi*).

[9]2012.02.16 email Enderton-Hayward.

[10]2012.02.17 email Enderton-Hayward [16].

[11]`http://www.cs.cmu.edu/~hde/hex/hexfaq/`

[12]White 2.d3 wins.

[13][74]

[14][74]

[15]This strategy uses 715 patterns [75].

[16]Jing Yang's center-opening 9×9 Hex applet can be downloaded at `http://www.mimuw.edu.pl/~pan/jang.zip`

[17][57, 58]

[18]David L. Silverman (1929-1978) was a fan of board games and word play. In addition to writing *Your Move*, from 1969 through 1975 Silverman edited the Kickshaws column of *Word Ways*, a magazine founded at the suggestion of Martin Gardner.

[19][63]

[20]2018.03.06 email Van Rijswijck-Hayward.

[21][8]

[22][32]

[23][38, 37]

[24][4]

[25][59]

[26][60]

[27]2013.7.10 email Pawlewicz-Hayward.

Chapter 12

Rise of bots

As soon as the Analytical Engine exists, it will necessarily guide the future course of the science.

Charles Babbage[1]

I am not depressed by [chess-playing machines getting so strong]. I am rooting for the machines!

Claude Shannon[2]

12.1 Shannon's circuit

In Hex — as in other games such as chess and Go — as computers got stronger, people started to write game-playing programs. Such programs typically combine evaluation and search: start from a game position, evaluate possible next moves, and — in the available time — explore continuations of the strongest-looking moves. Then pick the initial move with best minimax value.[3]

Shannon outlined how to build such a chess program using this evaluation scheme: give 1 point for each pawn, 3 points for each bishop or knight, 5 points for each rook and 9 points for the queen. The position's score is the sum of these points for the player minus the corresponding sum for the opponent.[4] This evaluation is easily computed and relatively accurate.

Shannon's Birdcage circuit plays Bridg-it well, and also gives an evaluation scheme for Hex. To win at Hex it suffices to block the opponent, so to evaluate a Hex position, construct the opponent's across-the-board circuit as shown in Figure 12.1. Include one node for each board cell and two terminal nodes for

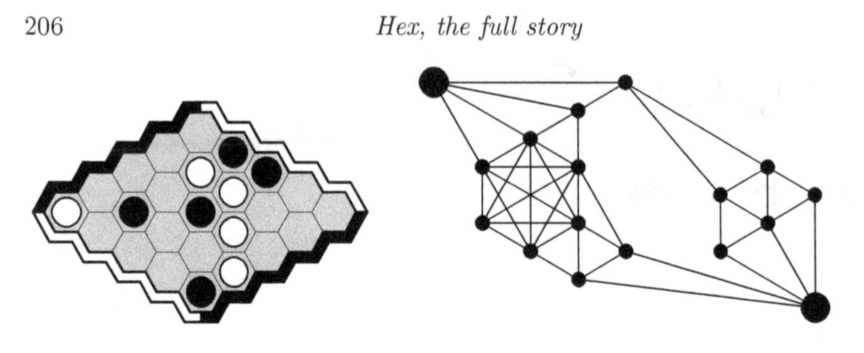

FIGURE 12.1: For this Hex position, the Shannon circuit for Black.

the opponent's sides.[5] Between each two adjacent nodes, put a 1-ohm resistor. (For a simpler circuit, omit resistors between adjacent cells both in the first or last row: these adjacencies are unnecessary, as a minimal side-to-side path has only one cell in each of these rows.) Delete nodes with player stones and contract nodes with opponent stones. Apply 1 volt across the two sides. The position's score is the resulting across-the-board current: the lower the opponent's current, the happier the player.[6]

Shannon's Hex circuit also ranks moves. The greater a node's voltage drop, the greater the interruption if the node is removed, so rank cells by voltage drop: the larger the drop, the better the move. Figure 12.2 shows voltage drops for the empty 6×6 board when 100.0 volts is applied across the white sides. All cells with a drop of at least 23.0 volts — shown in black — happen to win for Black, while all cells with a smaller drop are losing moves. For Hex, Shannon's circuit is not always this accurate, but it performs well in the early phase of a game.

12.2 Adding virtual connections

To play Hex well, you need to use virtual connections. Bert Enderton's Hex-playing program from the 1990s used some such connections. Later, Vadim Anshelevich— a consultant living in Texas — wrote Hexy, which adds virtual connections to Shannon's circuit, as we now explain.

To find virtual connections, start with adjacent cell pairs — which are trivially connected — and repeatedly apply combining rules, as in Berge's virtual connection example from an earlier chapter.[7] The number of connections found will grow rapidly as the game progresses — especially when the game is complicated — so terminate this process after a fixed period of time. Then update Shannon's circuit by adding a resistor between each pair of virtually connected cells.[8]

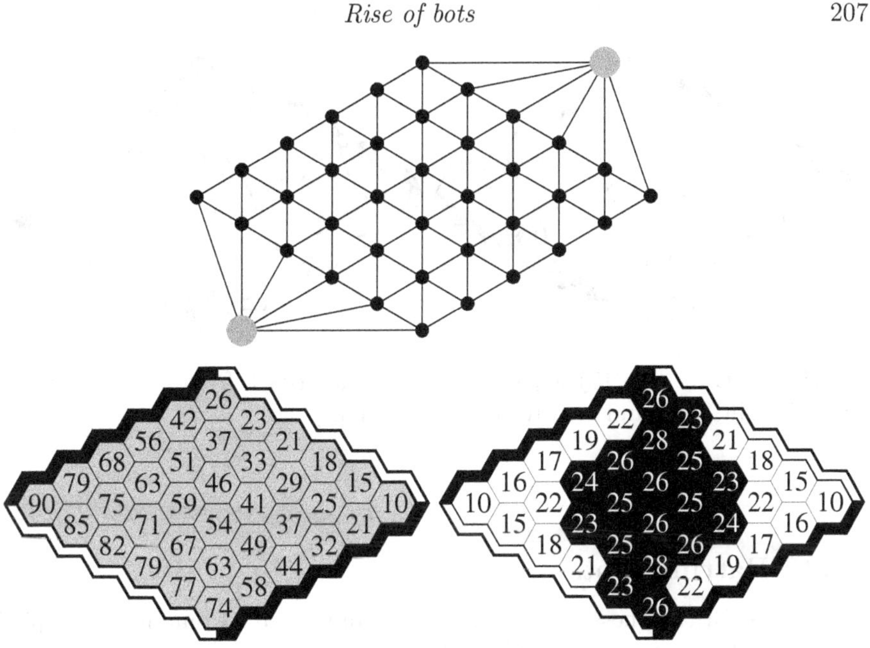

FIGURE 12.2: Shannon's circuit for Hex, for White. Rounded voltages (left) and voltage drops (right) when 100 volts is applied. Cells with a drop of at least 23 are winning Black moves.

While Anshelevich was writing Hexy, Jack van Rijswijck was writing Queenbee, a Hex-player based on distance. A simple distance measure is, for each cell, the sum of the shortest distance to each of the player's two sides. But as a Hex evaluation, this metric needs refining: on the empty $n \times n$ board, each cell has distance-sum $n - 1$, so all cells get the same score, which is not useful. In Hex, an opponent tries to block a player's progress, so Van Rijswijck defined a new measure: the *2-distance* from a cell to a side is zero if the cell touches the side; otherwise it is one plus the second-shortest 2-distance of all adjacent cells.

For example, in the left diagram of Figure 12.3, each cell on the bottom-right row touches that black side, so its 2-distance to that side is 0. In the next row, each cell that touches two cells with 2-distance 0 has 2-distance $1+ \max\{0, 0\} = 1 + 0 = 1$. One cell in that row touches one cell with 2-distance 0 and one cell with 2-distance 1, so its 2-distance is $1+ \max\{1, 0\} = 1 + 1 = 2$.

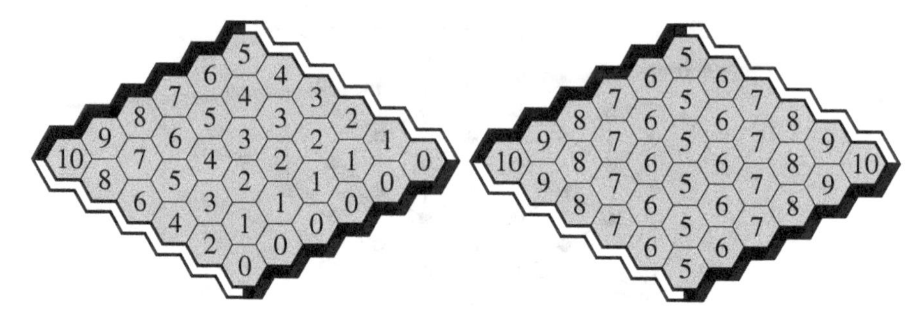

FIGURE 12.3: Van Rijswijck's 2-distances for Black from each cell to the bottom-right (left) and sum to both sides (right).

12.3 Battling bots

In 1989, British chess-master David Levy organized a Computer Olympiad with separate events for chess, Go and twelve other games.[9] In 2000, Hexy and Queenbee met at the 5th Computer Olympiad in London, the first Olympiad to feature a Hex competition. KillerBee by Emanuele Brasa also took part but lost all games, so the gold medal — played on the classic 11×11 board — was decided by the four-game Hexy-Queenbee match.

In the first game Queenbee opened at 1.b2 near the corner and Hexy swapped to take Black. The game is balanced until move 15,[10] when Queenbee jumps early, allowing Hexy to join its side with no positional advantage in exchange. Hexy then wins easily: after move 24, our Hex solver finds the win instantly. See Figure 12.4.

In the second game, the roles reversed: Hexy opened 1.a2 near the corner and Queenbee swapped to take Black. Hex played 3.f6, and the game was close until move 36, when Queenbee again jumped early instead of continuing to push. Our Hex solver shows that move 36 was a blunder: Queenbee had a winning move, although it is not easy to find. See Figure 12.5.

In the third game, Queenbee opened 1.a2 as Hexy had in the previous game. But Hexy played 2.f6, so the position was identical to that after move 3 of the previous game, with Queenbee (Black) near the corner and Hexy (White) in the middle. Both Hexy and Queenbee are deterministic, so moves 2 and following in this game were identical to moves 3 and following of the previous game. After fourteen more moves, Queenbee's operator resigned, and Hexy won gold.

Hex next appeared at the Computer Olympiad in 2003 in Graz, with two competitors similar in design to Hexy: Six by the Hungarian programmer Gábor Melis, and Mongoose by Yngvi Björnsson, Mike Johanson, Morgan Kan, Nathan Po and Hayward. Six won gold, 6 games to 2.[11] Figure 12.6

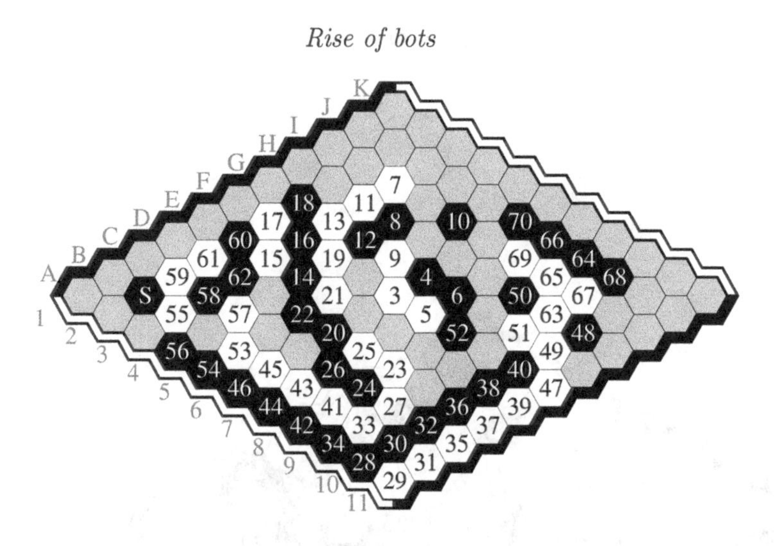

FIGURE 12.4: 2000 Computer Olympiad Queenbee-Hexy Game 1. Queenbee's move 15.e3 looks weak: MoHex prefers h5. Hexy — black after swap — is provably winning by move 23.

shows the start of a rare Mongoose victory. The same two programs met again in the 2004 Computer Olympiad, with Six winning 6-1.[12]

Three Hex programs took part in the 2006 Computer Olympiad in Turin: Six; Wolve, by a University of Alberta team that had added Broderick Arneson and Philip Henderson; and HexKriger by Rune Rasmussen, Cameron Browne, Auden Ellertsen, Ross Hayward and Frederic Maire from the University of Queensland. Six went undefeated and won gold again.[13] Figure 12.6 shows the deciding silver medal game, with Wolve defeating HexKriger. Both players erred in this game. Wolve blundered on move 13: today's solver finds a win there in 2 seconds. But HexKriger returned the favor, blundering on move 28.

12.4 Monte Carlo Tree Search

At the 2006 Computer Olympiad, a new kind of program — Crazystone by the French programmer Rémi Coulom — won the 9×9 Go tournament. Unlike chess or Hex, Go has no known quick-easy-accurate evaluation function. Earlier Go programs were hard to implement and easily defeated by intermediate strength humans, even with large handicaps. Coulom's program introduced an algorithm — Monte Carlo Tree Search (MCTS) — that evaluates a position by playing random moves to the end of the game and then recording the winner.

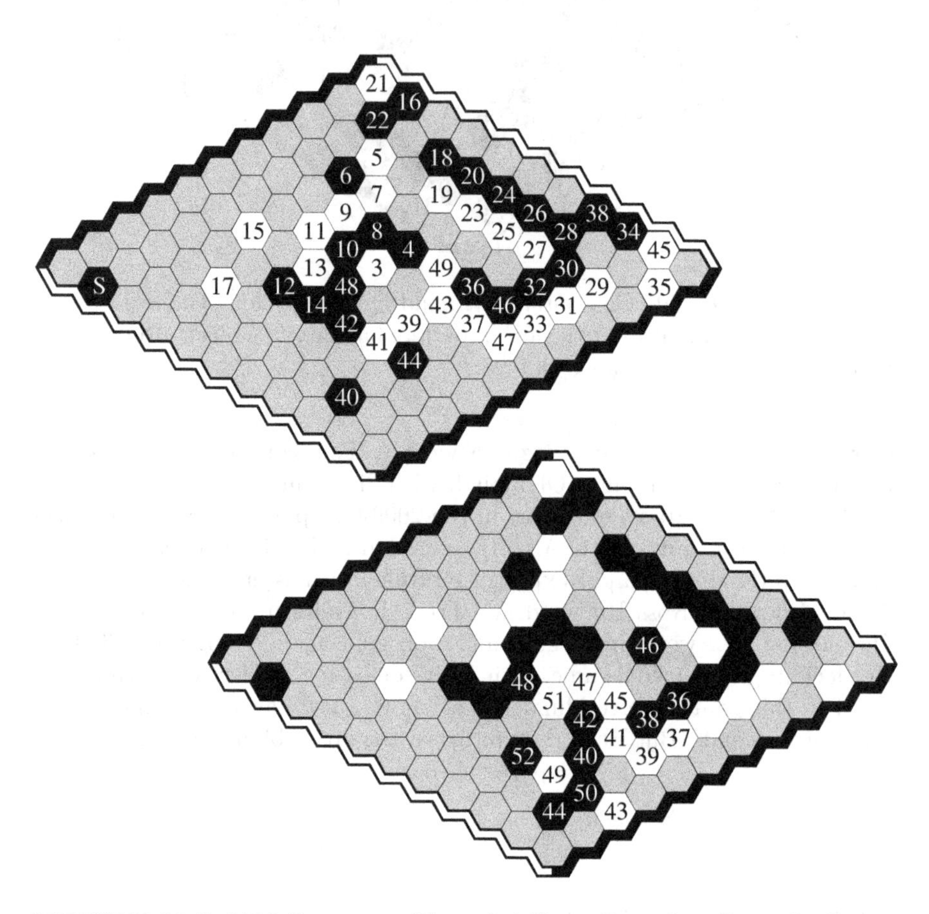

FIGURE 12.5: 2000 Computer Olympiad Hexy-Queenbee Game 2. Queenbee is Black after the swap. With move 36, Queenbee misses a narrow win: the variation (bottom) wins, but each of 36, 38, ..., 48 is the only winning move at that point.

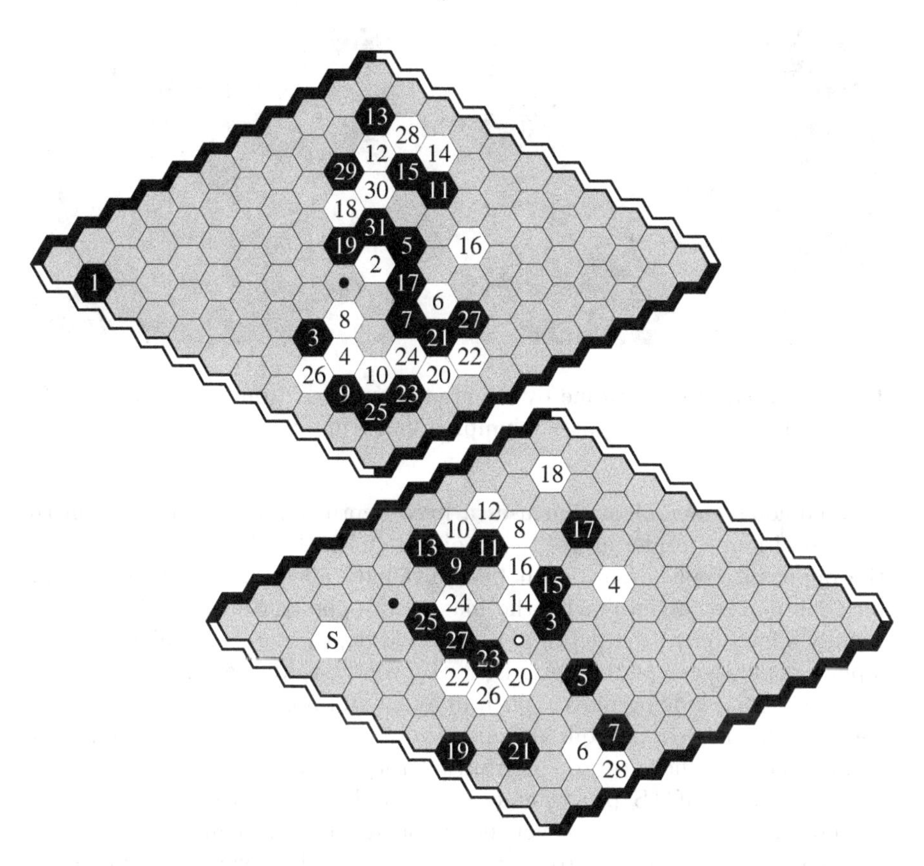

FIGURE 12.6: 2003 Computer Olympiad Six-Mongoose Game 5. Six errs on move 31: the dot wins. (bottom) 2006 Computer Olympiad Wolve-HexKriger Game 12. Wolve misplays move 13: the black dot wins. HexKriger errs on move 28: the white dot wins.

FIGURE 12.7: Wolve authors Hayward, Henderson, Arneson, Kan at 2006 Computer Olympiad in Turin.

FIGURE 12.8: Crazystone by Rémi Coulom (2nd from left) wins 9×9 Go gold medal at 2006 Computer Olympiad in Turin.

Shannon's chess algorithm uses a fixed-depth minimax search: from the given position, consider all continuations in which the player moves and then the opponent replies (so, depth two); evaluate all such continuations; and pick the move that minimizes the maximum opponent score. (This method usually allows only even-depth searches: odd-depth searches, which end with a player's move, can be too optimistic.) By contrast, Monte Carlo Tree Search explores strong-looking moves much more — and much more deeply — than weak-looking moves. Coulom also made sure that Crazystone explored weak-looking moves sufficiently to confirm that they are weak.

Soon after, MCTS programmers improved playout accuracy by adding local response patterns: whenever the opponent's last move matches a pattern, the player replies according to an associated rule.[14] For example, in Go and in Hex, if an opponent threatens a 'cut-connect' or angle connection, it is usually advantageous to reply and keep the stones locally connected. See Figure 12.9.

Requiring little game-specific knowledge, Crazystone and a new MCTS exploitation-exploration formula[15] inspired a new generation of game programs. At the University of Alberta, Martin Müller and Markus Enzenberger started the Fuego project, an open-source MCTS game independent Go library that includes a Go program. Henderson was quick to see the advantage of piggybacking: he and Arneson started writing the MCTS Hex-player MoHex on top of Fuego.

At the 2008 Beijing Computer Olympiad, four programs competed in the Hex event: Six, Wolve, MoHex, and another MCTS program Yopt by Abdallah Saffidine and Tristan Cazenave from France. Wolve narrowly squeezed past MoHex for gold. Wolve's improvements – including filling captured cells — finally allowed it to overtake bronze-winner Six. Yopt reached winning positions, but with no virtual connection computations suffered in the end-game. Figure 12.10 shows the deciding Wolve-MoHex and Six-Yopt games.

As Henderson had expected, MoHex — built on top of the Fuego library — got stronger as Müller and Enzenberger strengthened Fuego. In the 2009

FIGURE 12.9: The save-bridge pattern in Go (left) and Hex: when an opponent (here, White) attacks a bridge (angle connection), reply at the dot.

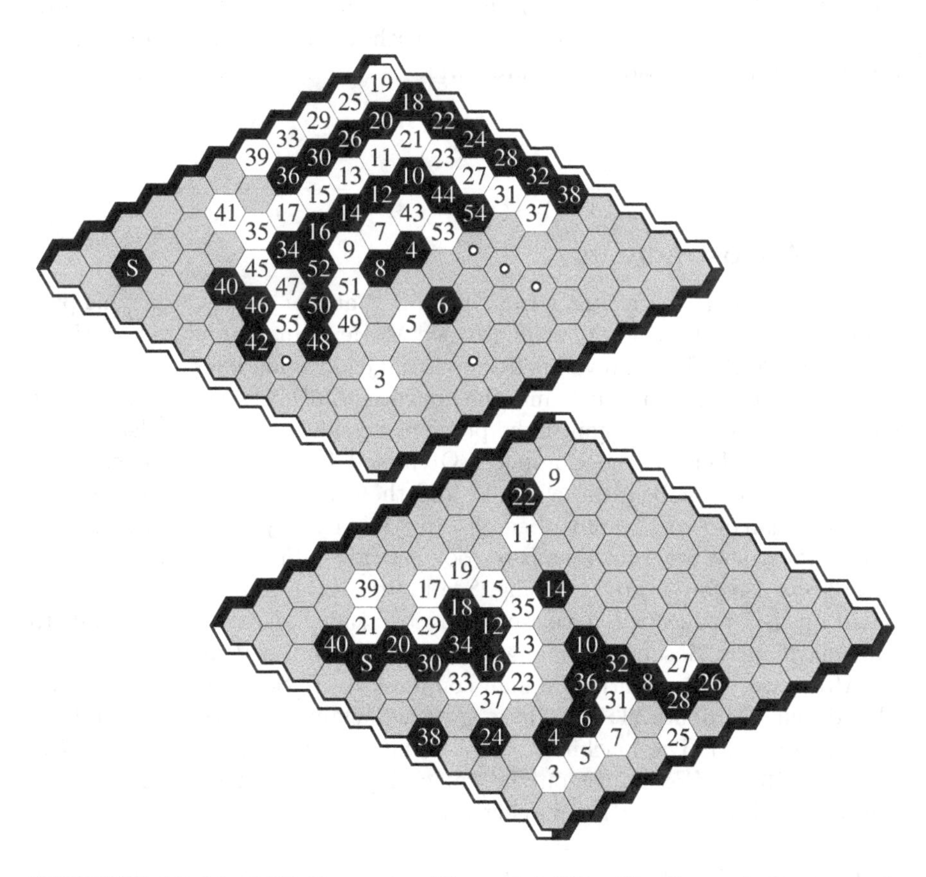

FIGURE 12.10: 2008 Computer Olympiad Yopt-Six Game 4. Post-match analysis shows that Yopt errs on move 55: any move to a dot wins. (bottom) MoHex-Wolve Game 3. Move 9 looks weak and Wolve — Black after swap — wins.

Computer Olympiad in Pamplona, Fuego and MoHex both ran parallel search processes, and MoHex dethroned single-process Wolve to take gold.

MoHex has kept its crown since, defeating several challengers over the years.[16] In 2009, students in an artificial intelligence course taught by Jakub Pawlewicz and Lukasz Lew at the University of Warsaw wrote MIMHex. Each week, the class worked on one part of the program. Assignments were graded by a tournament competition, with each week's winning code included in the final version. Unfortunately the course finished before virtual connections could be added, so — like Yopt — MIMHex is strong in opening play but suffers at the end. See Figure 12.11.

At the 2011 Computer Olympiad in Tilburg, Panoramex — by Fabien Teytaud, Tristan Cazenave and Nicolas Jouandeau from France — competed on an 18-node cluster of 4-core machines. In its first game it quickly reached a winning position against MoHex but — with no virtual connection computation — eventually lost. See Figure 12.11.

12.5 Almost human

In 2011, postdocs Aja Huang from Taiwan — author of the Go-playing program Erica — and Pawlewicz strengthened MoHex: Huang updated MoHex's rollouts while Pawlewicz revamped the virtual connection finder. At the 2013 Computer Olympiad in Yokohama, the improved MoHex — which wins about 90 per cent of its games against the previous version — went undefeated.

In 2015 in Leiden, the Computer Olympiad for the first time had two Hex events, one for board size 11×11 and the other for 13×13. Based on a generalization of Proof Number Search, a method usually used to solve games, Pawlewicz's new program DeepHex took MoHex to a playoff round on the 13×13 board before narrowly losing. See Figure 12.11. Ezo,a minimax player based on new connection measures by Kei Takada, Masaya Honjo, Makoto Mitsuhashi, and Masahito Yamamoto from Japan, finished third.

Afterwards, in an informal human-computer exhibition, Tony van der Valk played four 30-minute 11×11 games, two each against DeepHex and MoHex. See Figure 12.13. The machines won 4-0 and Van der Valk — then the 5th strongest 13×13 Hex player on the game-playing site Little Golem — was impressed: Hex-bots — Hex-playing programs — were approaching human strength.

Go-bots — Go-playing programs — were also improving. In 2014 at the University of Electro-Communications Cup in Japan — the world's biggest annual computer Go competition — the top two programs, Zen by Yoji Ojima and Hideki Kato and Crazystone by Coulom, won the right to each play a game against human professional Norimoto Yoda. To balance the game, the computers had a four-stone handicap: a black stone is placed near each

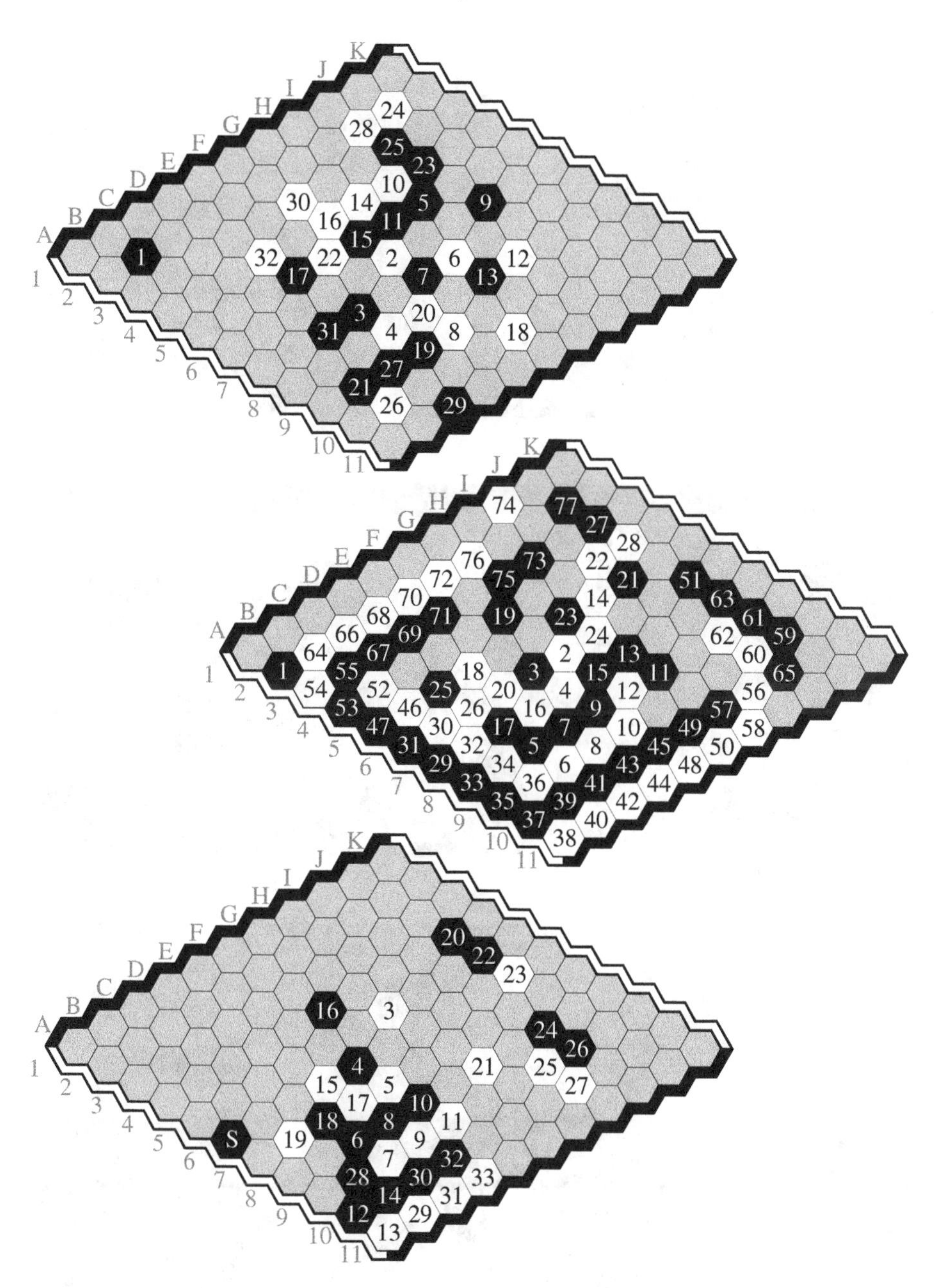

FIGURE 12.11: Computer Olympiad games. (top) 2010 MIMHex-MoHex. MIMHex errs with 29.c11: c7, f8, g6, h3 all win. 2011 MoHex-Panoramex. Panoramex errs with 56.h10: g9, g10 win. 2015 MoHex-DeepHex. DeepHex considered 22.i8, leading to positions hard for MoHex, but played 22.j5, which loses.

FIGURE 12.12: (top) Pawlewicz, Hayward and Huang at the 2013 Computer Olympiad in Yokohama. (middle) Van der Valk, Pawlewicz, Takada, Hiroyuki, Yasumasa at the 2015 Olympiad in Leiden. (bottom) Weninger, Young, Yamamoto, Takada, Zhang at the 2016 Olympiad in Leiden.

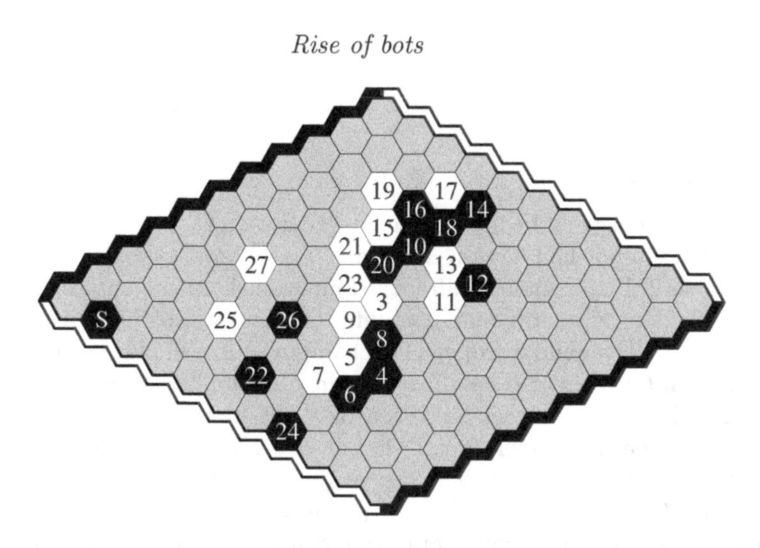

FIGURE 12.13: Exhibition game MoHex-Van der Valk. MoHex (White after swap) sees the win by move 21.

four corners (row 4, column 4) before Yoda makes his first move. With this handicap, Zen lost but Crazystone won.

So, in 2014, on the 19×19 board, top Go programs were about four stones behind top humans. Coulom was asked how long it would be until computers caught up to humans:[17]

I think maybe ten years. But I do not like to make predictions.

Coulom's hesitancy was understandable: strange things were happening in the world of computer Go. In 2012, scientists led by University of Toronto professor Geoff Hinton figured out how to train a deep convolutional neural net (a large computer network that models the connectivity of neurons in the human brain) to solve image classification problems (look at a picture and classify it: bird? plane? squirrel?). And by 2014, two different groups — Christopher Clark and Amos Storkey at the University of Edinburgh; Chris Maddison, Huang (who had worked on MoHex), Ilya Sutskever and David Silver from Google DeepMind — had independently trained neural nets to answer this question: given a Go position, where would a human move next?

No one had found an effective evaluation function for Go, but these neural nets played intermediate level Go without needing any search. Could they somehow be used in a next-generation Go program?

12.6 AlphaGo

Yes, neural nets could be used in a next-generation Go program. Construction of such a Go-bot had already started.

In October 2015, just months after the four-game human-computer Hex exhibition in Leiden, a secret human-computer Go contest took place in London: professional Go player and European Go champion Fan Hui played a five-game match on even terms (so, no handicap) against a computer program.

DeepMind's Go neural net had been only the first step of an ambitious project: build the first superhuman Go-bot. DeepMind, bought in 2014 by Google for over $500 million, had the people and the necessary computational power. CEO Demis Hassabis, project leader Silver, and lead programmer Huang had blended Monte Carlo Tree Search, neural nets and new machine learning ideas into this new program AlphaGo.

On January 27 2016, the journal Nature revealed their new ideas and the match result: AlphaGo had defeated Fan Hui 5-0.[18]

They also announced that AlphaGo would play the legendary Lee Sedol — for many years the world's strongest Go player — in a five-game match at a luxury hotel in Seoul in March 2016. According to Go experts, AlphaGo and Fan Hui had each made errors in the October match that Lee would be unlikely to make. But AlphaGo had learned Go via self-play: would the AlphaGo that played Lee in March 2016 be much stronger than the version that had played Fan Hui?

On March 9 2016, with tens of millions of people watching around the world, the first game started. Hoping to create a position unfamiliar to AlphaGo, Lee made some unusual early moves. But AlphaGo adapted as a human master might and went on to win the game.

On March 10 in game 2, after making move 36, Lee left the room for a cigarette break. While he was out of the room, AlphaGo played move 37 to reach the position in Figure 12.14. When he returned, Lee was stunned: such a 'shoulder hit' — one diagonal from the opponent, allowing the opponent stone to extend horizontally or vertically but not both — is often played on the fourth line from the edge, but in a professional game is almost never played on the fifth line. Lee studied the board a long time before making his next move.

The era of human domination over Go-bots arguably ended in this game: AlphaGo went on to find more strong moves unpredicted by humans. Two games later, Lee brilliantly escaped from a dire position to claim one win, but AlphaGo won the match 4-1.

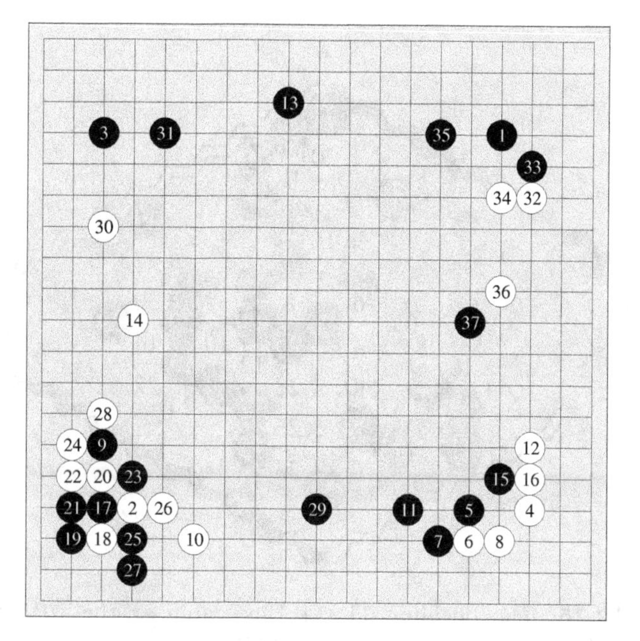

FIGURE 12.14: AlphaGo-Lee Sedol game 2 move 37.

12.7 Neural net Hex-bots

The methods of AlphaGo were immediately picked up by programmers of other games, including Hex. At the 2016 Computer Olympiad, the three competitors had ordinary MCTS algorithms, but by 2017 both Ezo and Mo-Hex had added neural nets. While the computational power available to these research groups is dwarfed by that of Google, some progress has been made. MoHex-CNN, developed by doctoral student Chao Gao and operated by Noah Weninger, won the 13×13 event with Ezo-CNN taking silver, while ordinary MoHex 2.0 edged Ezo-CNN on 11×11. See Figure 12.15.

Hex programs have come far since 2000. Hexy and Six lose easily to original MoHex which in turn loses easily (90 per cent of the time) to MoHex 2.0, while neural net players are even stronger. On the 11×11 board today's Hex-bots still make errors, although they often play perfectly after the first 20 moves.

To illustrate this progress, we close this chapter with an analysis of the first Computer Olympiad Hex gold-medal games by MoHex 2.0. See Figure 12.17.[19]

In Game 1, Queenbee opens at b2 and Hexy swaps. First-generation Hex-bots play better as first-player, perhaps because they cannot see far enough ahead to allow a solid defense. While near the corner, b2 captures a1,a2,b1,c1, so this black move is a relatively bold opening: Hexy's swap looks reasonable.

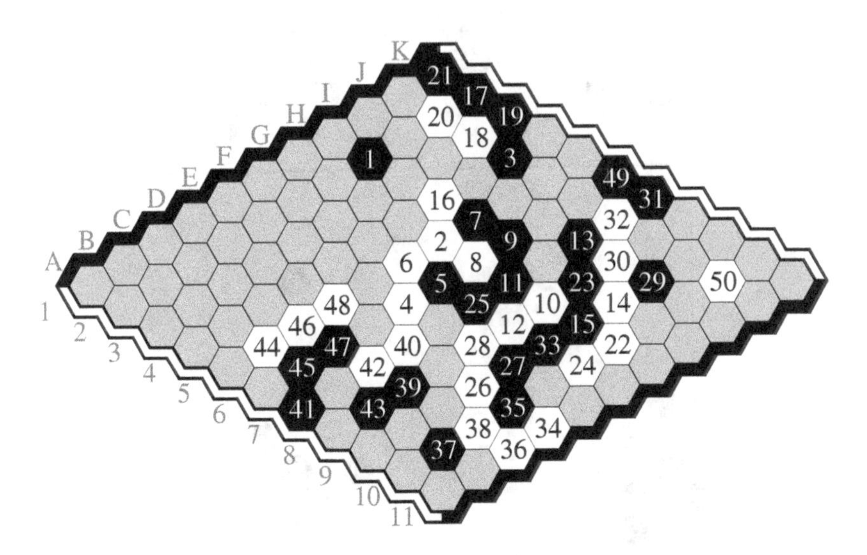

FIGURE 12.15: The deciding 11×11 Hex game at the 2017 Computer Olympiad. MoHex 2.0 (White) defeats Ezo-CNN.

FIGURE 12.16: Takada, Gao, Hayward at the 2018 Computer Olympiad in New Taipei City.

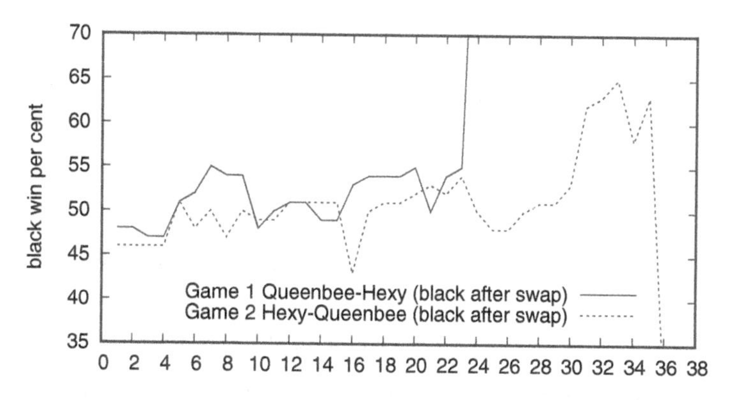

FIGURE 12.17: After each move, MoHex 2.0's estimated Black win rate (60 seconds/move) for the 2000 Computer Olympiad gold-medal games.

By contrast, after 60 seconds, MoHex (version 2.0) builds a search tree with almost 2 000 000 nodes, in the process performing about 350 000 game playouts, with the first black move winning .48 of the time. So, rather than swap, MoHex prefers to reply with white: it has found the best reply to in the center, f6.

On move 5 the score jumps from .47 to .51: perhaps Queenbee (White after swap) erred on move 4: MoHex prefers 5.g5 to Queenbee's 5.f7. After Hexy's 10.i5 the score drops from .54 to .48: MoHex prefers 10.h5. The next drop is after move 15, when Queenbee jumps to e3 instead of pushing to f4. Hexy's 20.d6 looks weak, but Queenbee fails to exploit it, replying 21.e5 instead of 21.d8. Now Queenbee is in trouble. After 23.d8, MoHex instantly finds the winning 24.c8, which Hexy also finds. Queebee could have prolonged the game with 23.c9, but this move also looks to be losing.

In Game 2, Hexy opens at a2 and Queenbee then swaps. A2 is a weaker opening than b2: MoHex scores it .46 and so would not swap. But Black's scores slowly grow, and soon the game is virtually even. Queenbee stumbles with 16.k2: MoHex expects 16.c4 and Black's score tumbles. But Hexy fails to exploit this and Queenbee recovers. By move 31, Queenbee's score is .62.

With move 32, human players can see Black's obvious line of play: extend a ladder i9,h9,g9 . . . , and at the right moment, jump from the ladder to establish a winning virtual connection. But when and where to jump?

With no virtual connection computation, Queenbee has no chance to make this decision correctly, and blunders with 36.g8: MoHex expects 36.g9. The correct ladder is i9,h9,g9,f9 and then jump to d9 — and then the jump is the only winning move! Almost twenty years after Hexy and Queenbee played this game, MoHex 2.0 finishes it correctly.

Notes

[1]Charles Babbage, *Passages from the life of a philosopher*, London, 1864.

[2][50]

[3]A player's minimax strategy maximizes — over all possible player strategies — the minimum score achieved by that strategy, over all possible opponent strategies.

[4][66]

[5]Modern implementations also build the player's circuit and combine the scores [2, 3].

[6]By Kirchoff's circuit laws, in a uniform-resistor circuit, each node's voltage is the average of that of its neighbours. This identity yields a linear system of equations, and solving the system gives node voltages. Also — if you do not want to implement an exact linear system solver — this system can be solved approximately: set the voltages of the two terminal nodes, set other voltages to any intermediate value, and repeatedly iterate over non-terminal nodes, setting each voltage to the neighbours' average. This process converges quickly.

[7]Here is one combining rule: if cells a and b are virtually connected via a strategy that uses cell set X, and b and c are virtually connected via set Y, and the player has a stone at b, and X and Y do not intersect, then a and c are virtually connected via a strategy that uses the union of X and Y. Also, if cells d and e are semi-connected by various strategies with respected cell sets R, S, T, and the intersection of these three sets is empty, then d and e are virtually connected by a strategy that uses the union of these sets. See [2] for the full set of combining rules.

[8][2]

[9]The 1989 ICGA Computer Olympiad in London included awari, backgammon, bridge, checkers, chess, Chinese chess, Connect Four, dominoes, draughts, Go (9×9 and 19×19), gomoku, Othello, renju, and Scrabble.

[10]We analyzed this game using our Hex solver.

[11][52]

[12]With the outcome decided, the operators did not play the last two scheduled games.

[13][28]

[14]Bruno Bouzy used 3×3 patterns in his program Indigo. Later the program MoGo used 3×3 patterns centered on the previous move.

[15]In 2006 Levente Kocsis and Csaba Szepesvári introduced a new exploitation-exploration formula — upper confidence bound applied to trees (UCB) — that was quickly adopted by MCTS programmers.

[16]MoHex competitors after 2009: MIMHex (by Jakub Pawlewicz, Lukasz Lew and 11 students from Poland) in 2010 at Kanazawa, Panoramex (by Fabien Teytaud, Tristan Cazenave and Nicolas Jouandeau from France) in 2011 at Tilburg, Ezo (by Kei Takada, Masaya Honjo, Makoto Mitsuhashi and Masahito Yamamoto from Japan) and Jhinenox (by Jinno Masatoshi from Japan) in 2013 at Yokohama, DeepHex (by Pawlewicz) in 2015 in Leiden, Hexamaze (by Tianli Zhang from China) in 2016 in Leiden, and Hexcited (by Ma Shengjie from China) and Hexcellent (by Wu Tong from China) in 2017 in Leiden.

[17][48]

[18][68]

[19]MoHex-CNN is stronger than MoHex 2.0 but more erratic, so we used the latter for our analysis. MoHex 2.0 scores computed April 30 2018, 60 seconds/move, on a 2007 iMac.

Epilogue

> *I believe that the future belongs to [artificial intelligence].*
>
> Ke Jie, after playing AlphaGo.

Sometime around 1986, Ryan Hayward asked Claude Berge, do you think computers will ever beat humans at Hex? Never, he replied. If we could ask him today, he might change his mind.

After the 2016 Lee Sedol match, DeepMind announced a three-game match to be played in March 2017 between AlphaGo and Ke Jie, the world's top Go player. Starting in December 2016, on an online Go server in China an unknown new player started beating top professionals. DeepMind soon confirmed that this was a (constantly improving) version of AlphaGo being tested in preparation for the Ke Jie match. AlphaGo won all 60 online games and defeated Ke Jie 3-0. And in October 2017, DeepMind announced in an article in Nature that they had developed a zero-knowledge version of AlphaGo, one that had learned to play only from the rules, so without any human game records.

So what about Hex? With MoHex-CNN by Chao Gao et al. at the University of Alberta and Ezo-CNN by Kei Takada at Hokkaido University, the process of applying AlphaGo methods to Hex has begun. But progress is slow. The AlphaGo project presumably cost DeepMind millions of dollars for people-power and tens of millions of dollars for computer-power: few research groups have that kind of money. MoHex-CNN loses to humans on the 13×13 and larger boards. The story of Hex continues.

☼ ☼ ☼

For those who like to ponder problems, we include in the appendix some open questions. For those who like to write computer programs, why not write your own Hex-bot? And for those who just want to play, good luck in your next game! Remember, to win, it suffices to block your opponent ...

Chronology

1939 (or earlier) Piet Hein starts the design of Polygon [Hex].

1942 Hein finishes the design of Polygon. On September 30 Hein orders 25 50-sheet 11×11 Polygon game pads, and 500 more sheets on October 31. He signs a contract with Politiken on November 16, applies for a Norwegian patent on November 20, offers a Christmas goose for Polygon puzzles on December 4 and speaks to the student club *Parentesen* around December 14. On December 26 Politiken publishes the first Polygon column.

1943 Politiken hosts Polygon salons with Hein and Lindhard on February 1 and 10. On August 11 the last Politiken puzzle, number 49, appears.

1949 (late winter, perhaps February) Nash tells Gale of a game with a first-player-wins strategy, Gale makes a 14×14 board for Fine Hall.

1950 Parker Brothers market 11×11 Hex. Gale invents the game now called Bridg-it or Birdcage.

1951 Shannon builds a Birdcage machine.

1952 Nash writes a Rand Corporation memo on Hex, *Some Games and Machines for Playing Them.*

1953 Shannon and Moore describe their Hex-playing machine in *Computers and Automata.*

1957 In January Martin Gardner starts research on Hex for his new monthly *Scientific American* Mathematical Games column. *Concerning the game of Hex, which may be played on the tiles of the bathroom floor* appears in July. An endnote to Gardner's October column credits Nash with independently inventing Hex.

1958 Gardner's October column mentions Gale's game [Bridg-it].

1960 Hasbro markets 5×5 Bridg-it.

1961 Pierce's book *Symbols, Signals and Noise* includes a Hex no-draw proof. Gardner's July *Scientific American* column has Gross's Bridg-it solution.

1964 Lehman's paper *Solution of the Shannon Edge-Switching Game* appears.

1969 Beck et al. text *Excursions into Mathematics* has a proof that the acute corner is a losing Hex opening.

1974 Evans paper *A Winning Opening in Reverse Hex* shows that on even boards the acute corner is a winning Rex opening.

1975 Titus and Schensted's book *Mudcrack Y & Poly-Y* appears. Even and Tarjan show Shannon node-switching is P-space-complete. Gardner's June *Scientific American* column discusses Rex.

1977 Berge writes the manuscript *L'Art Subtil du Hex.*

1979 Gale's paper *The Game of Hex and Brouwer's Fixed-Point Theorem* appears.

1981 Reisch shows Hex is P-space-complete.

1991 Enderton speaks on search techniques for solving Hex, Yang solves the 7×7 center Hex opening.

1995 Enderton's bot solves all 6×6 and three 7×7 Hex openings.

2000 In August, Hexy defeats Queenbee at the London Computer Olympiad. Anshelevich's paper *Game of Hex: An Automatic Theorem Proving Approach to Game Programming*, Van Rijswijck's MSc thesis *Computer Hex: Are Bees better than Fruitflies?* and Browne's book *Hex Strategy: Making the Right Connections* appear.

2001 Van Rijswijck's bot solves all 6×6 2-move Hex openings.

2003 Six defeats Mongoose at the Graz Computer Olympiad. A bot by Björnsson et al. solves all 7×7 Hex openings. Yang solves the centermost 8×8 and 9×9 Hex openings.

2004 Maarup finds Hein's Polygon manuscripts. Noshita's *Union-Connections and a Simple Readable Winning Way in 7×7 Hex* appears. Six defeats Mongoose at the Bar-Ilan Computer Olympiad.

2005 Maarup's MSc thesis *Everything you always wanted to know about Hex but were afraid to ask* and Browne's book *Connection Games: Variations on a Theme* appear.

2006 Van Rijswijck's PhD thesis *Set Colouring Games* appears. Six defeats Wolve and HexKriger at Turin Computer Olympiad.

2008 Henderson et al. bot solves all Hex 8×8 openings. Wolve defeats MoHex, Six and Yopt at the Beijing Computer Olympiad.

2009 MoHex defeats Wolve, Six and Yopt at the Pamplona Computer Olympiad.

2010 Henderson's PhD thesis *Playing and Solving Hex* appears. MoHex defeats MIMHex, Wolve, Yopt at the Kanazawa Computer Olympiad.

2011 MoHex defeats Wolve and Panoramex at the Tilburg Computer Olympiad.

2012 Pawlewicz et al. bot solves all Hex 9×9 openings.

2013 Pawlewicz et al. bot solves two Hex 10×10 openings. MoHex defeats Ezo and Jhinenox at the Yokohama Computer Olympiad.

2015 Hayward, Pawlewicz and Toft find more Hein Hex documents. MoHex defeats DeepHex and Ezo at the Leiden Computer Olympiad.

2016 AlphaGo defeats Lee Sedol 4-1. MoHex defeats Ezo and Hexamaze at the Leiden Computer Olympiad.

2017 Hayward and Toft identify Hein's Polygon call-for-puzzles, Toft finds Lindhard's Polygon documents. At the Leiden Computer Olympiad, MoHex defeats Ezo-CNN and Hexcited in 11×11 Hex, MoHex-CNN defeats Ezo-CNN and Hexcellent in 13×13 Hex.

2018 MoHex-3HNN defeats DeepEzo at the New Taipei City Computer Olympiad.

Appendix A

Politiken Polygon puzzles

In order to promote his new game Polygon — now called Hex — Piet Hein wrote a series of columns for the Danish newspaper Politiken, many with a puzzle and/or opening sequence. Here are all 49 puzzles and all 9 openings.

A.1 Politiken Polygon puzzles

Hein hand-drew each puzzle and opening, but contracted out their creation. Hein perhaps composed Puzzle 1, and Emil Christensen perhaps composed Puzzle 26. But most — if not all — other puzzles were composed by Jens Lindhard.

Each puzzle is White to play and was intended to have exactly one winning move. Puzzle 45 has no winning move: White's best move leaves Black with two winning replies. Puzzle 38 originally appeared with a marker missing: we have added it back in.

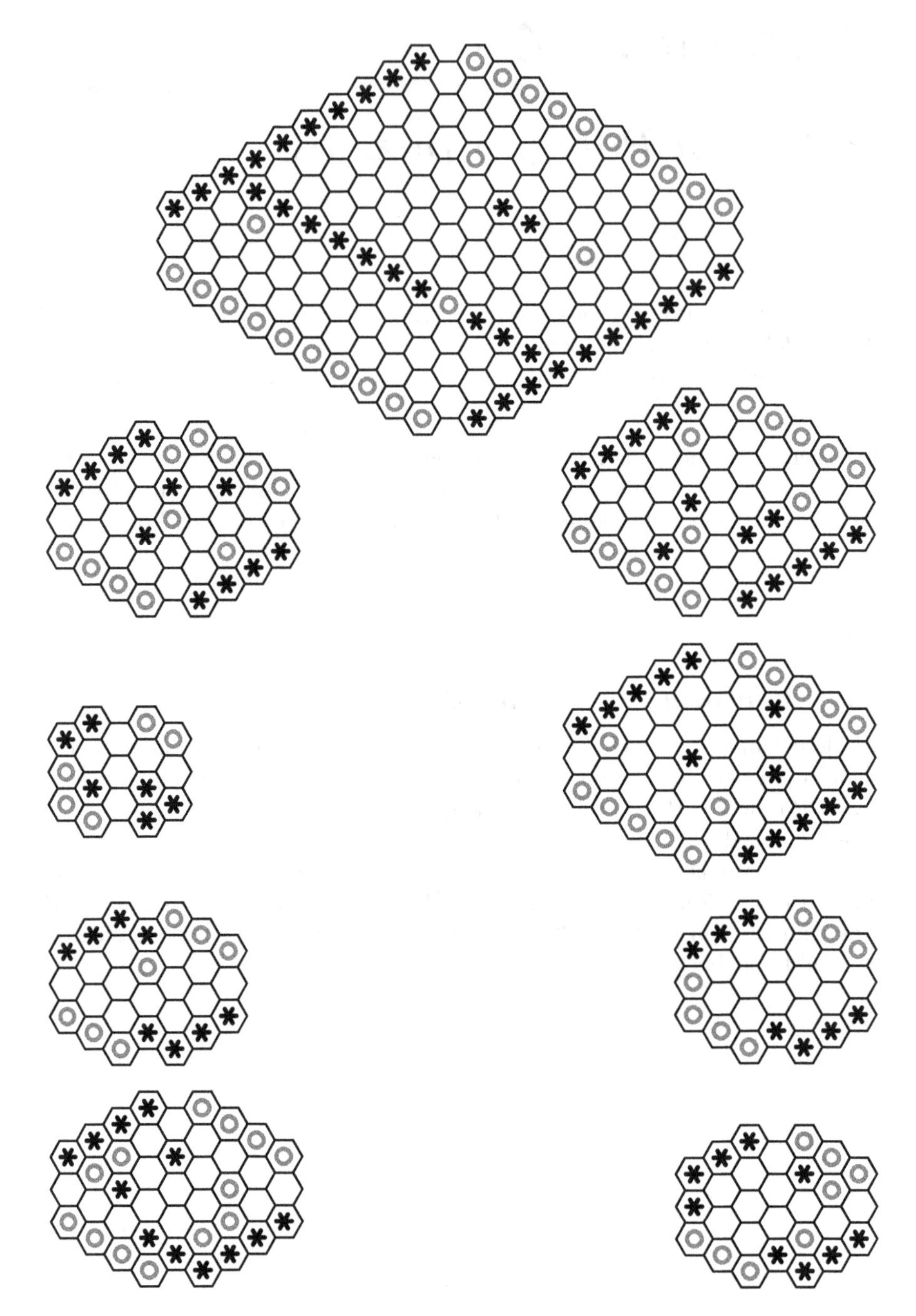

FIGURE A.1: Politiken puzzles 1-9.

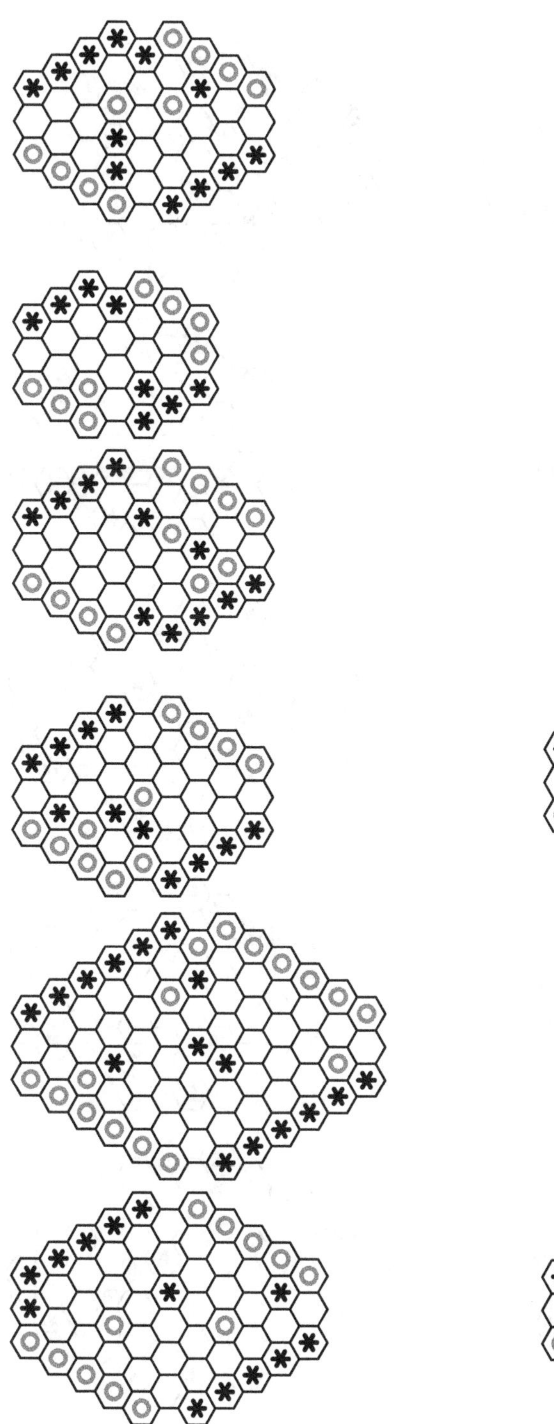

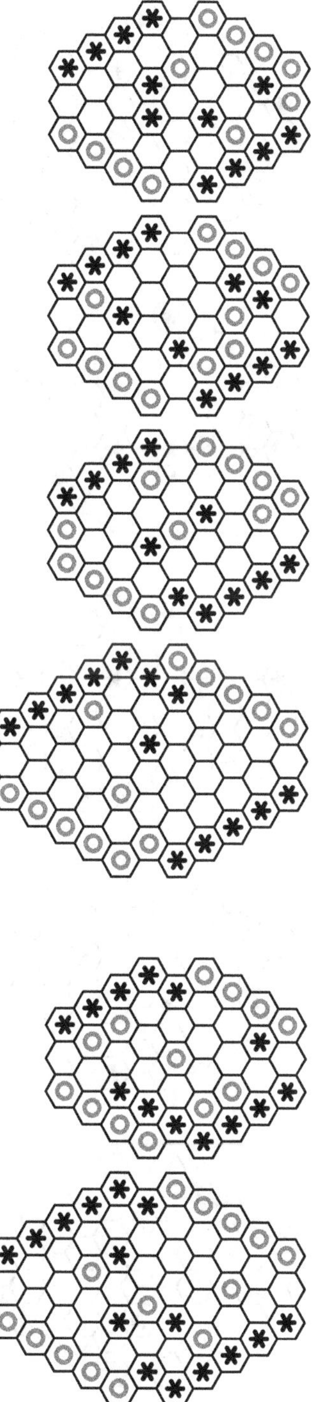

FIGURE A.2: Politiken puzzles 10-21.

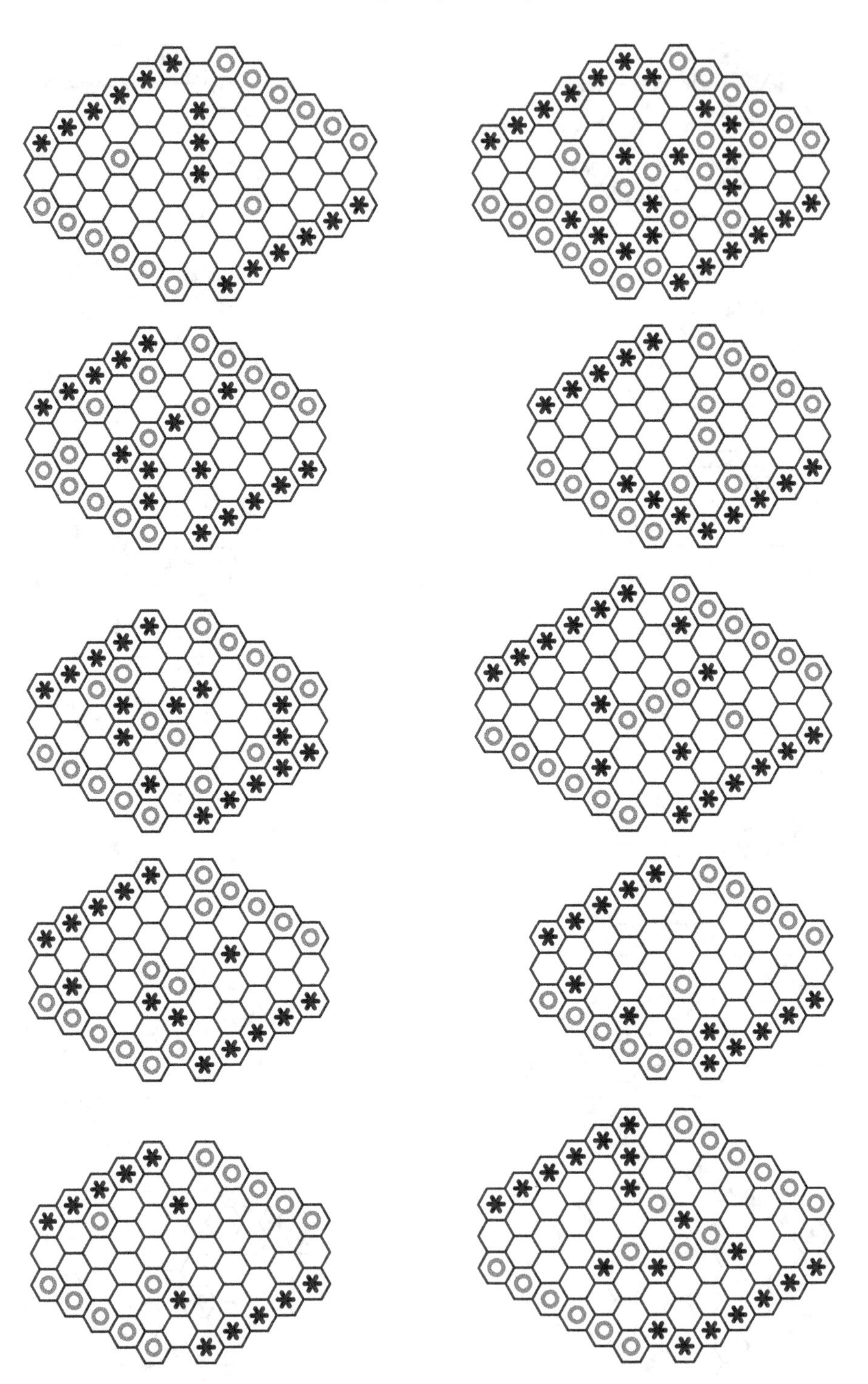

FIGURE A.3: Politiken puzzles 22-31.

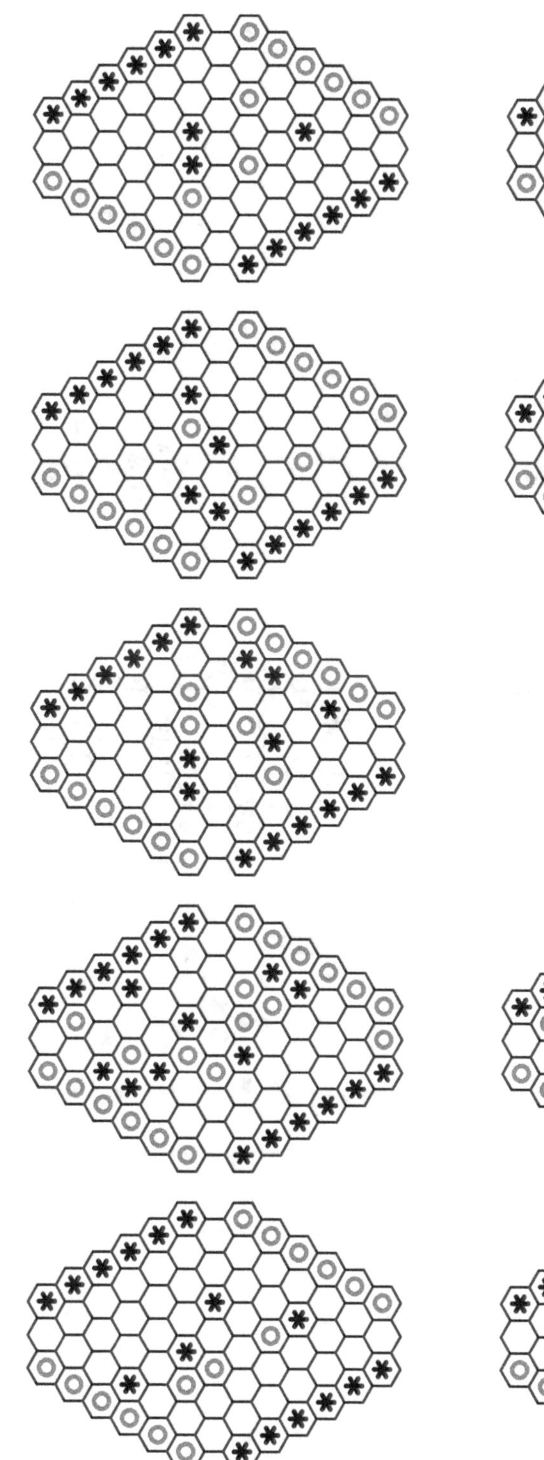

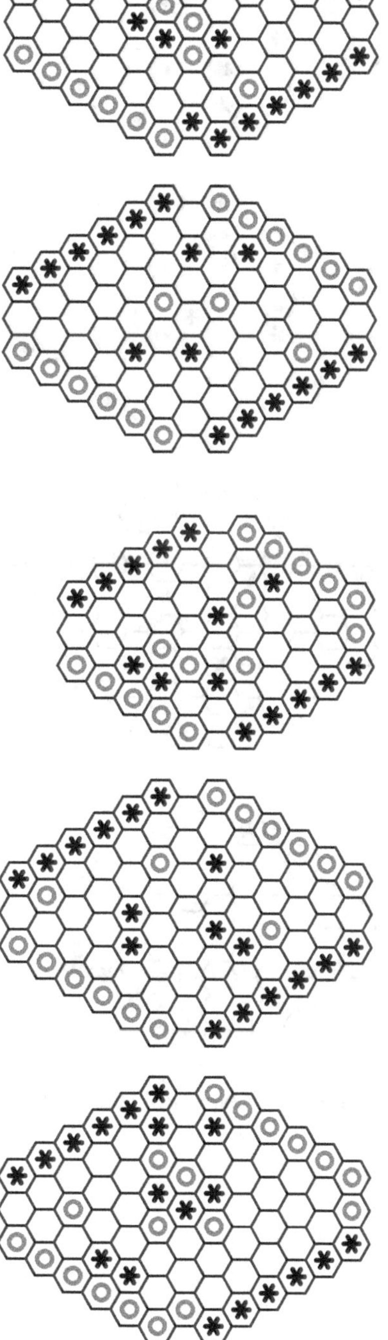

FIGURE A.4: Politiken puzzles 32-41.

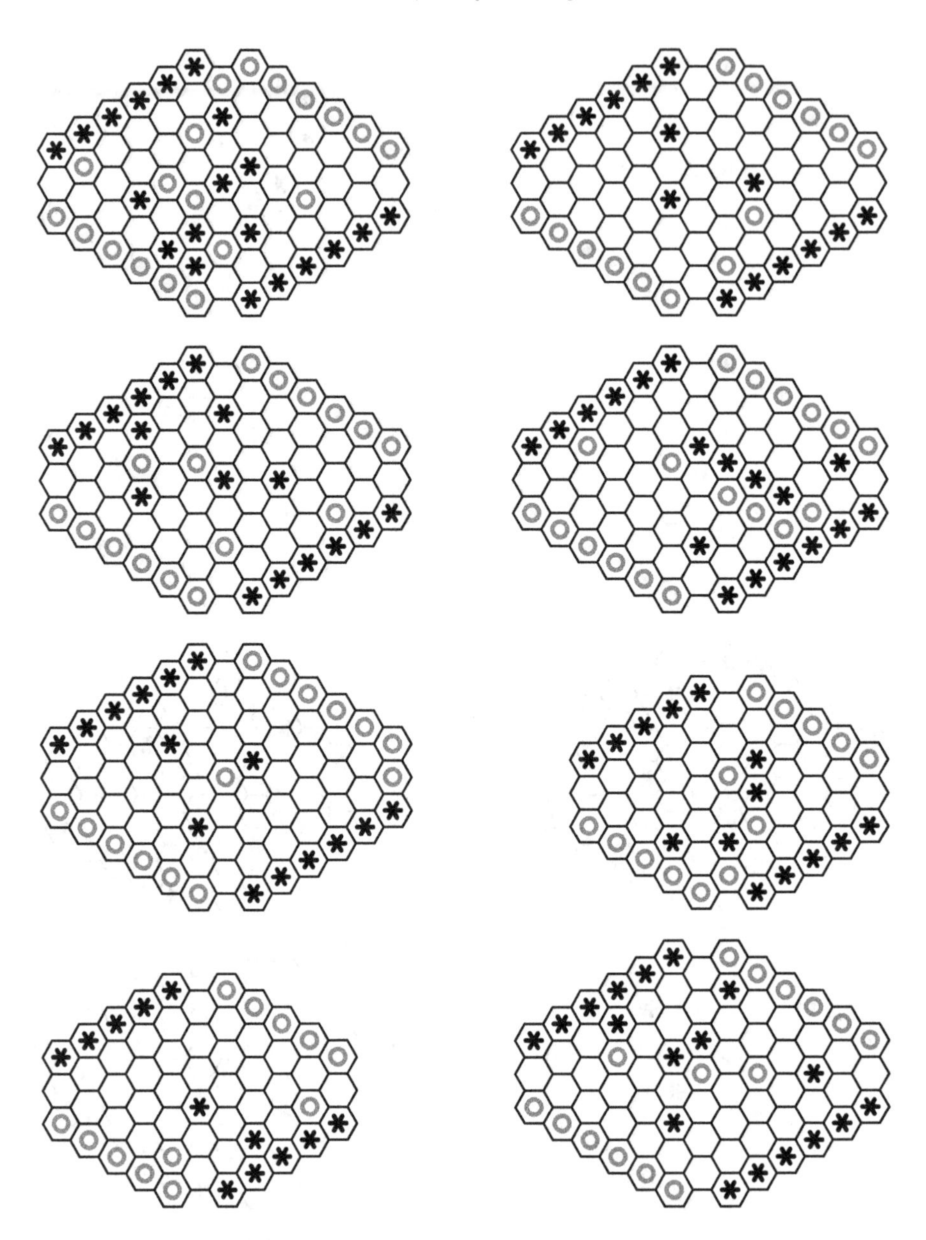

FIGURE A.5: Politiken puzzles 42-49. Puzzle 45 has no winning move.

A.2 Politiken Polygon openings

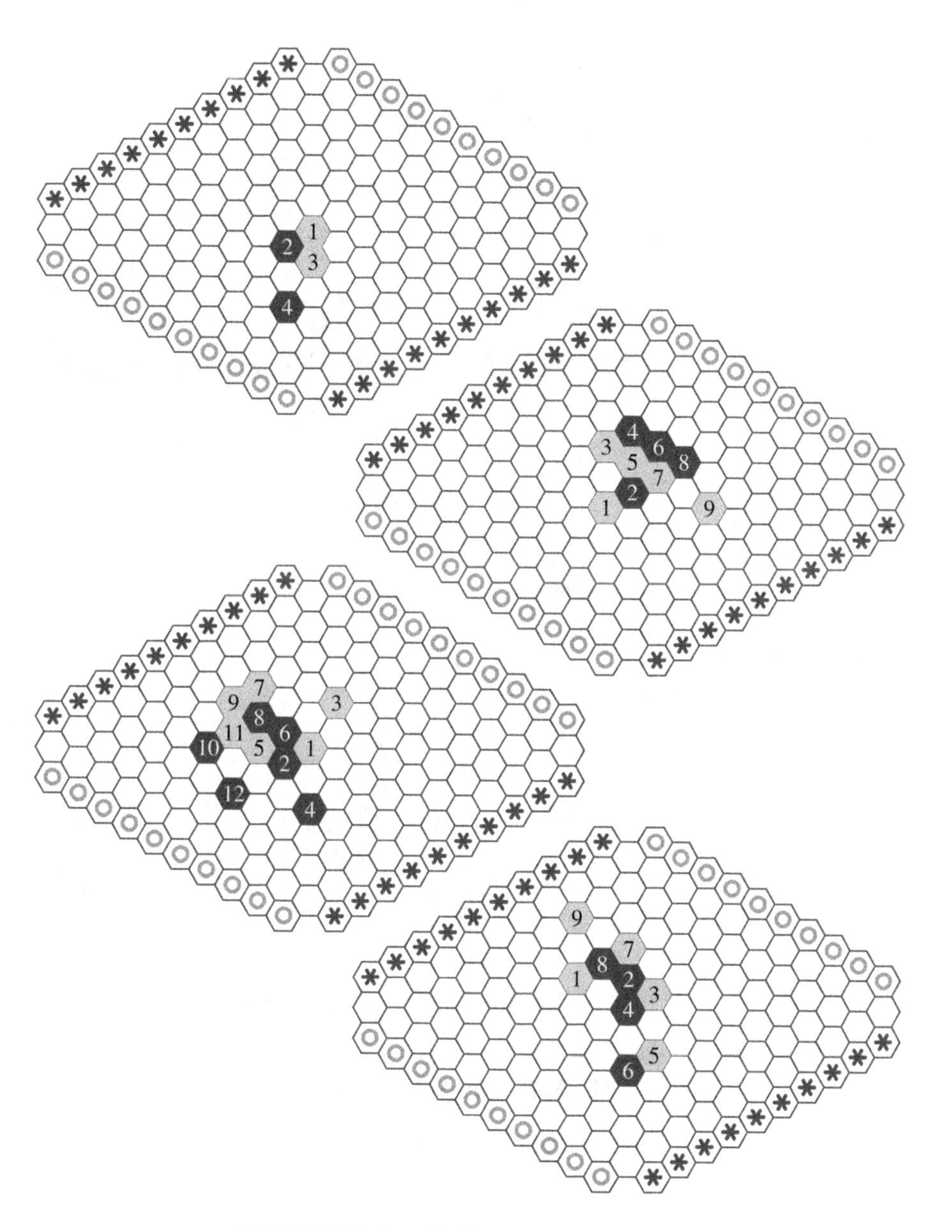

FIGURE A.6: Politiken openings 1-4.

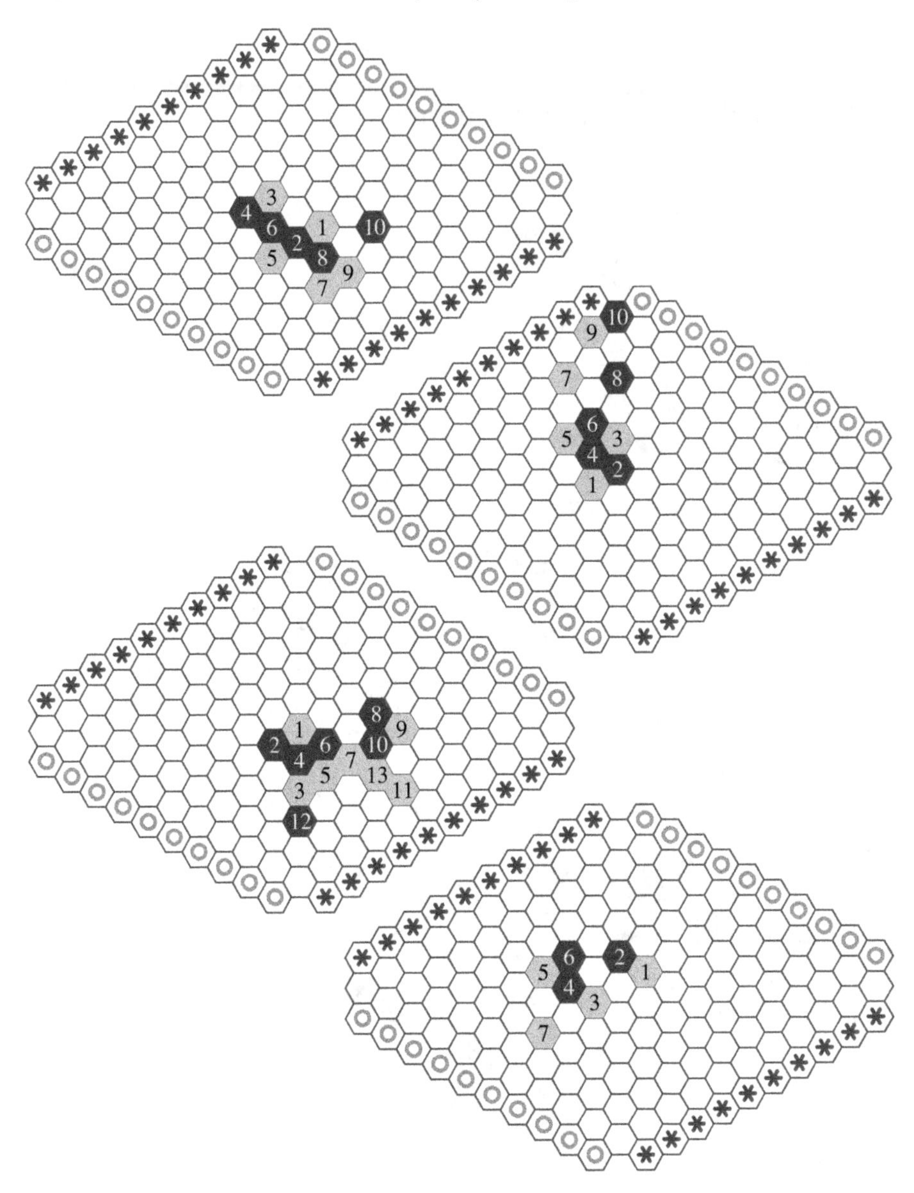

FIGURE A.7: Politiken openings 5-8.

A.3 Politiken Polygon solutions

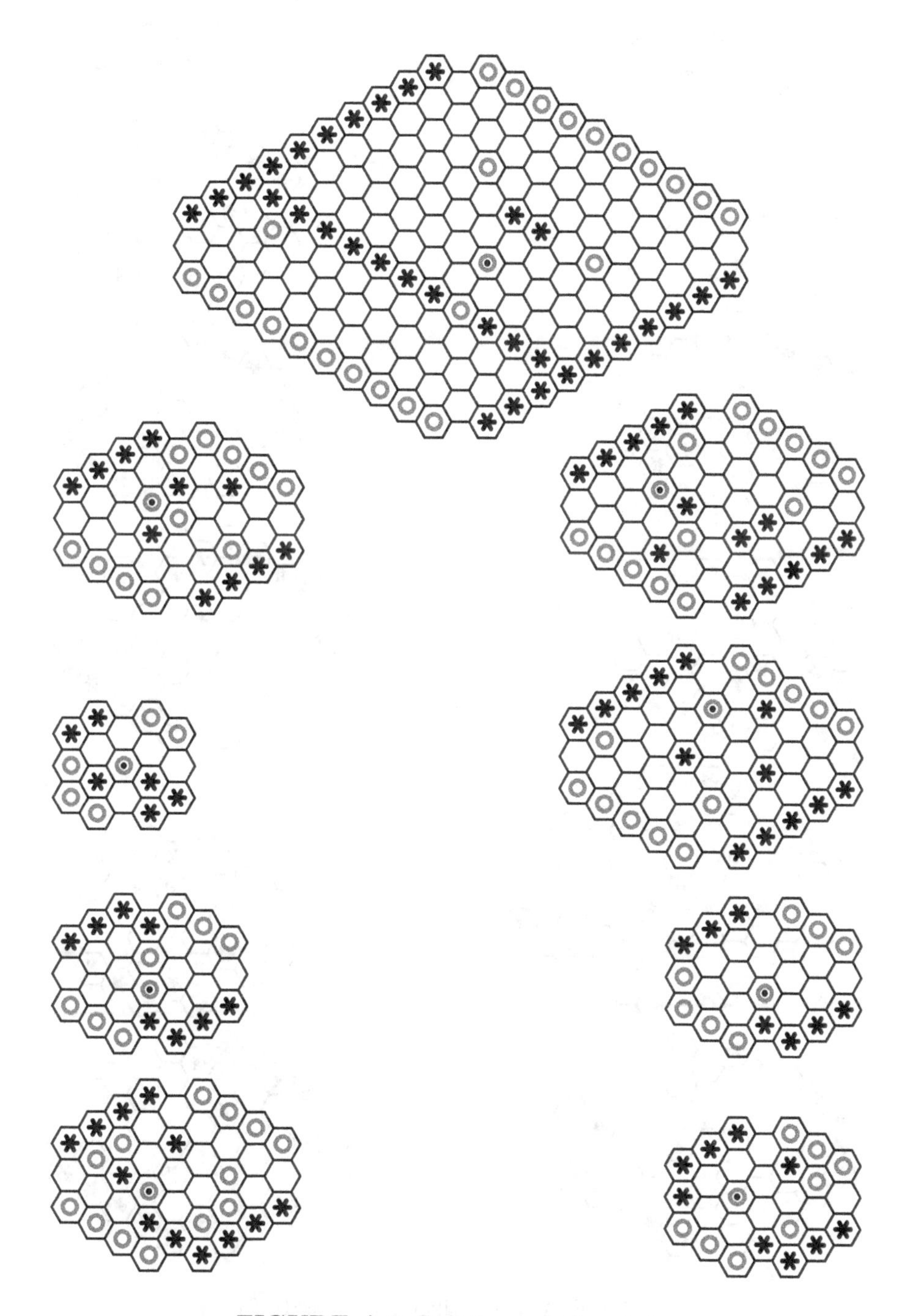

FIGURE A.8: Politiken solutions 1-9.

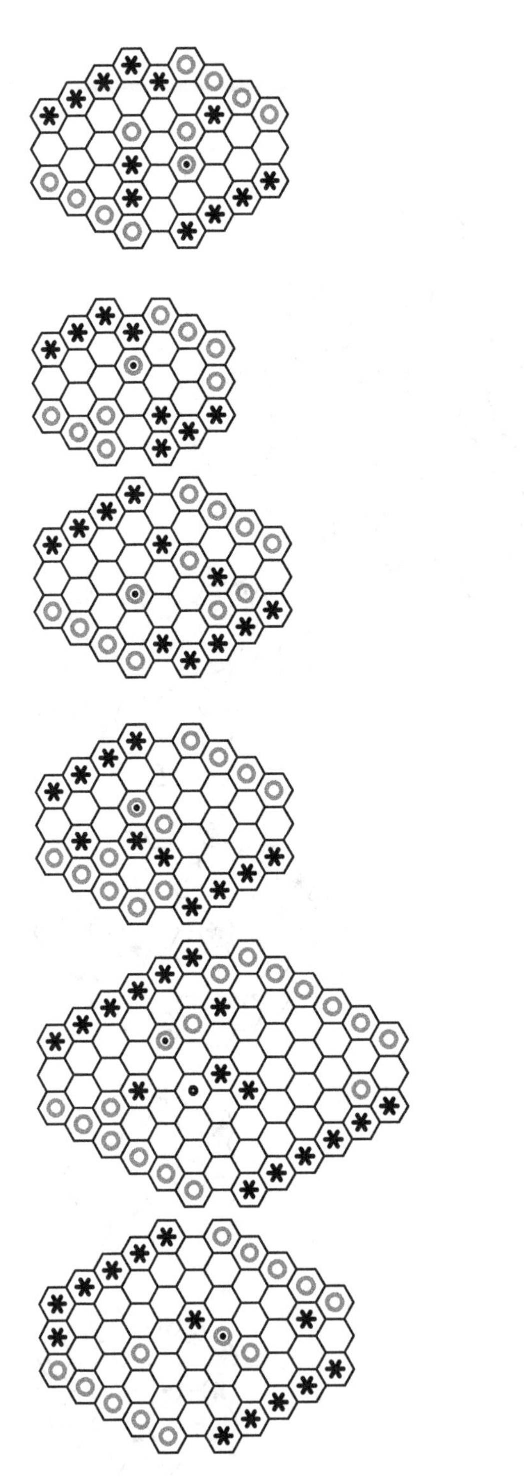

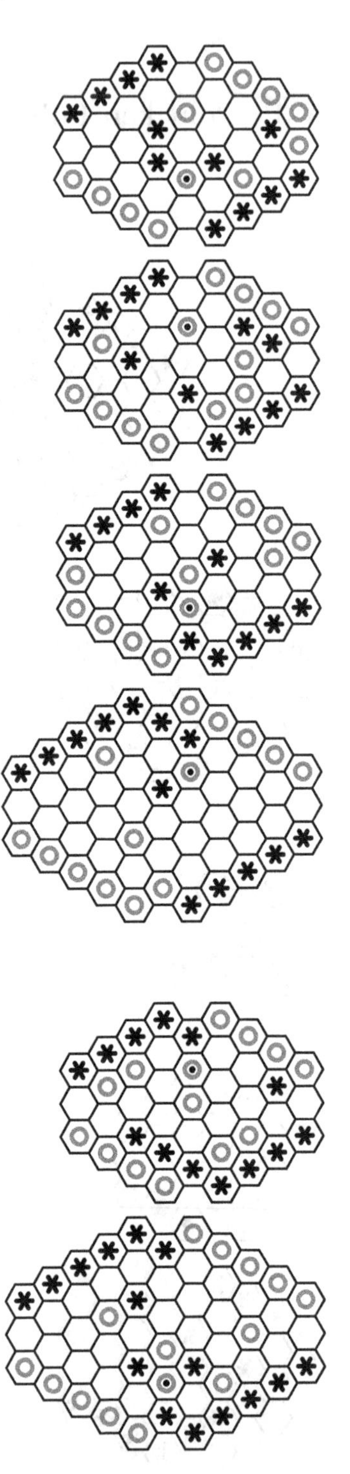

FIGURE A.9: Politiken solutions 10-21.

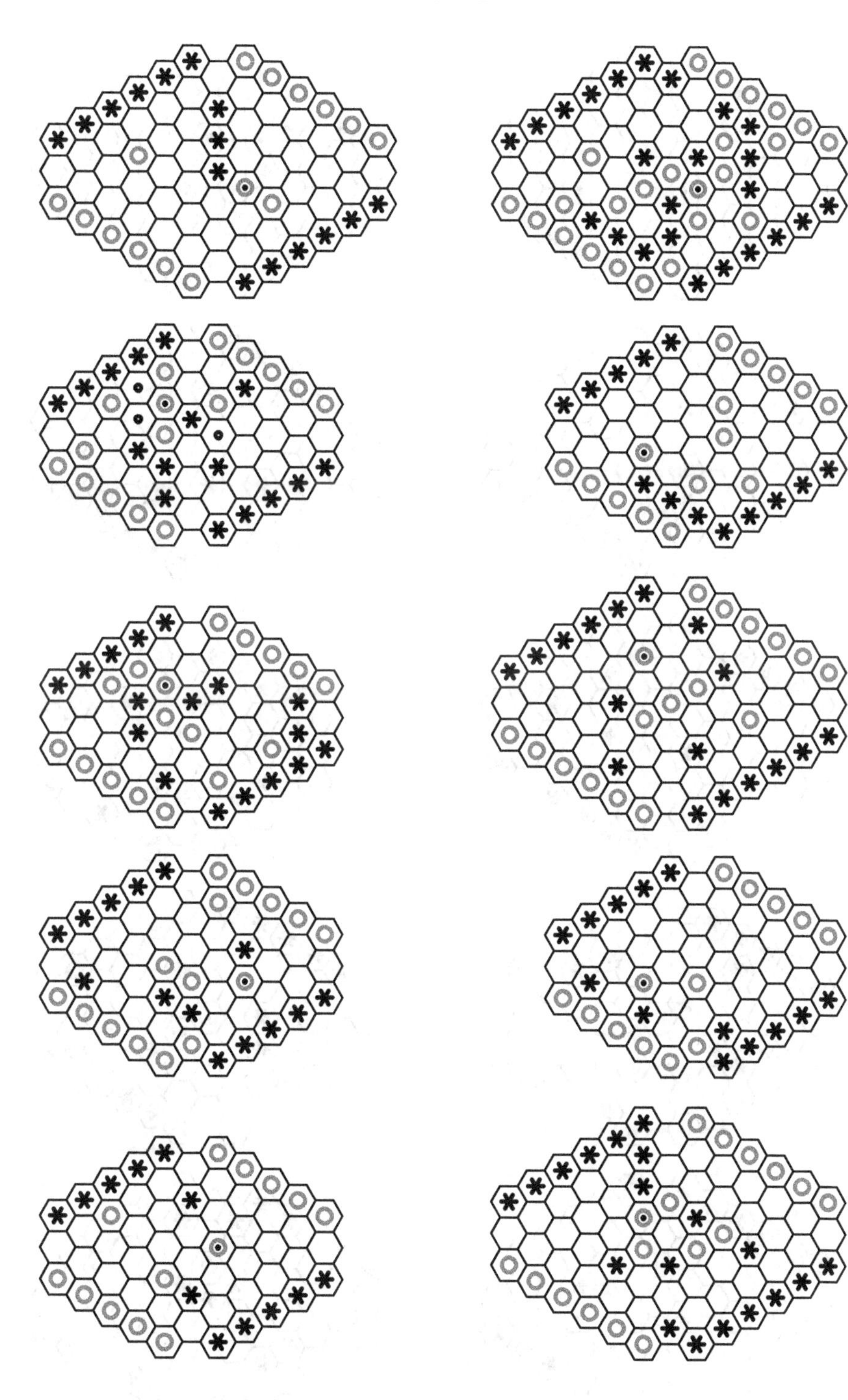

FIGURE A.10: Politiken solutions 22-31.

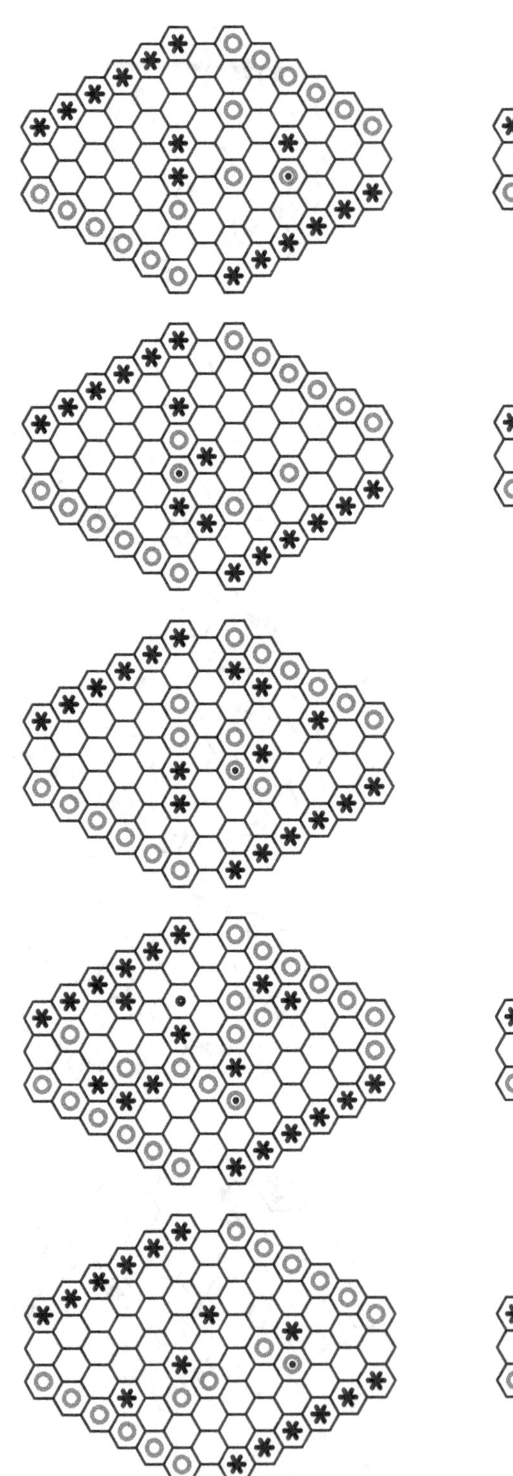

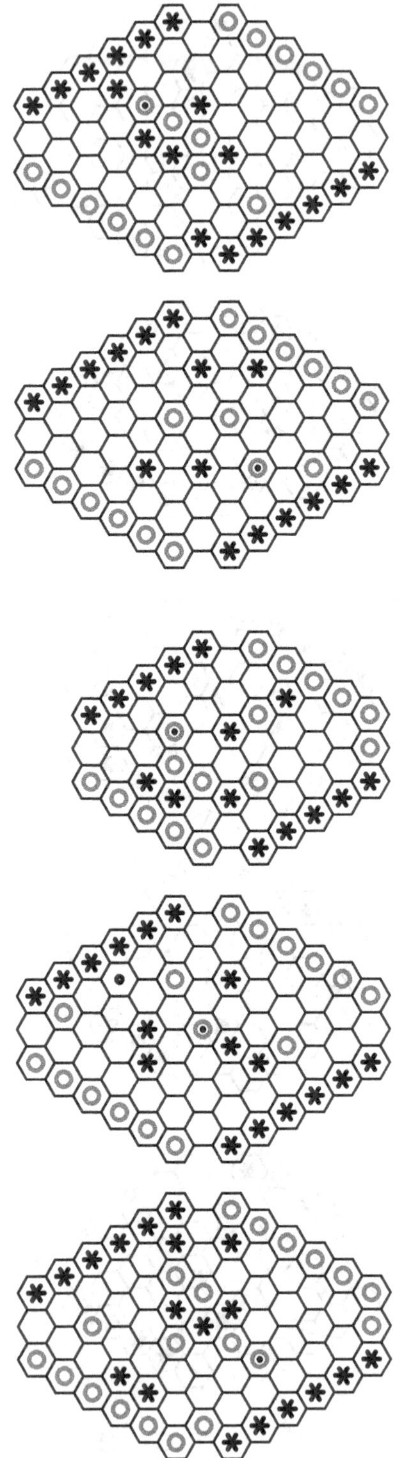

FIGURE A.11: Politiken solutions 32-41. Each dot also wins.

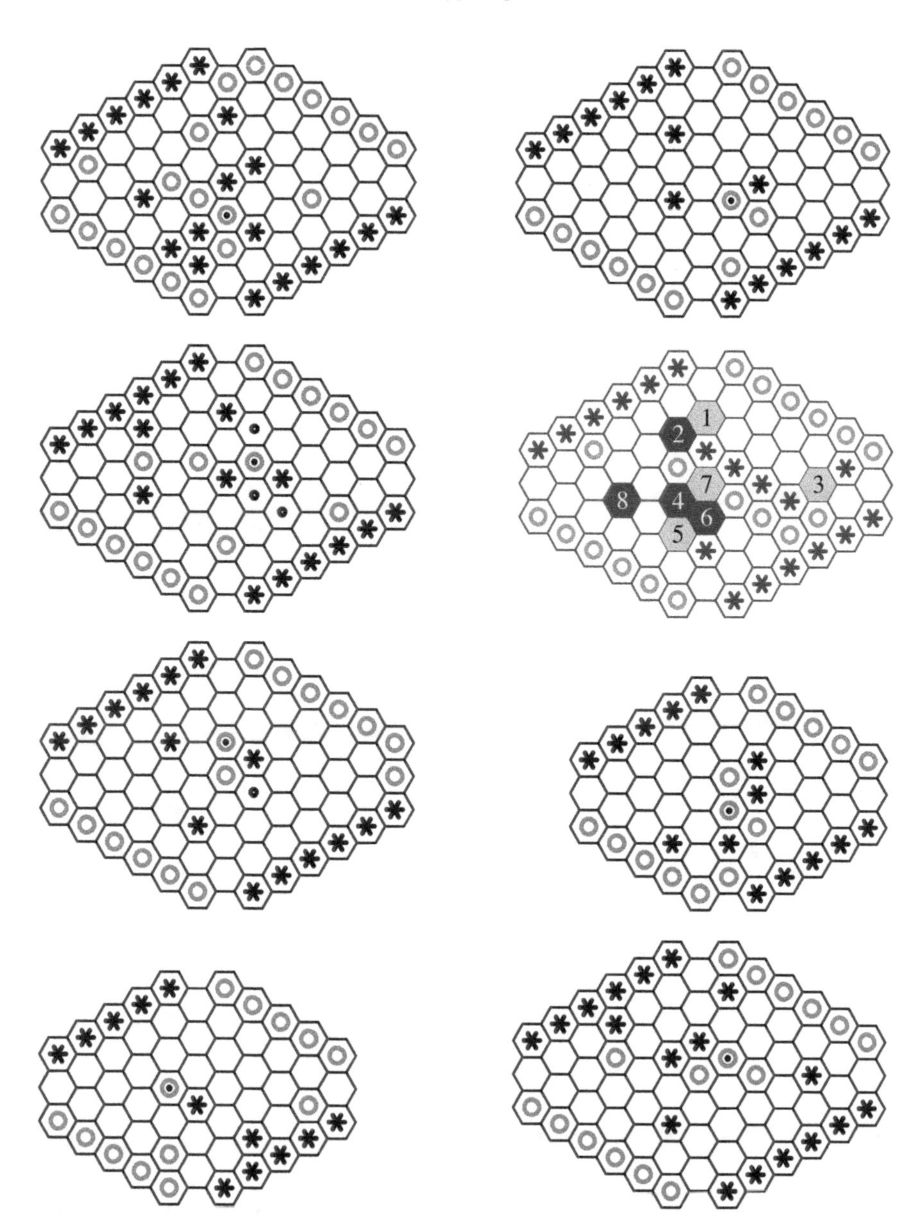

FIGURE A.12: Politiken solutions 42-49. Each dot also wins. Puzzle 45 has no winning move, we show White's best defense again Black's winning line of play.

Appendix B

Unpublished Lindhard puzzles

As we discovered in the fall of 2017, Jens Lindhard was Hein's principal Polygon puzzle composer. From the Lindhard archive at Aarhus University, here are all unpublished Polygon puzzle drafts and openings.

B.1 Lindhard puzzles

Lindhard's unpublished puzzle drafts were not numbered: we labelled them arbitrarily. Some drafts seem preliminary: in Drafts 13, 14 and 22, each move — including the proposed move — loses. So, in Drafts 13 and 14 we added Lindhard's proposed move and in Draft 22 we added a white stone at b4, which makes Lindhard's proposed solution correct. After these edits, each draft except Puzzle 24 has a winning move.

For each puzzle, find the best move. Who wins?

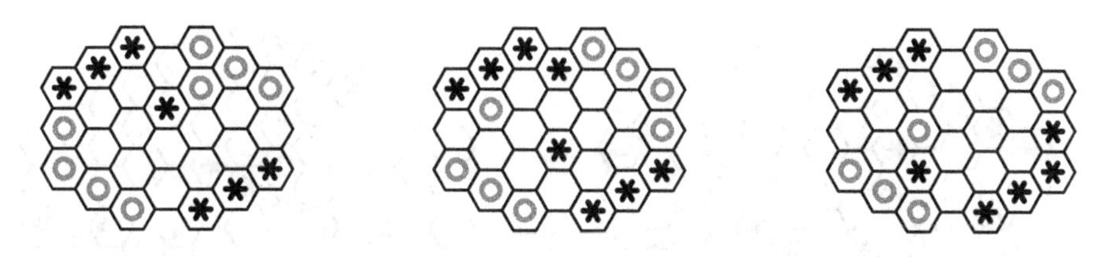

FIGURE B.1: Lindhard puzzles 1-3. White to play.

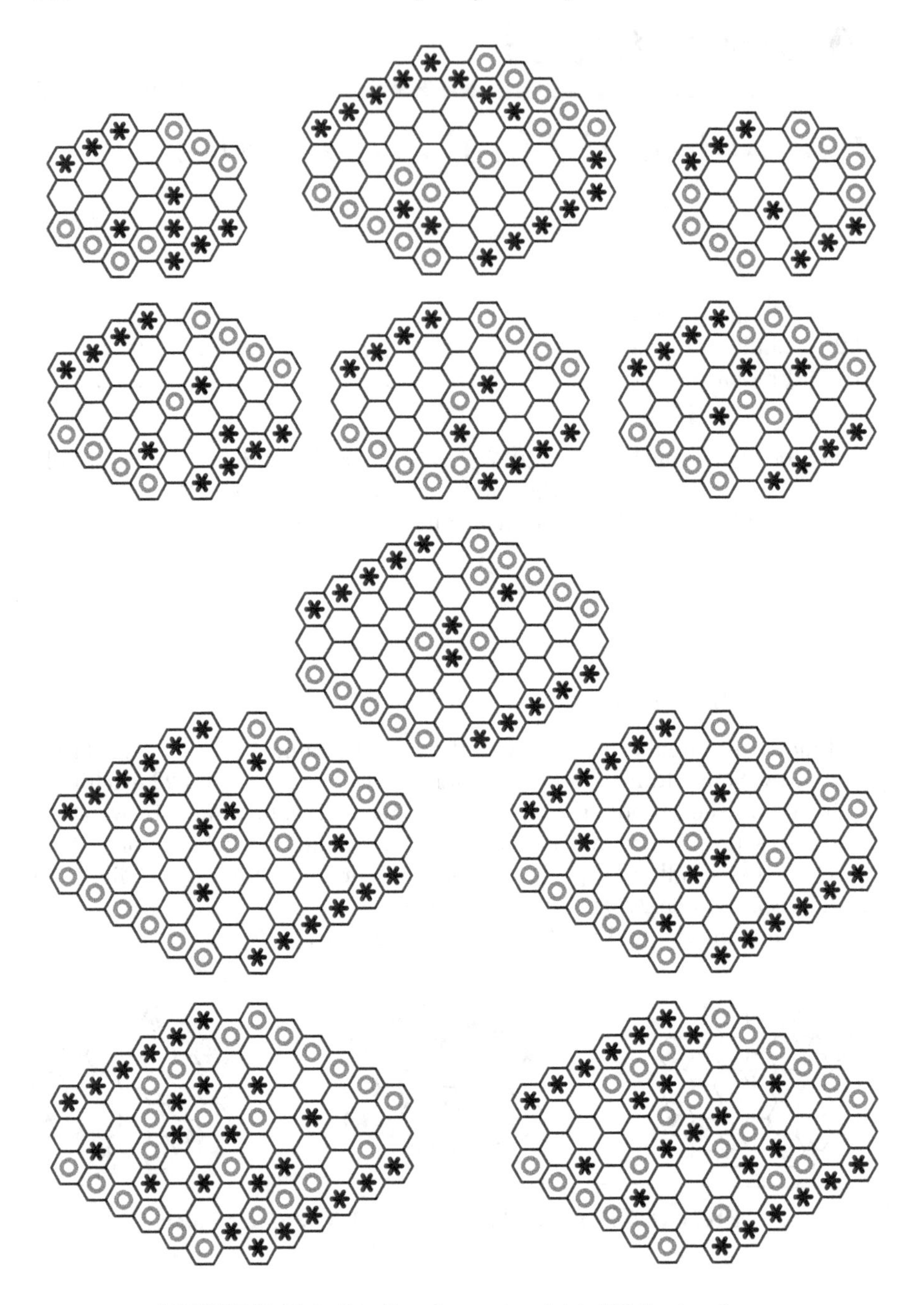

FIGURE B.2: Lindhard puzzles 4-14. White to play.

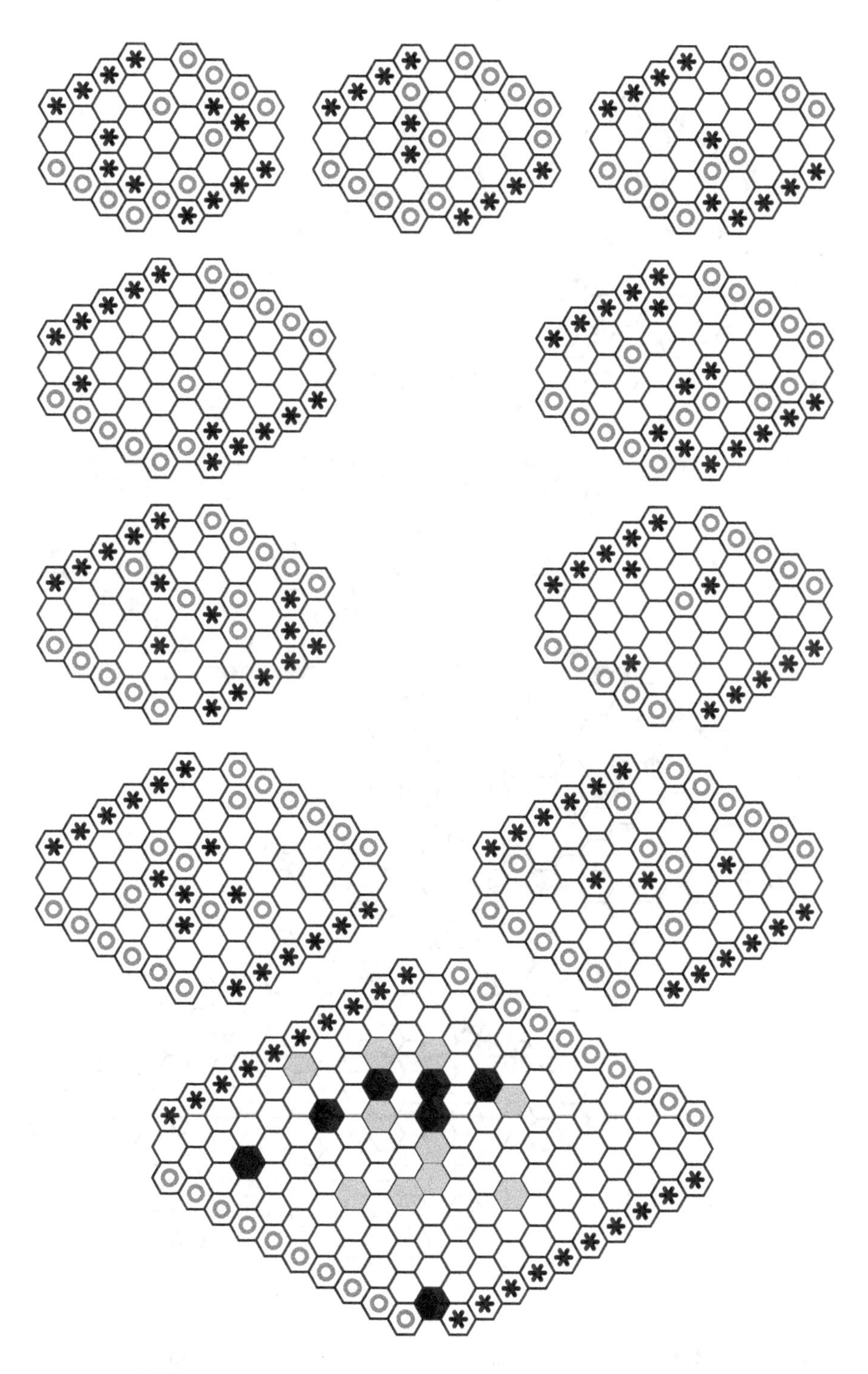

FIGURE B.3: Lindhard puzzles 15-24. Black to play.

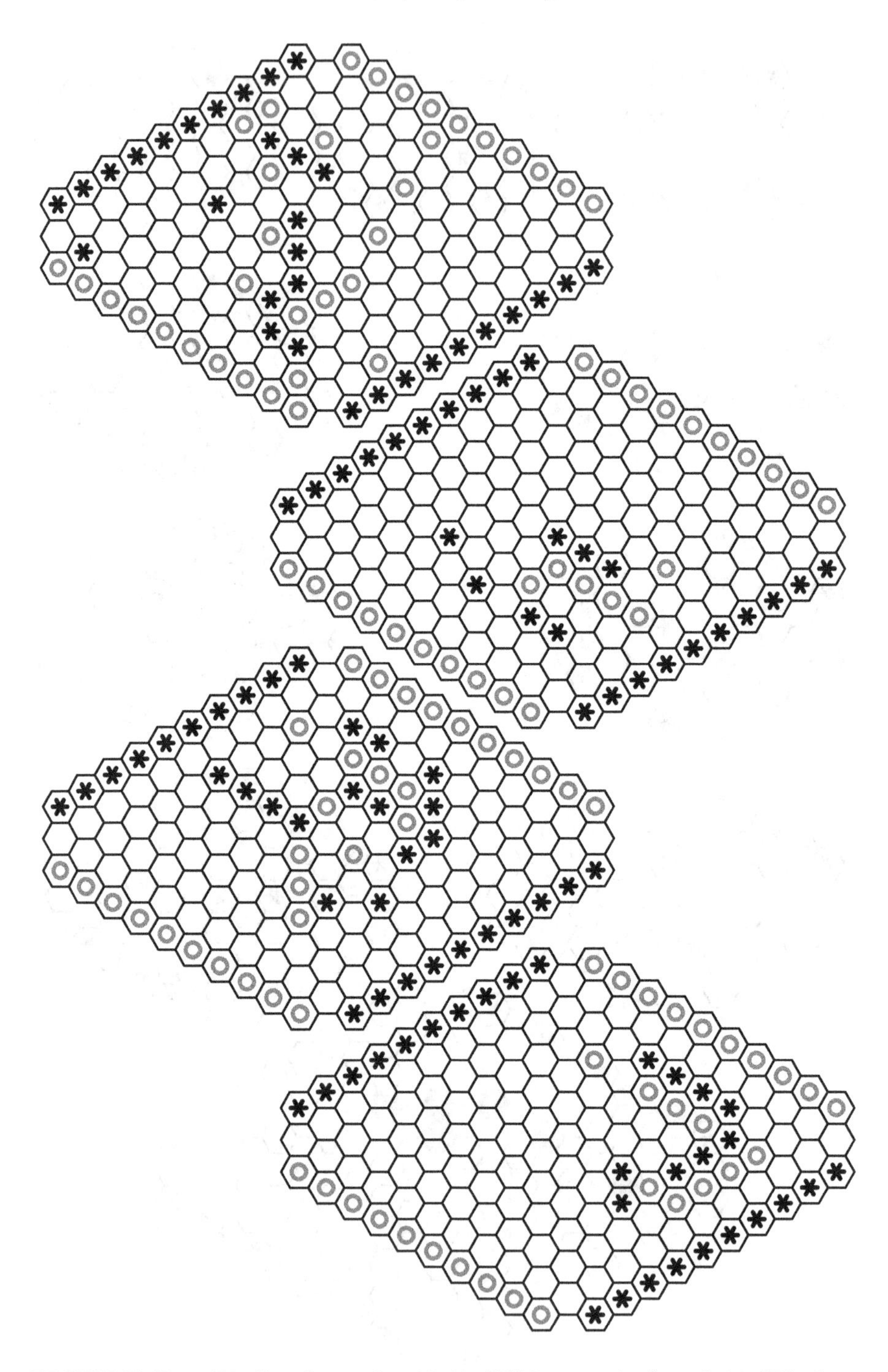

FIGURE B.4: Lindhard puzzles 25-27 (White to play) and 28 (Black to play). Puzzle 28 might be a rubbing-shoulders diagram.

B.2 Lindhard openings

> *The opening moves in any decent [11×11 Polygon] game happen more or less at random.*
>
> Jens Lindhard, from the Lindhard archive

Openings 1-10, numbered by Lindhard and in polished form, were perhaps candidates for the planned Polygon booklet. The other openings were roughly drawn, perhaps intended only as sketches or experimental continuations.

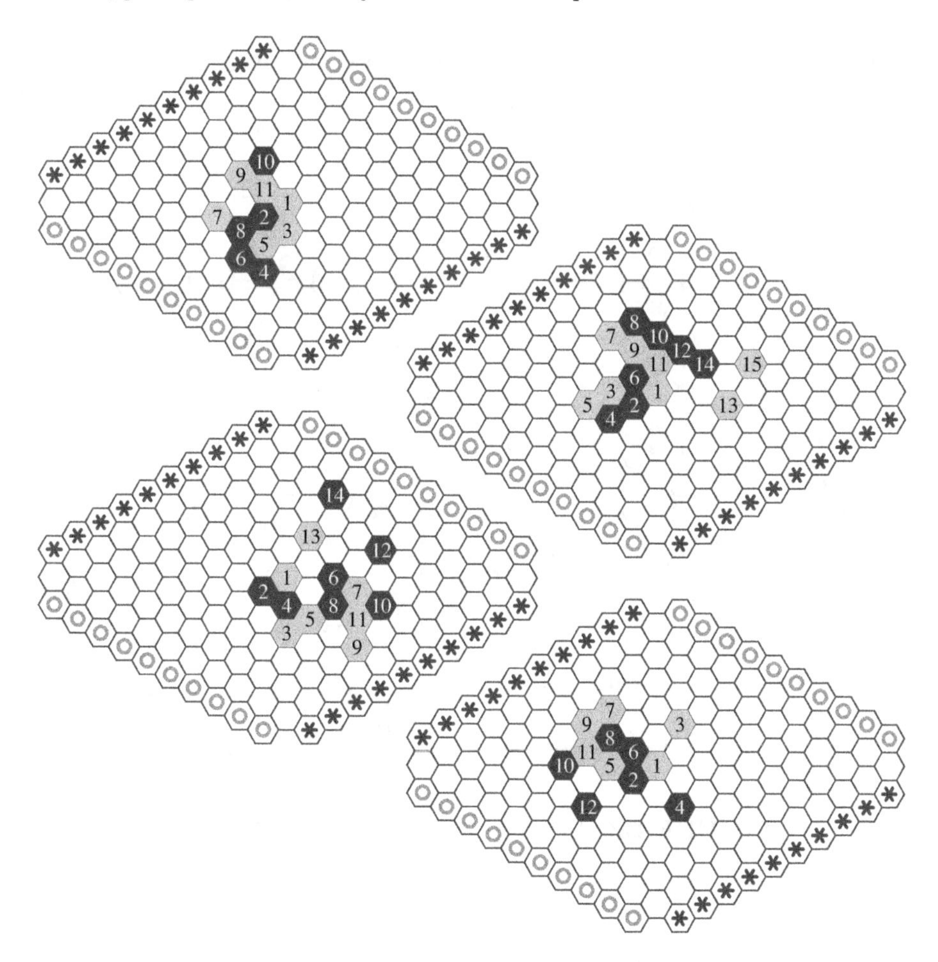

FIGURE B.5: Lindhard openings 1, 1a, 1b, 1c.

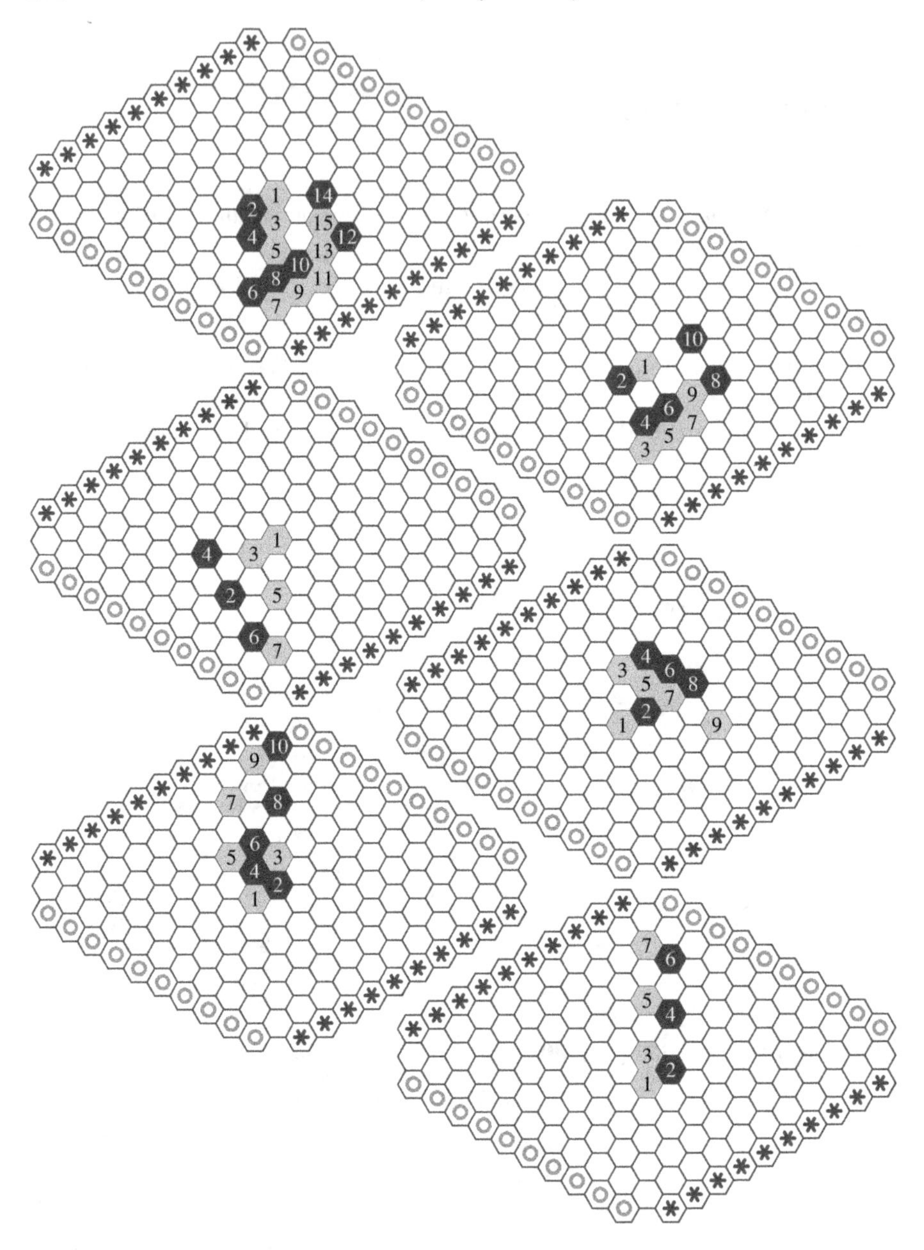

FIGURE B.6: Lindhard openings 1d, 2, 3, 4, 4a, 5.

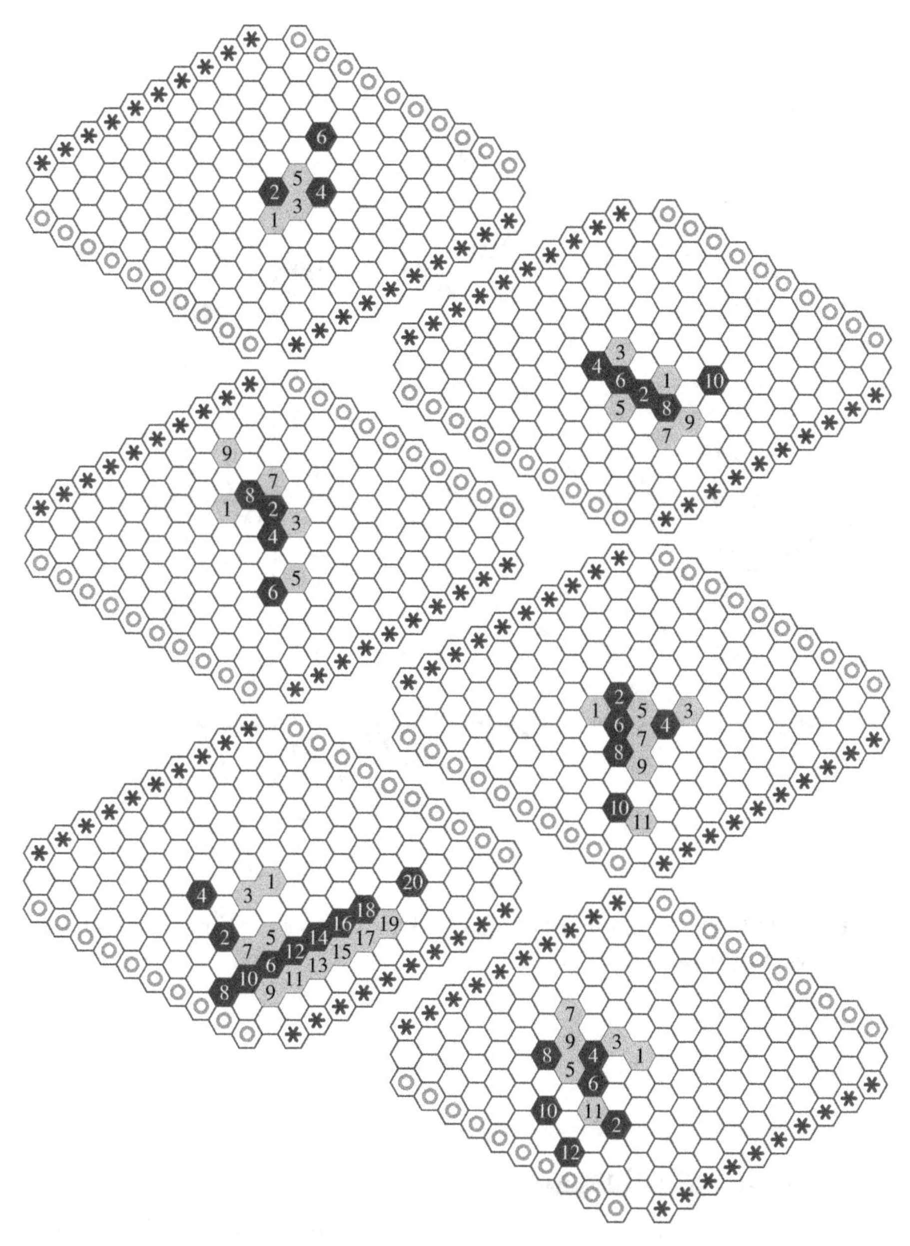

FIGURE B.7: Lindhard openings 5a, 6-10.

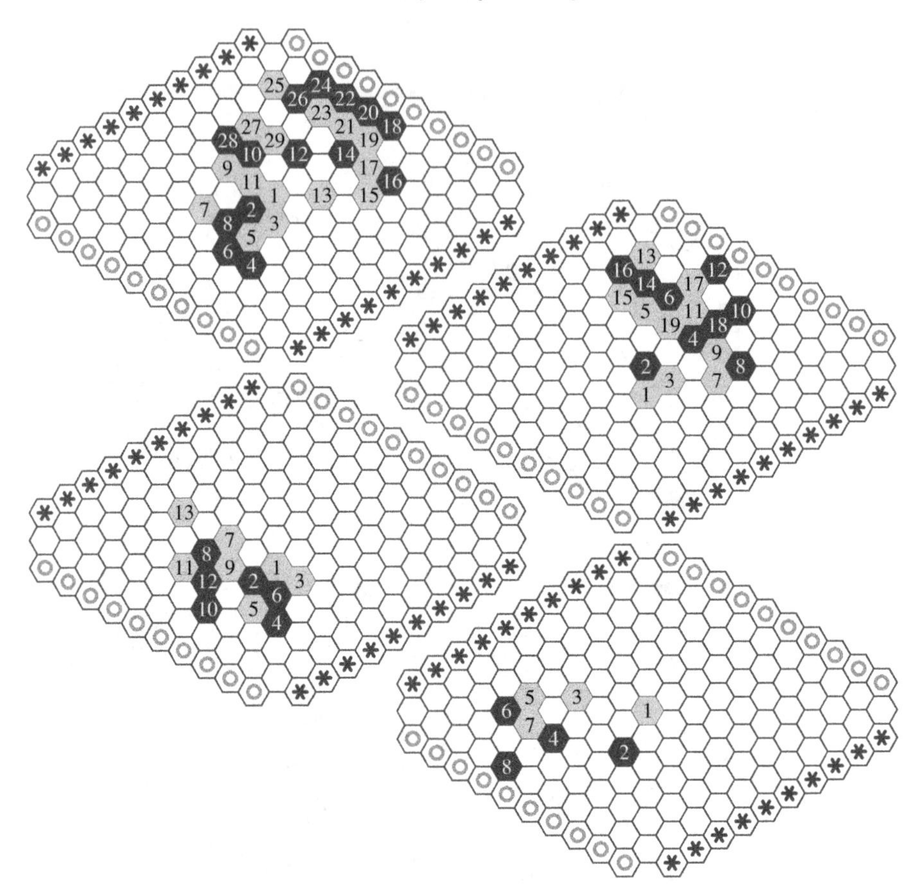

FIGURE B.8: Other Lindhard openings. The topmost continues from Opening 1, with White winning.

B.3 Lindhard solutions

Except for the 11×11 puzzles, we show all winning moves. Except for Puzzle 24, the player-to-move can win.

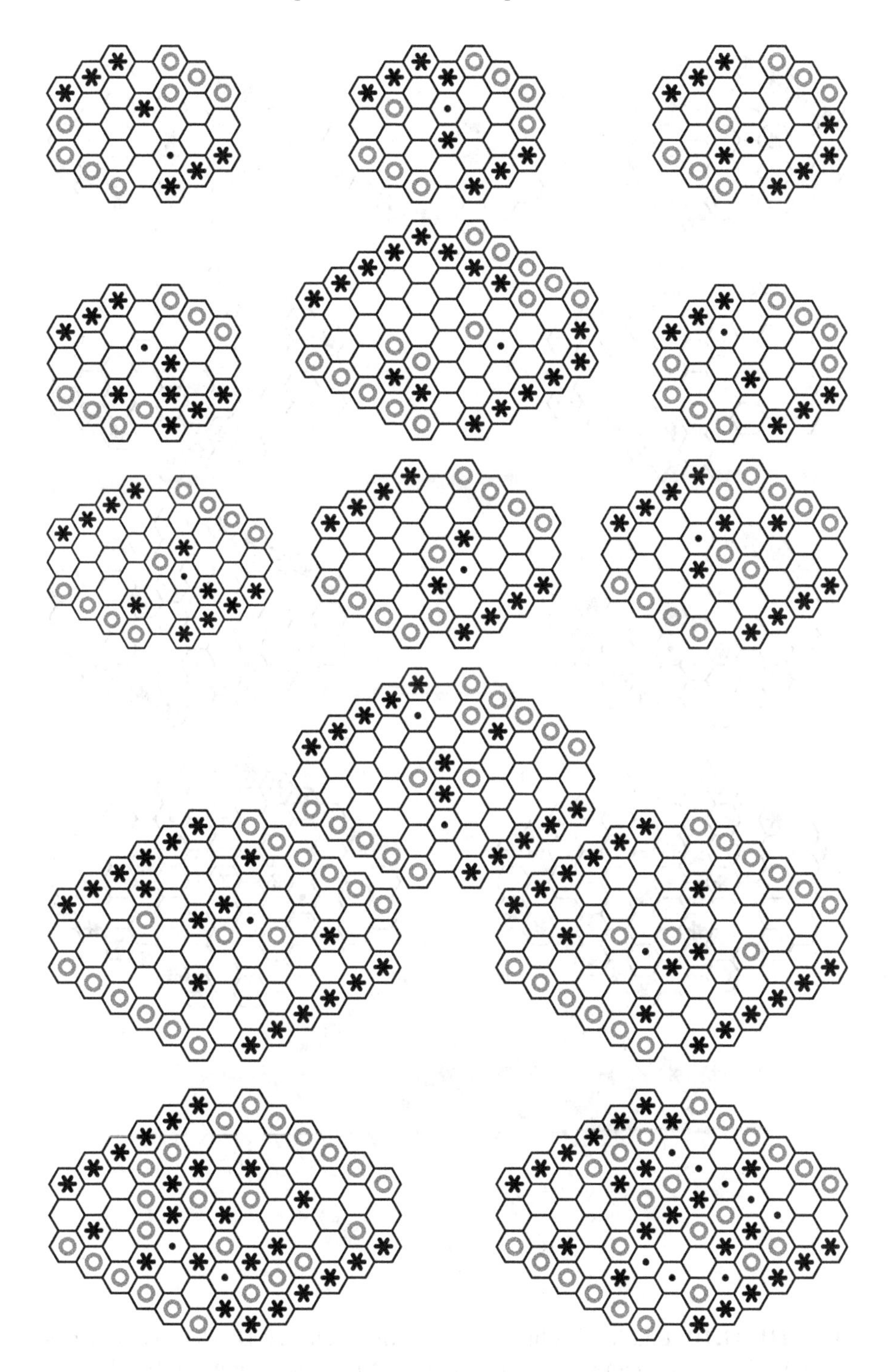

FIGURE B.9: Lindhard solutions 1-14.

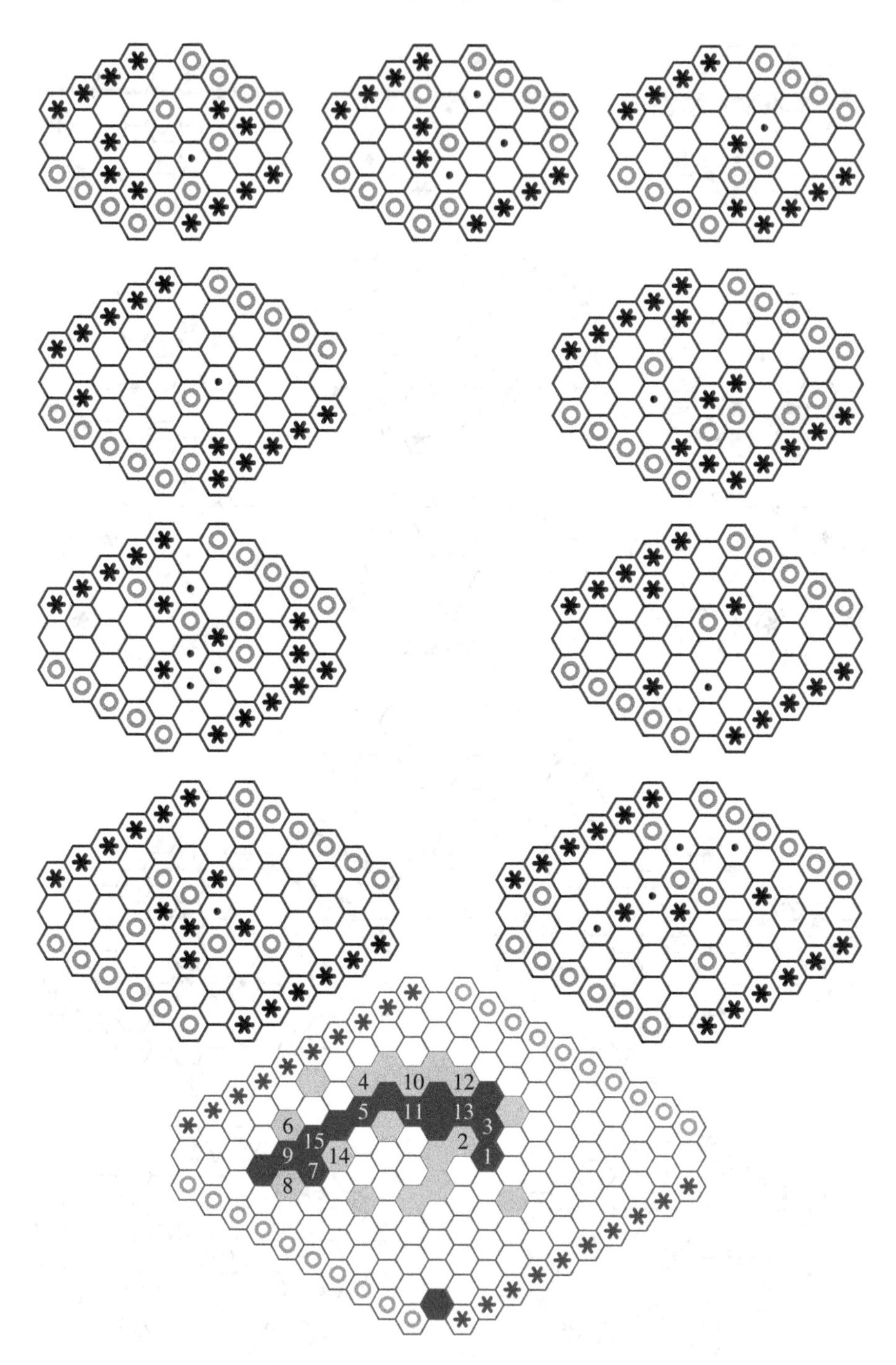

FIGURE B.10: Lindhard solutions 15-24. In Puzzle 24, Black's best move is 1, but White 2 wins: can you find the unique White winning move 16? Answer in Figure B.11.

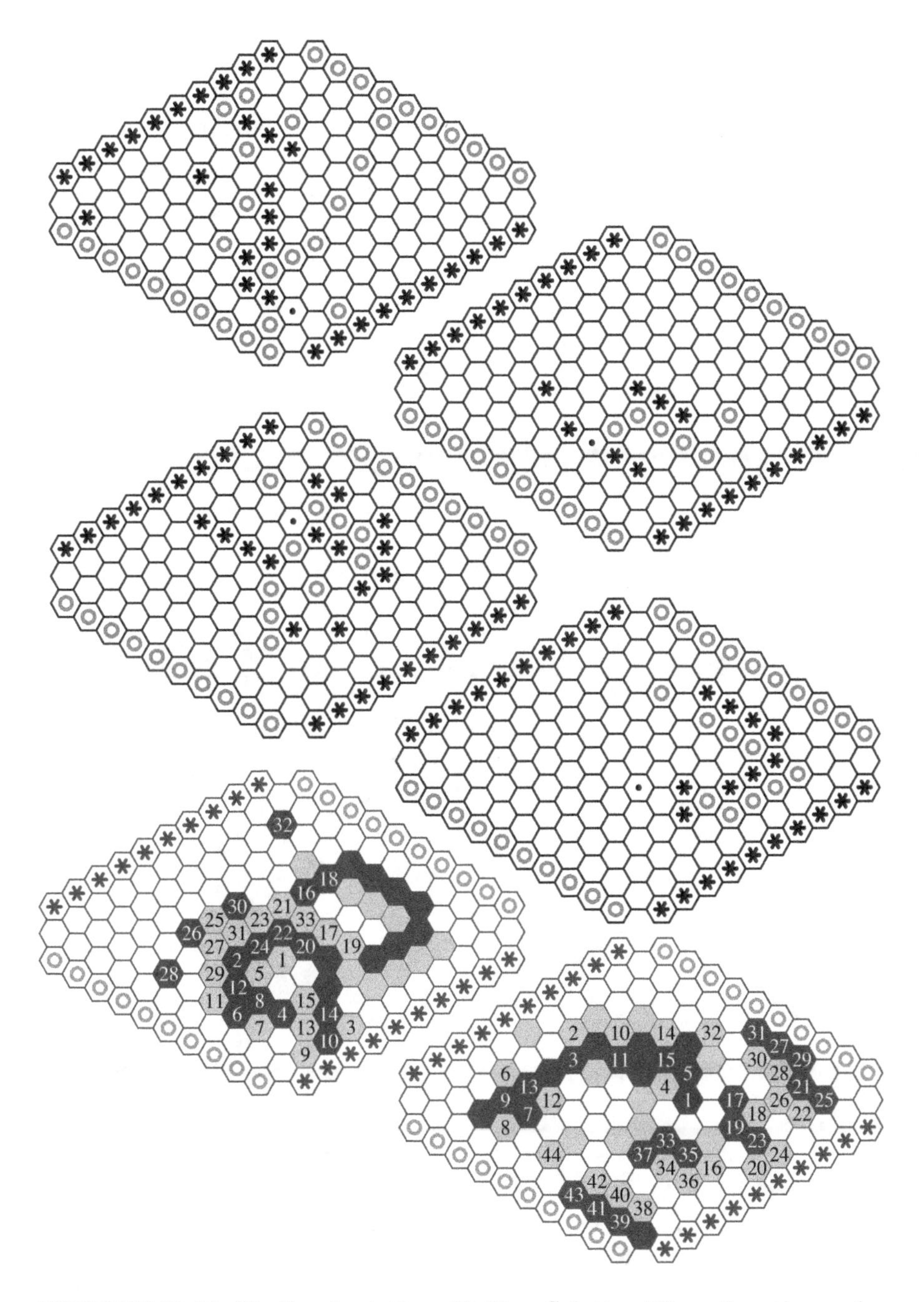

FIGURE B.11: Lindhard solutions 25-28, a Solution 28 continuation and a Solution 24 continuation that answers the question in Figure B.10.

Appendix C

Henderson Hex puzzles

Here are some new Hex puzzles collected by Henderson. Solver verified these and other Hex solutions in this book.

C.1 Henderson puzzles

FIGURE C.1: Henderson puzzles B1-B9. Black to play.

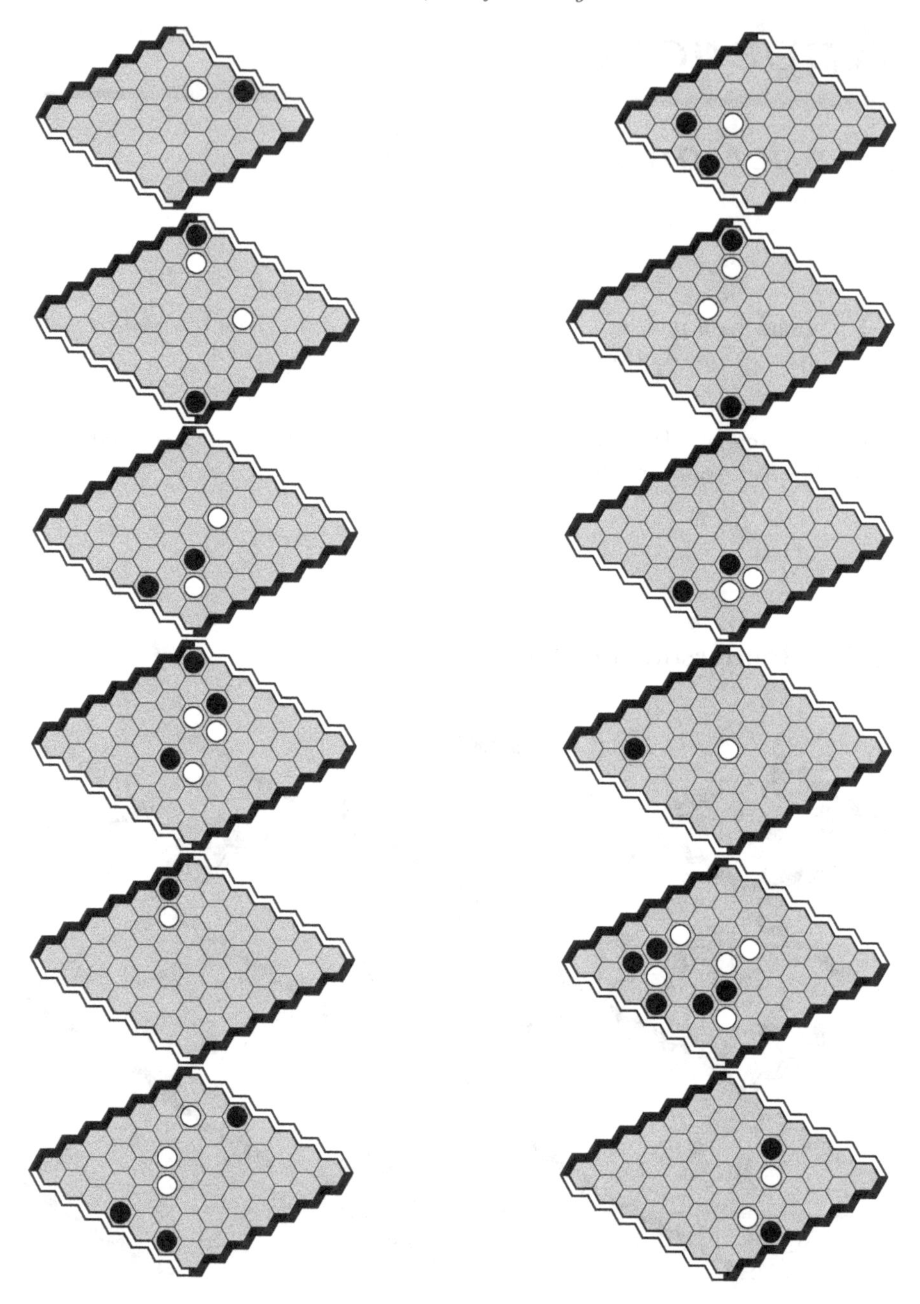

FIGURE C.2: Henderson puzzles B10-B21. Black to play.

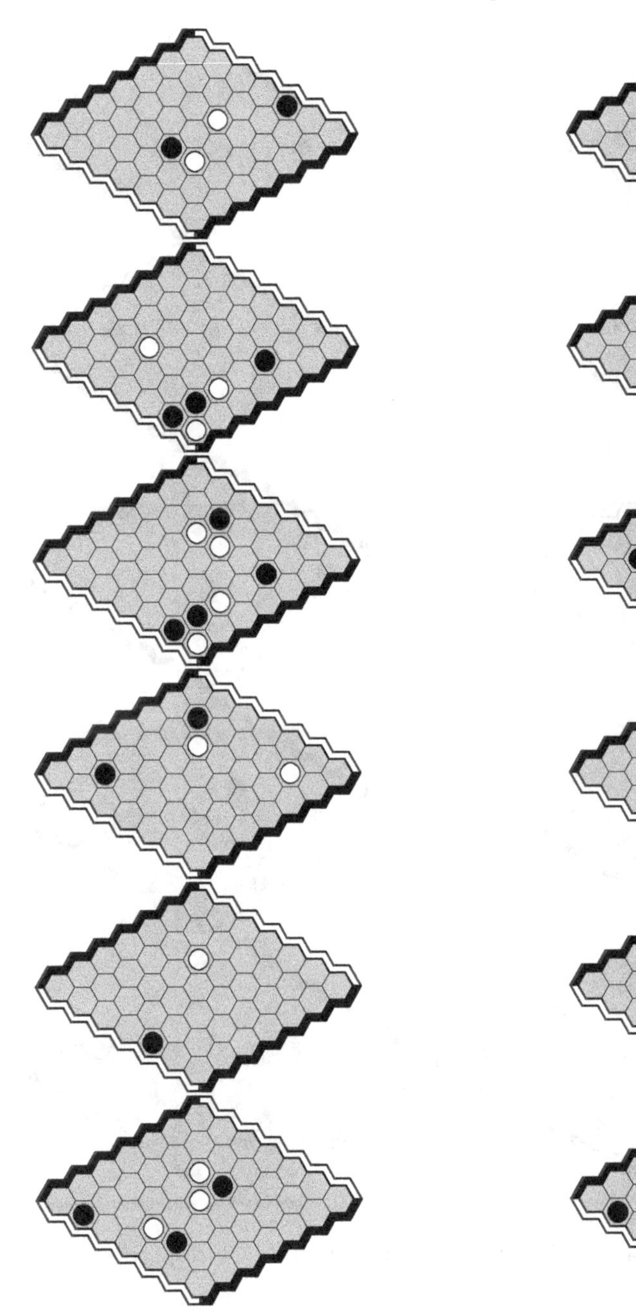

FIGURE C.3: Henderson puzzles B22-B33. Black to play.

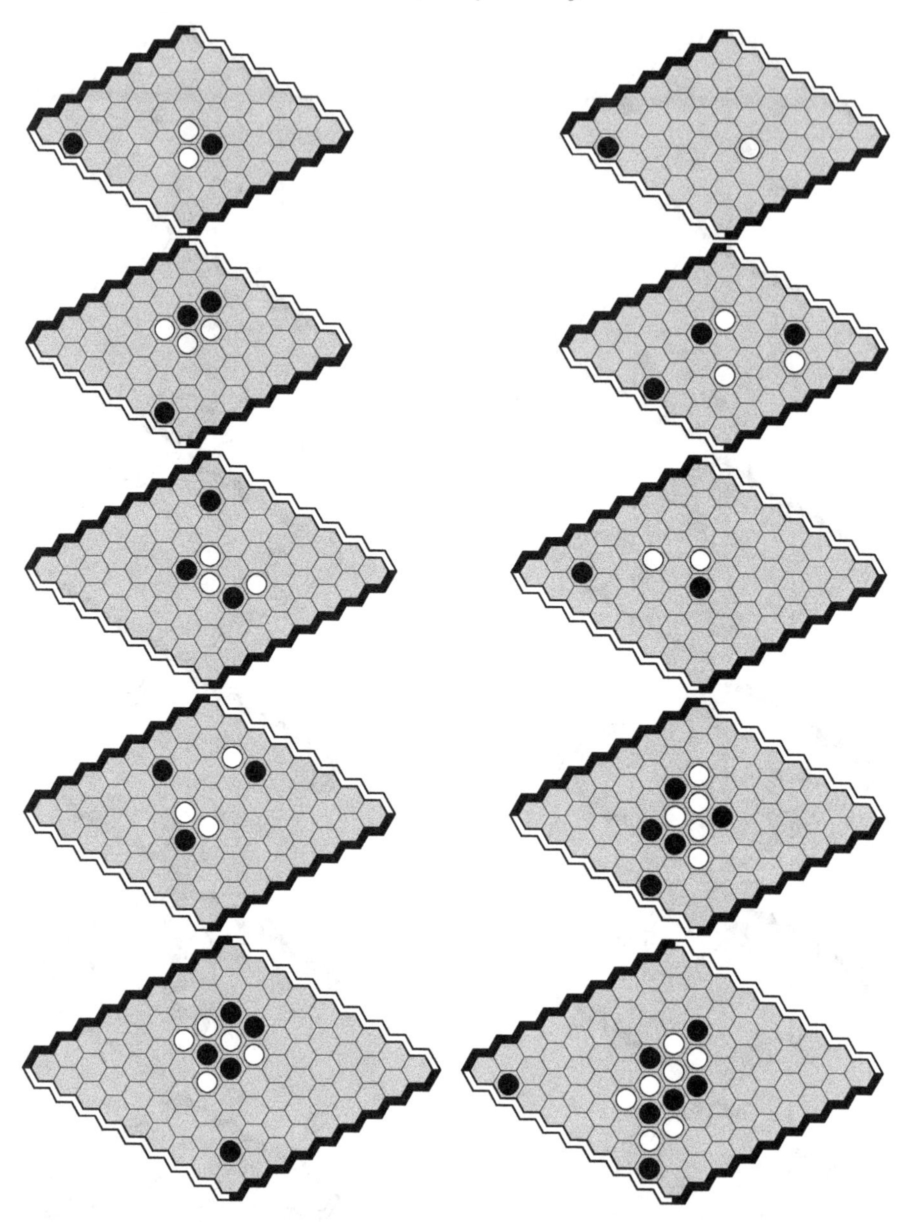

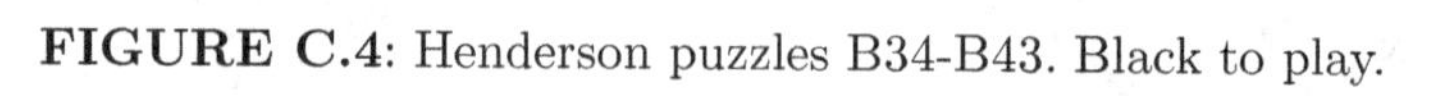

FIGURE C.4: Henderson puzzles B34-B43. Black to play.

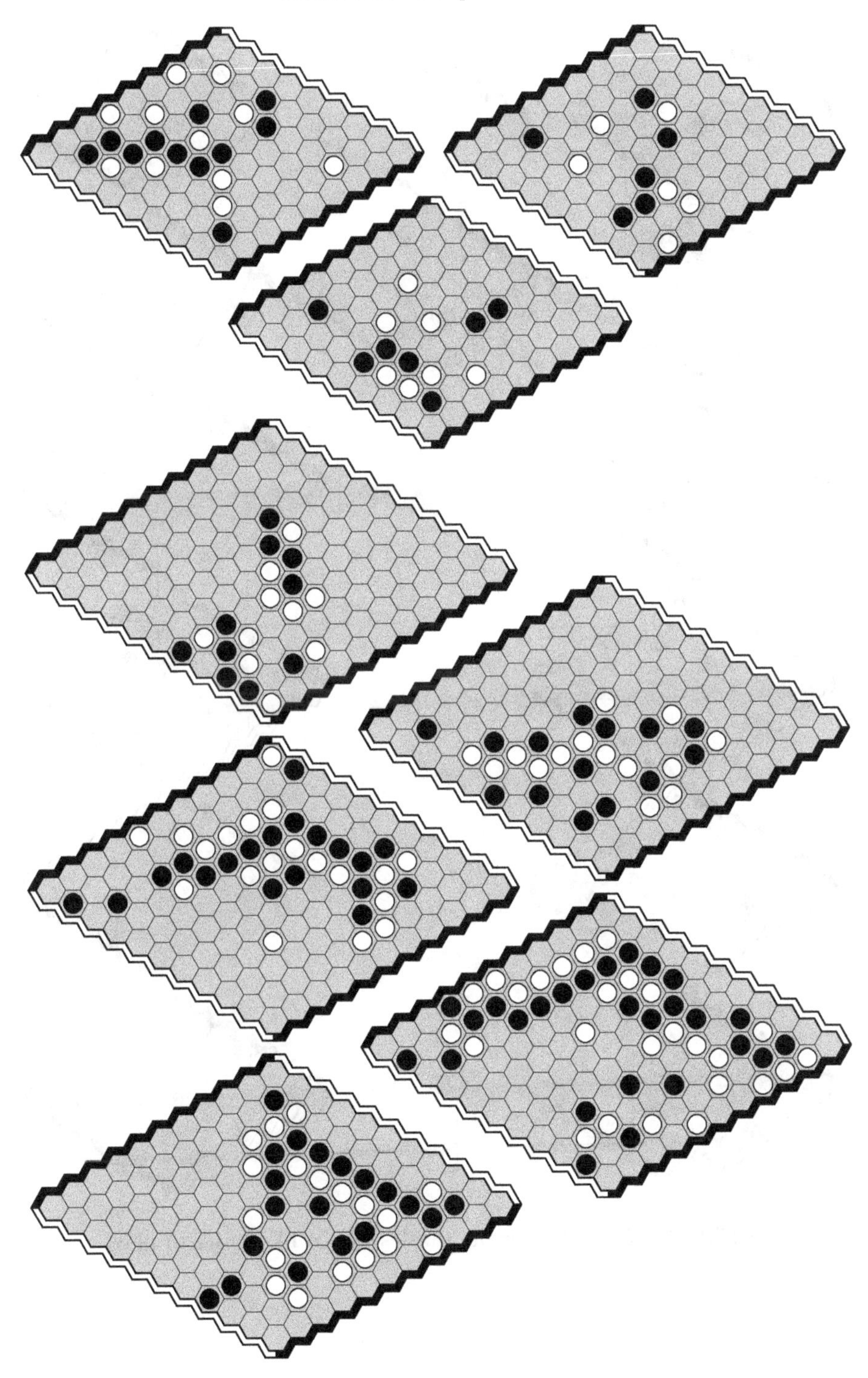

FIGURE C.5: Henderson puzzles B44-B51. Black to play.

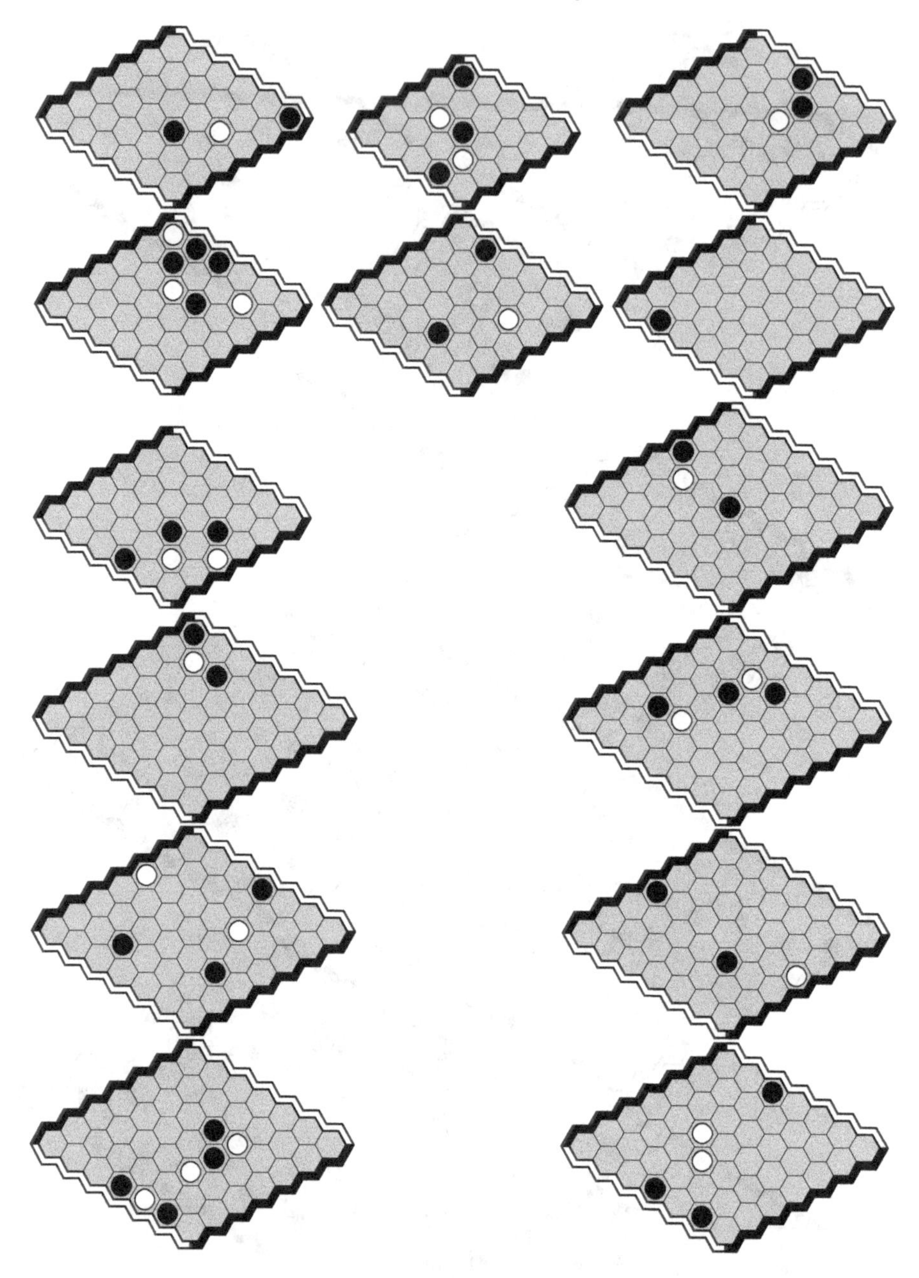

FIGURE C.6: Henderson puzzles W1-W14. White to play.

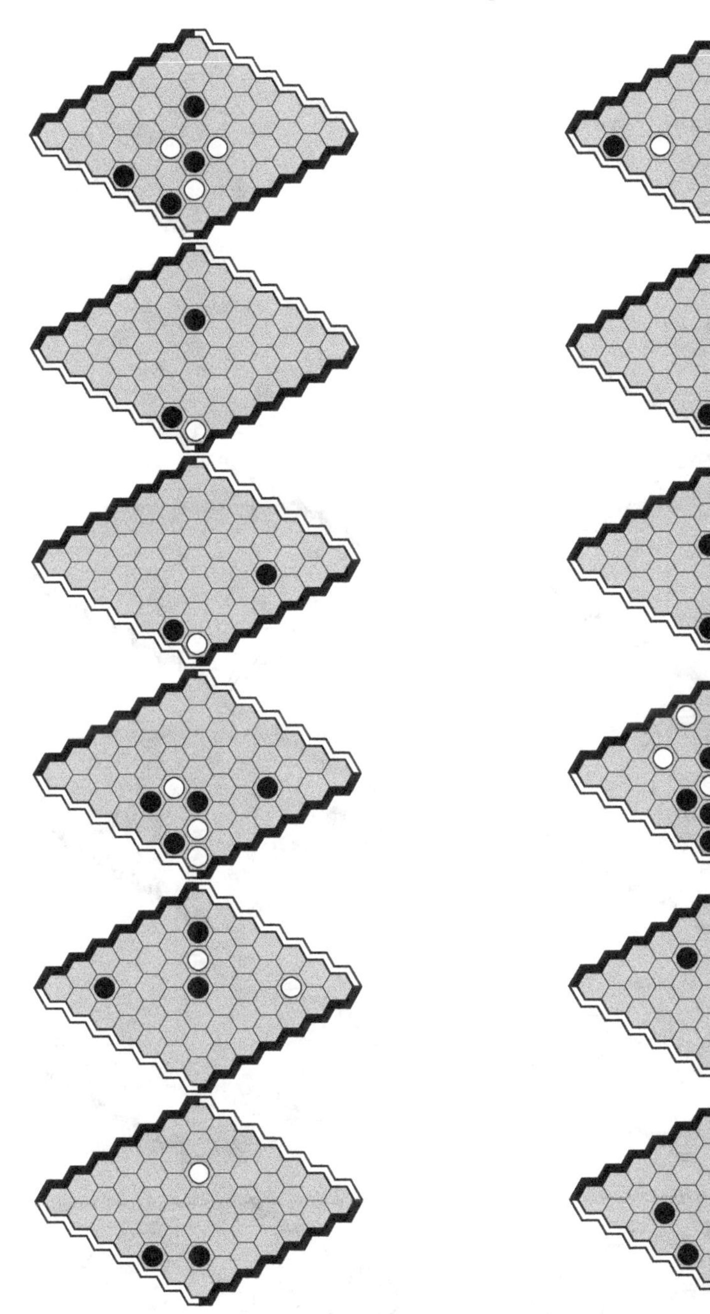

FIGURE C.7: Henderson puzzles W15-W26. White to play.

FIGURE C.8: Henderson puzzles W27-W38. White to play.

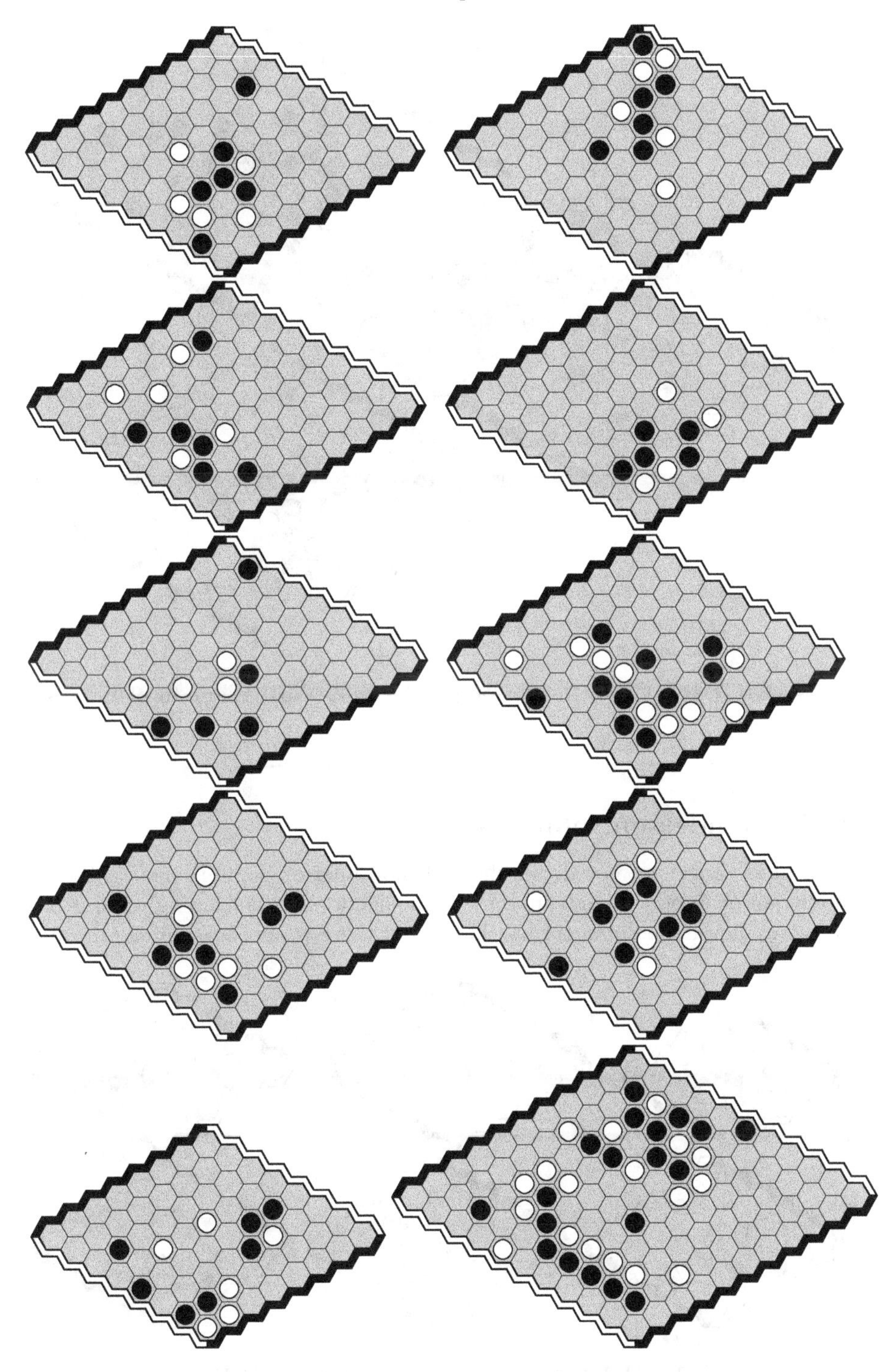

FIGURE C.9: Henderson puzzles W39-W48. White to play.

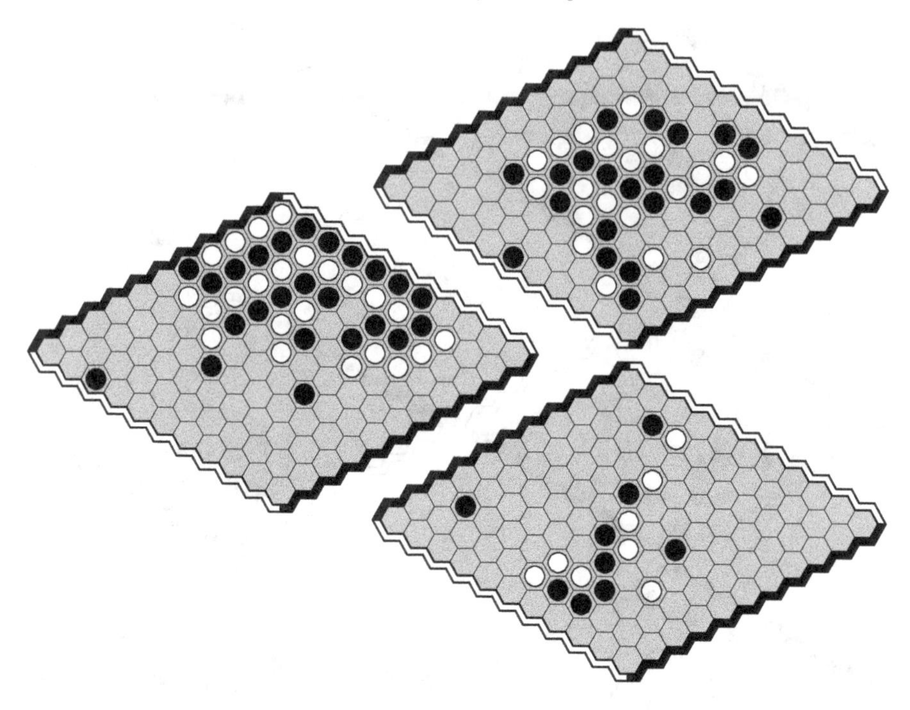

FIGURE C.10: Henderson puzzles W49-W51. White to play.

C.2 Henderson solutions

Unless otherwise noted, we show all non-inferior winning moves. We do not examine opponent-captured moves: if such a move is winning, then it is a delaying move, the opponent can kill it and there is another winning move.

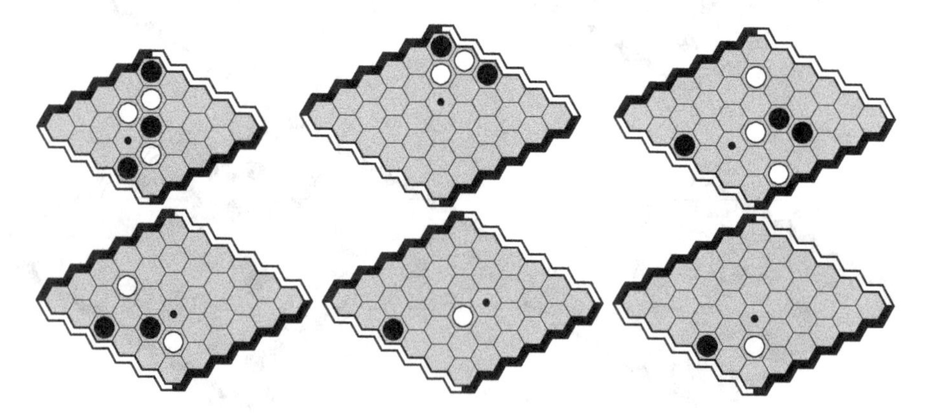

FIGURE C.11: Henderson solutions B1-B6.

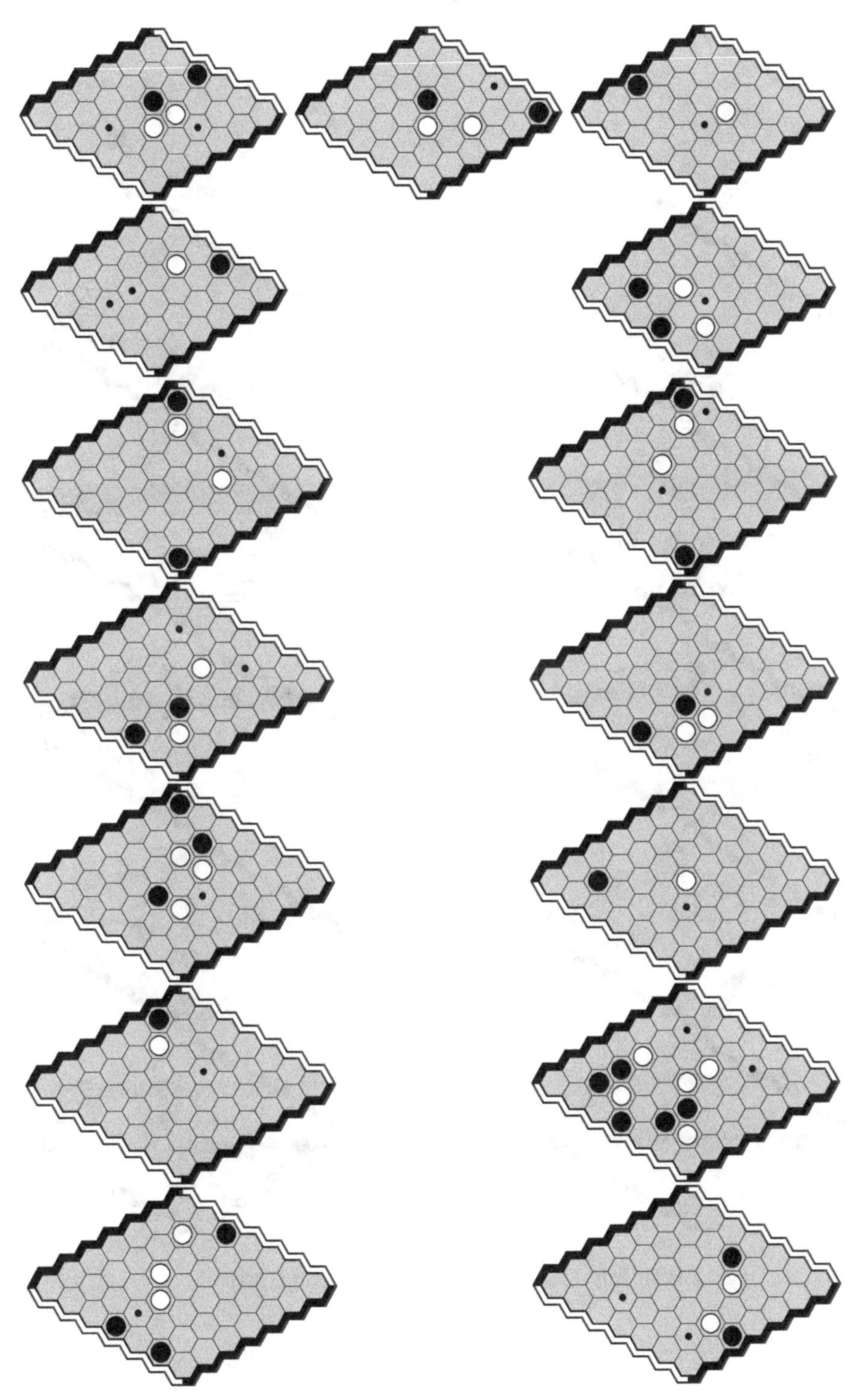

FIGURE C.12: Henderson solutions B7-B21.

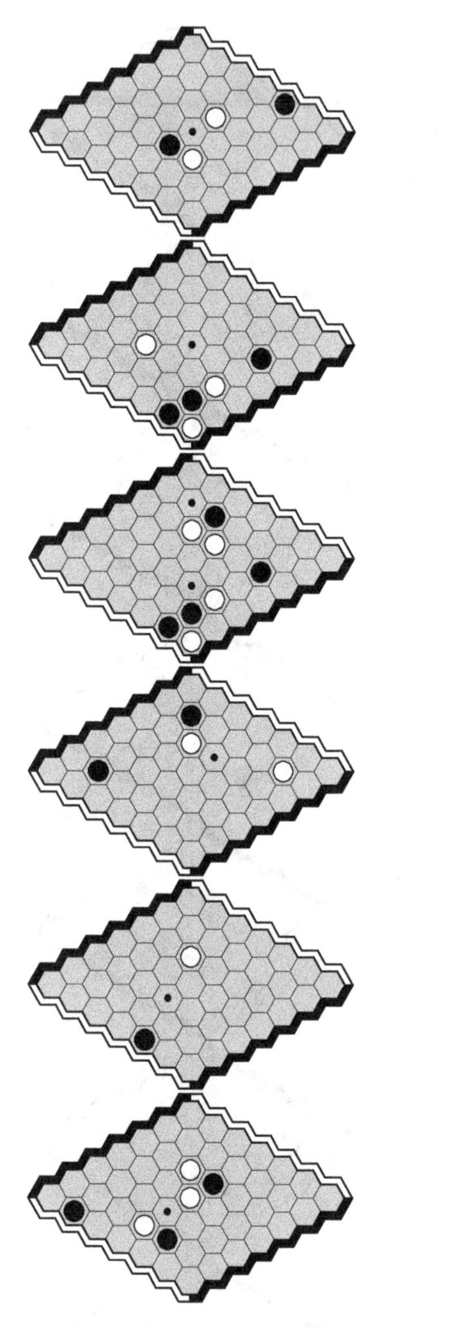

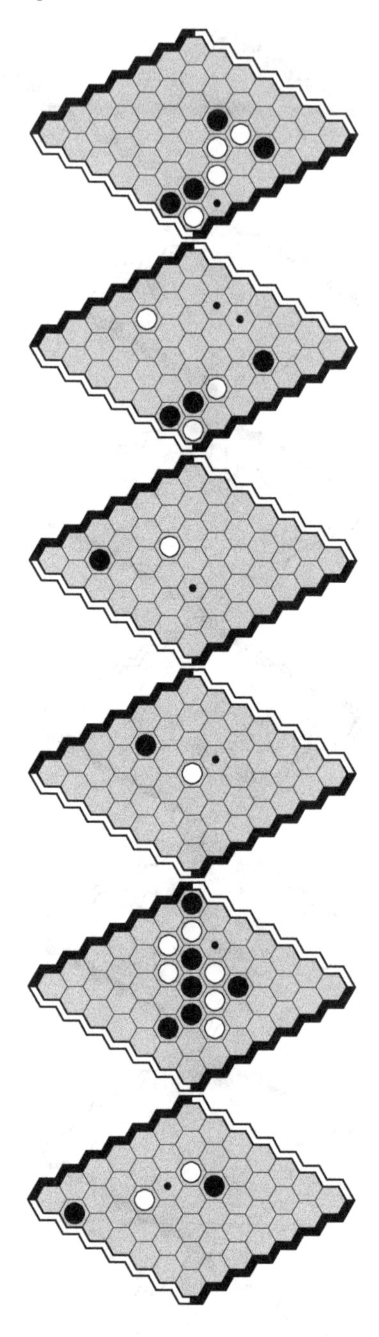

FIGURE C.13: Henderson solutions B22-B33.

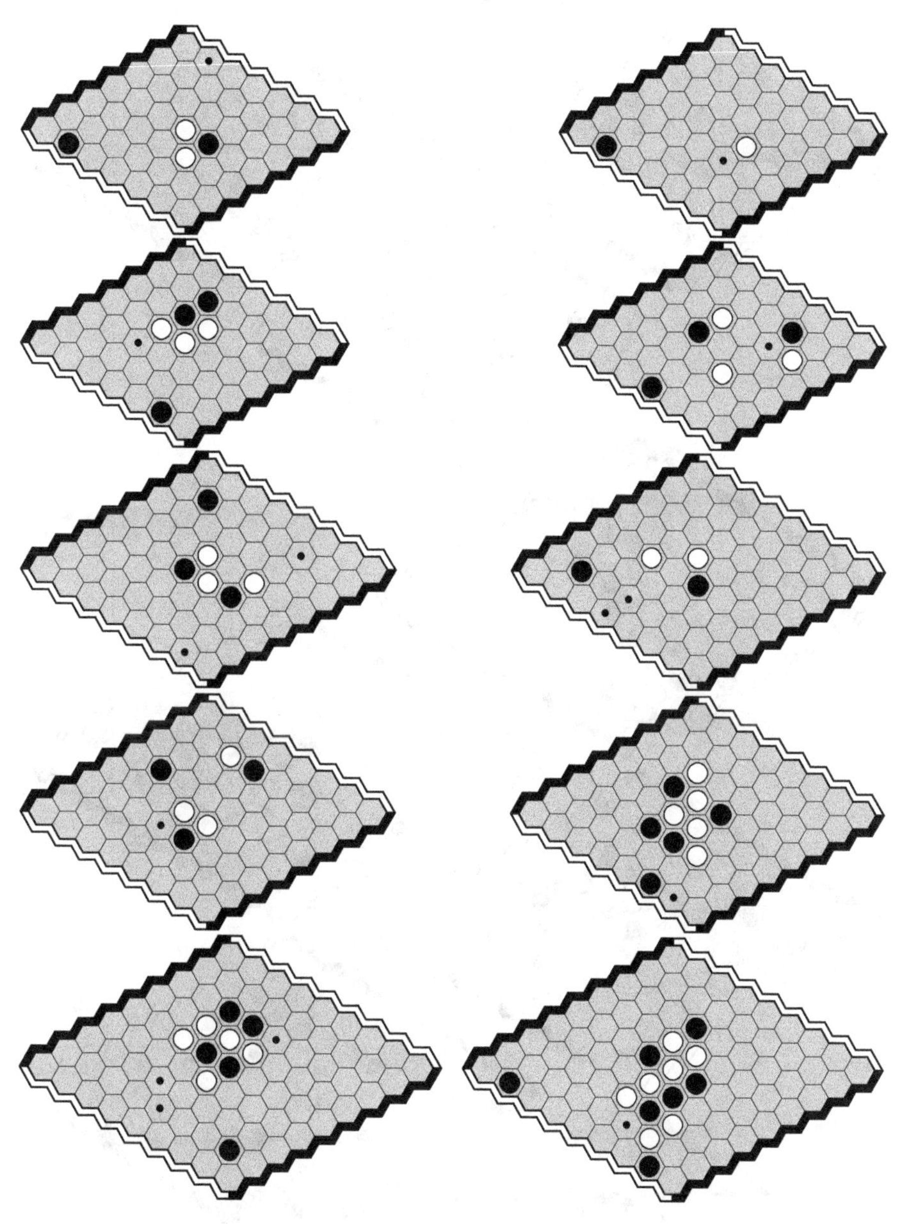

FIGURE C.14: Henderson solutions B34-B43.

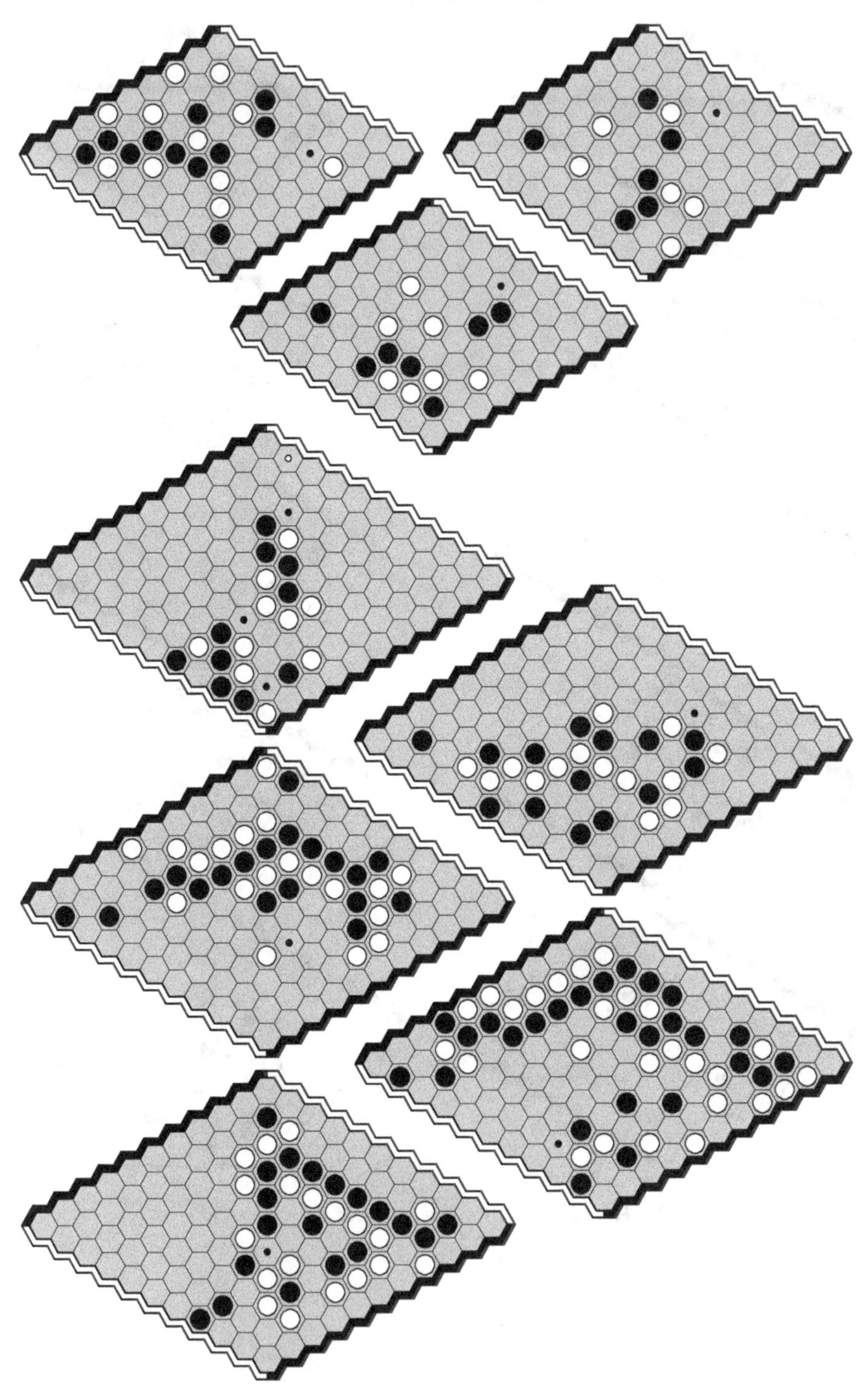

FIGURE C.15: Henderson solutions B44-B51.

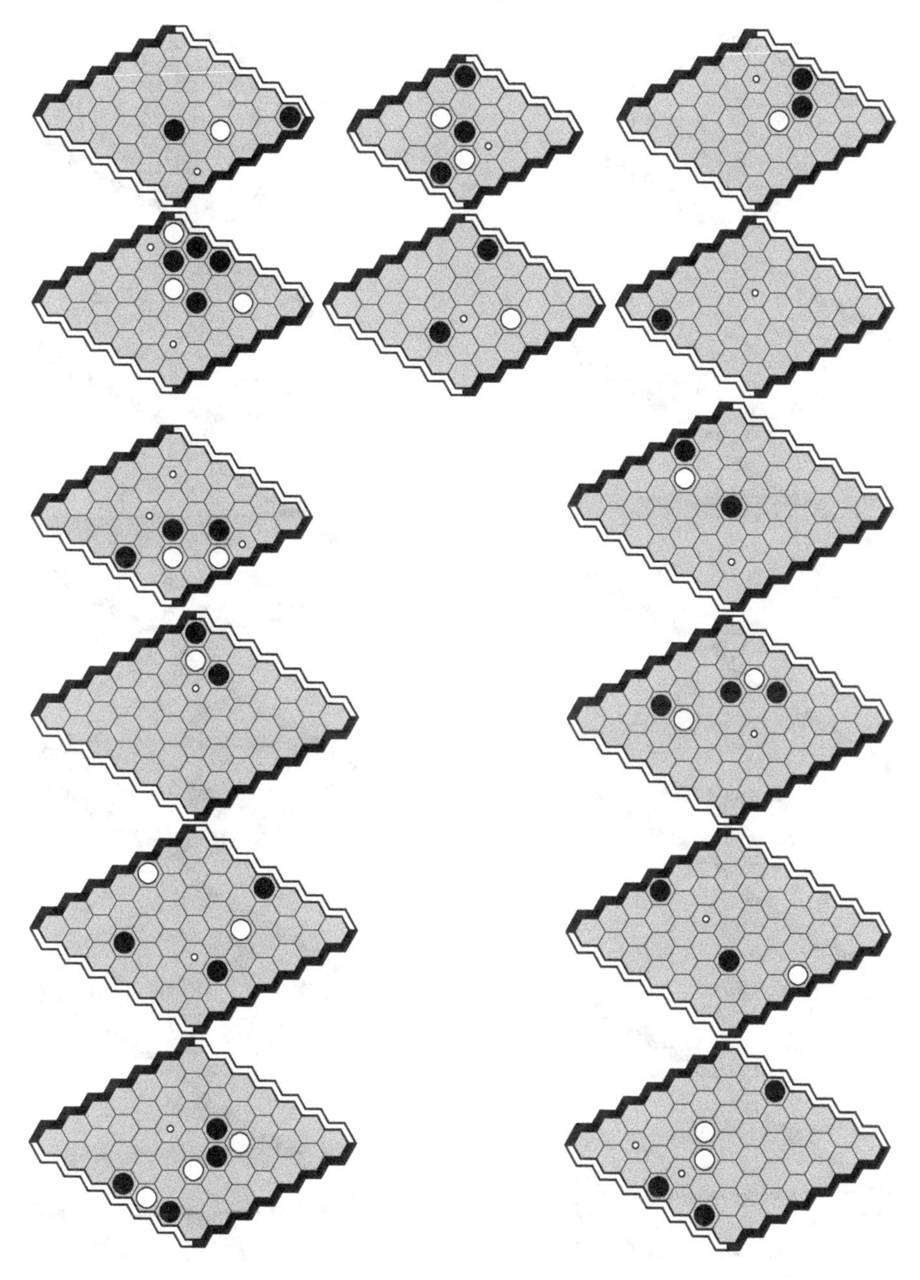

FIGURE C.16: Henderson solutions W1-W14.

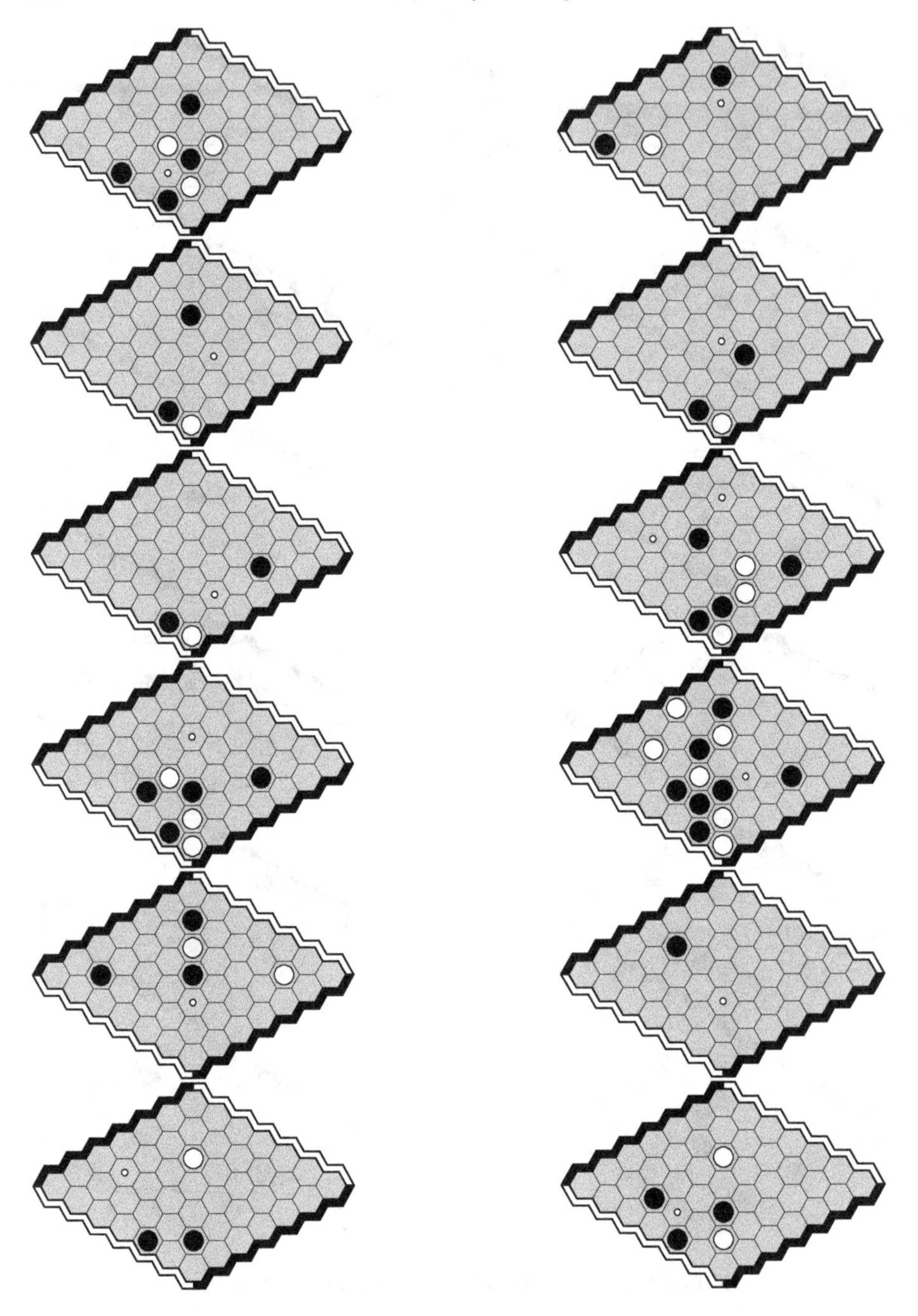

FIGURE C.17: Henderson solutions W15-W26.

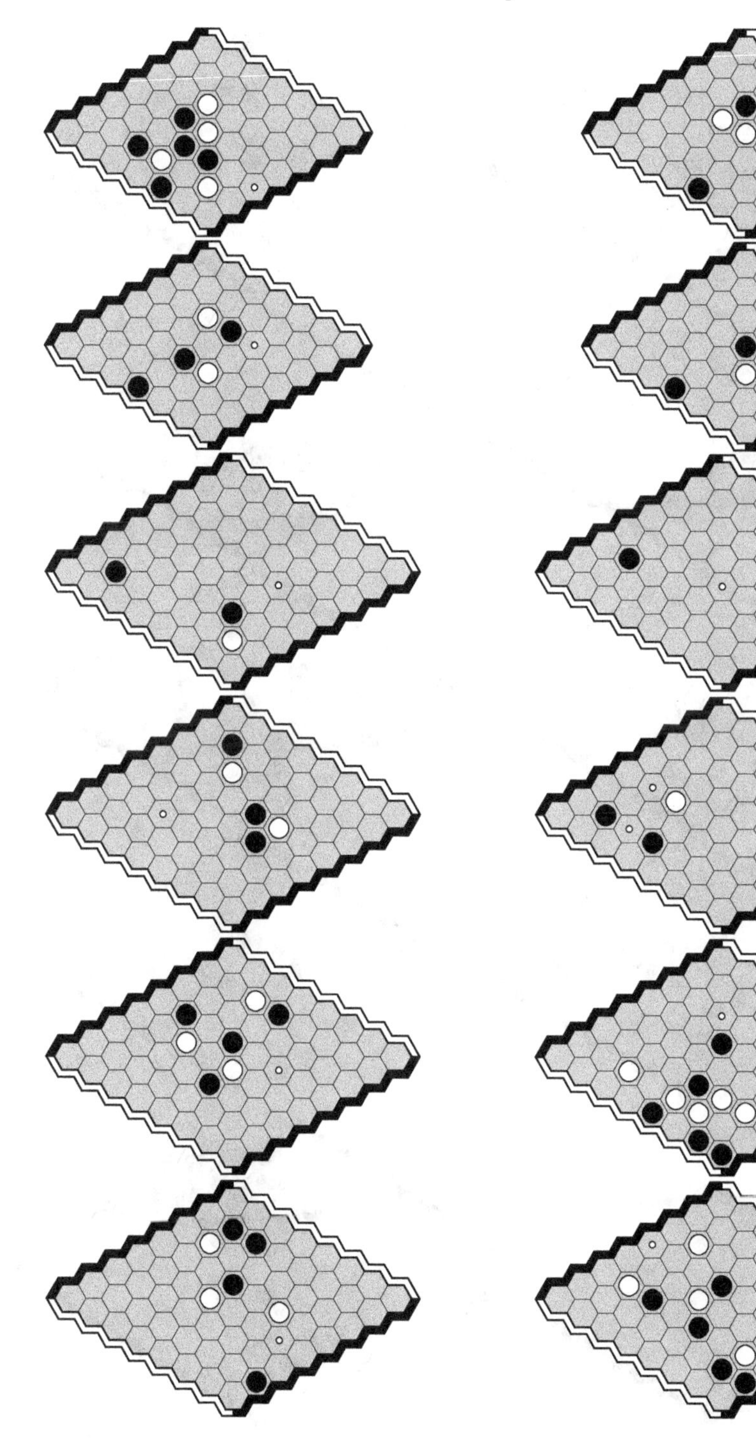

FIGURE C.18: Henderson solutions W27-W38.

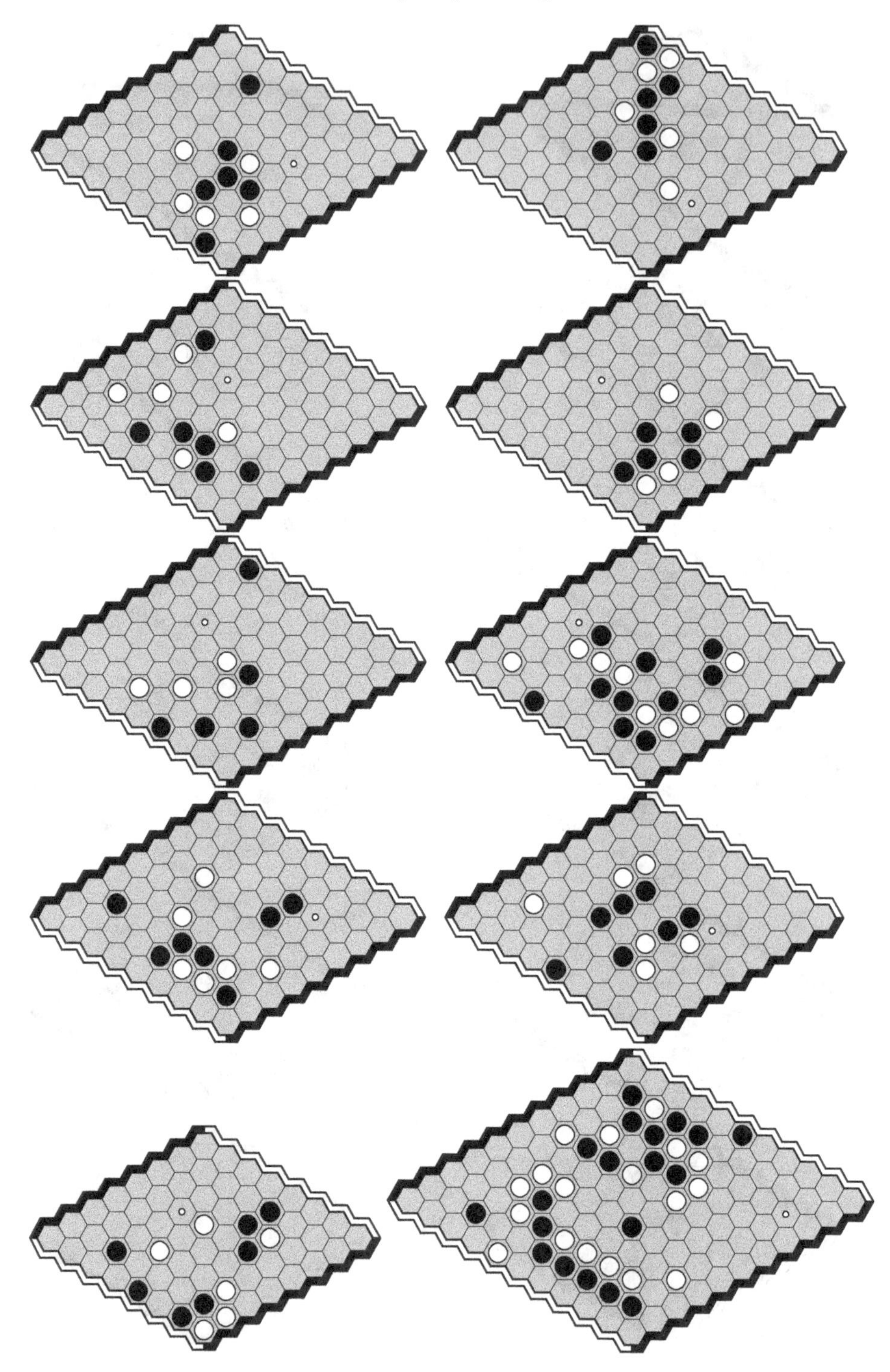

FIGURE C.19: Henderson solutions W39-W48.

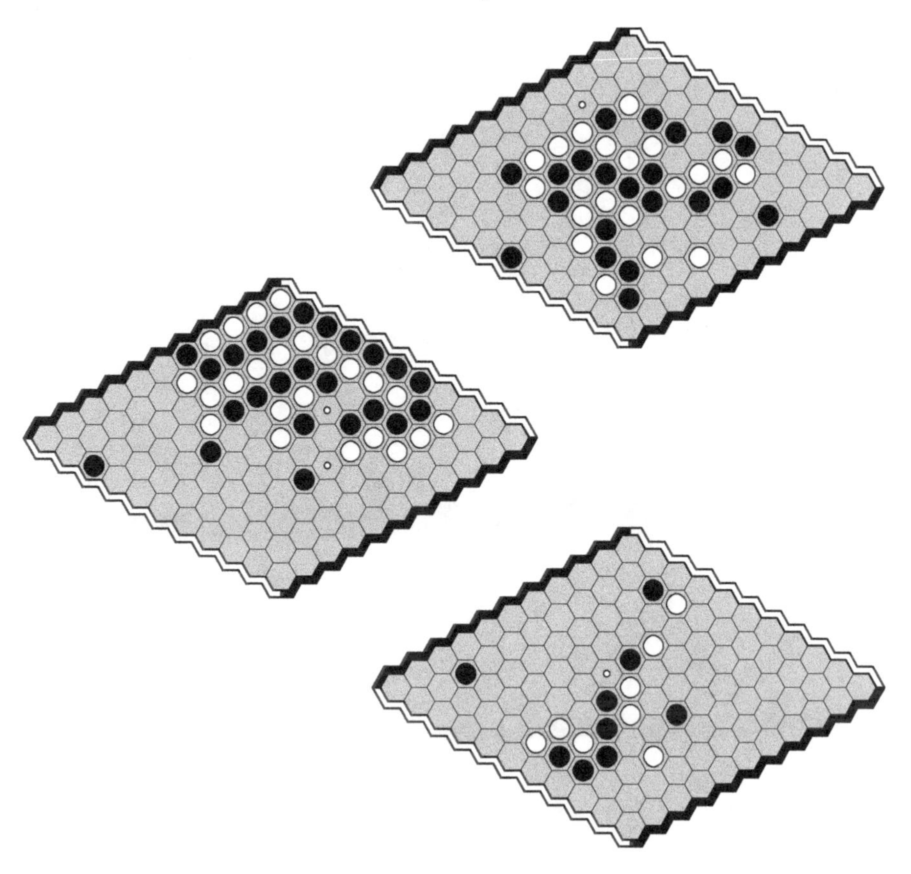

FIGURE C.20: Henderson solutions W49-W51.

Appendix D

Rex puzzles

D.1 Rex puzzles

Here are some new Rex puzzles composed by Ryan Hayward using the Rex solving software written by Kenny Young. Each puzzle is Black to play and has at least one winning move. Can you find all winning moves?

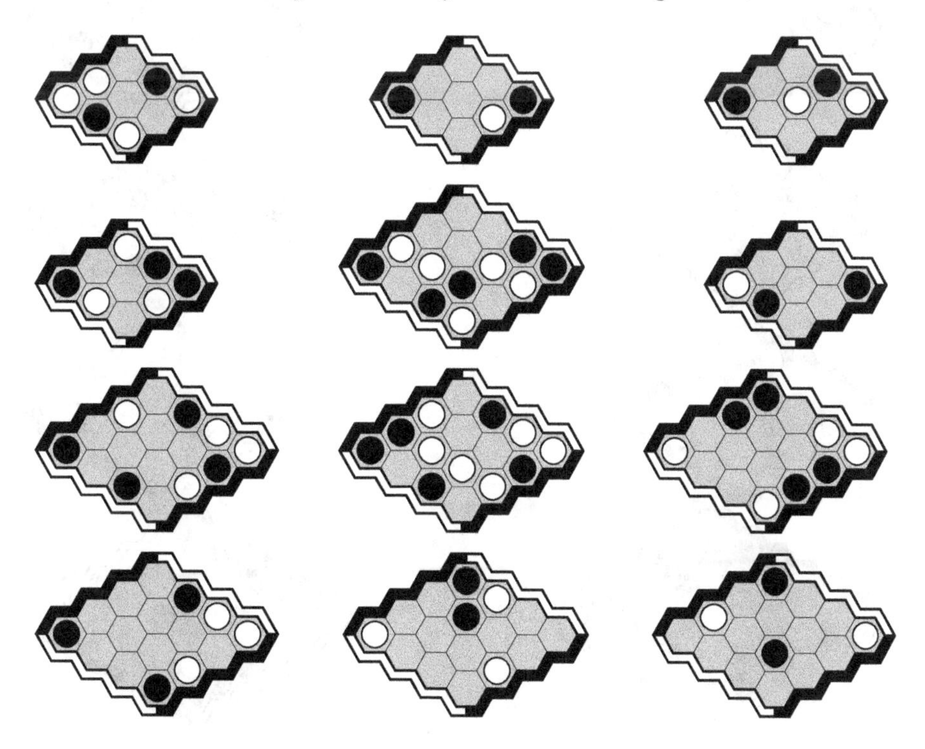

FIGURE D.1: Rex puzzles 1-12. Black to play.

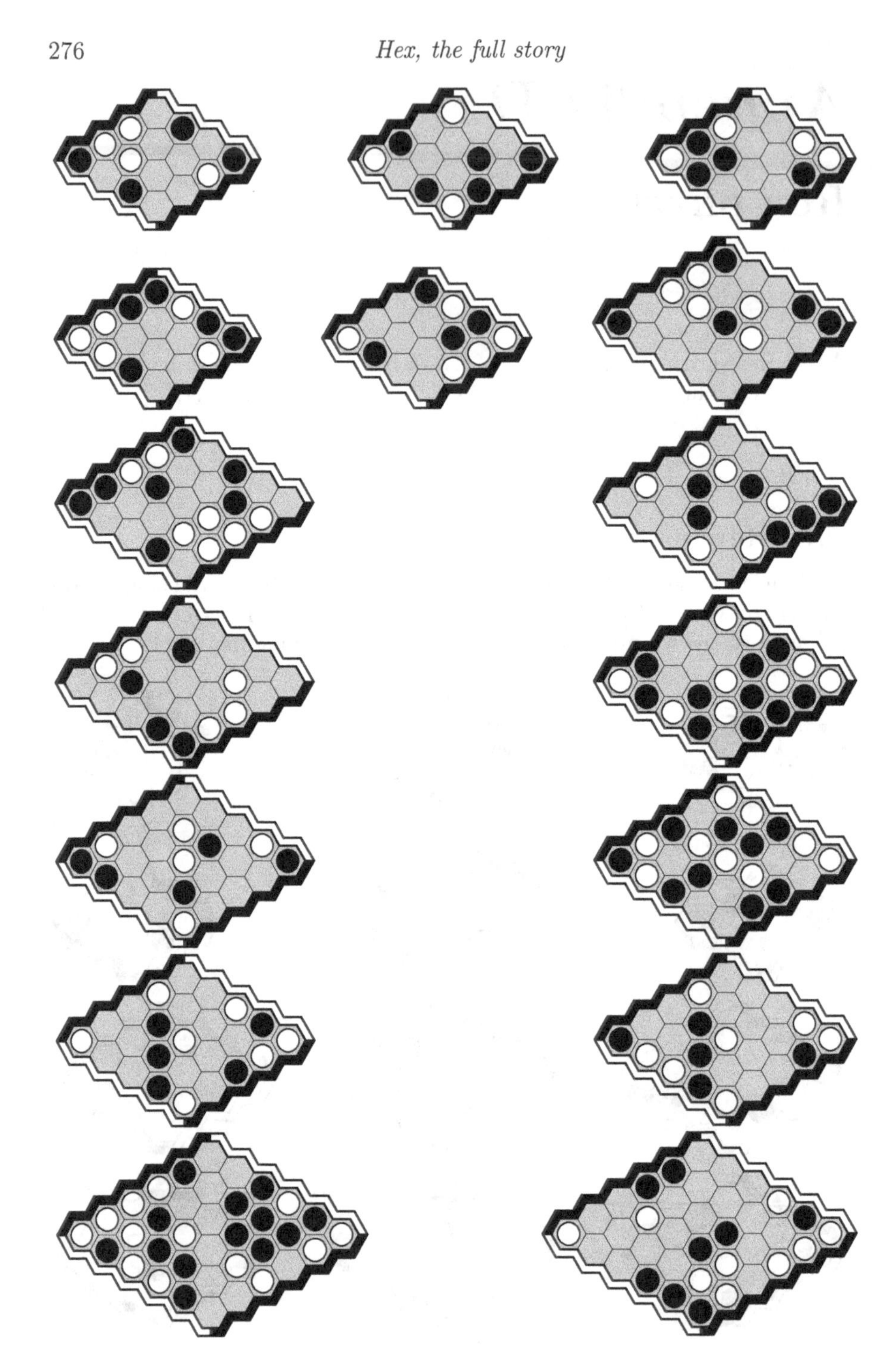

FIGURE D.2: Rex puzzles 13-28. Black to play.

D.2 Rex solutions

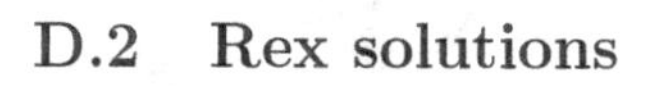

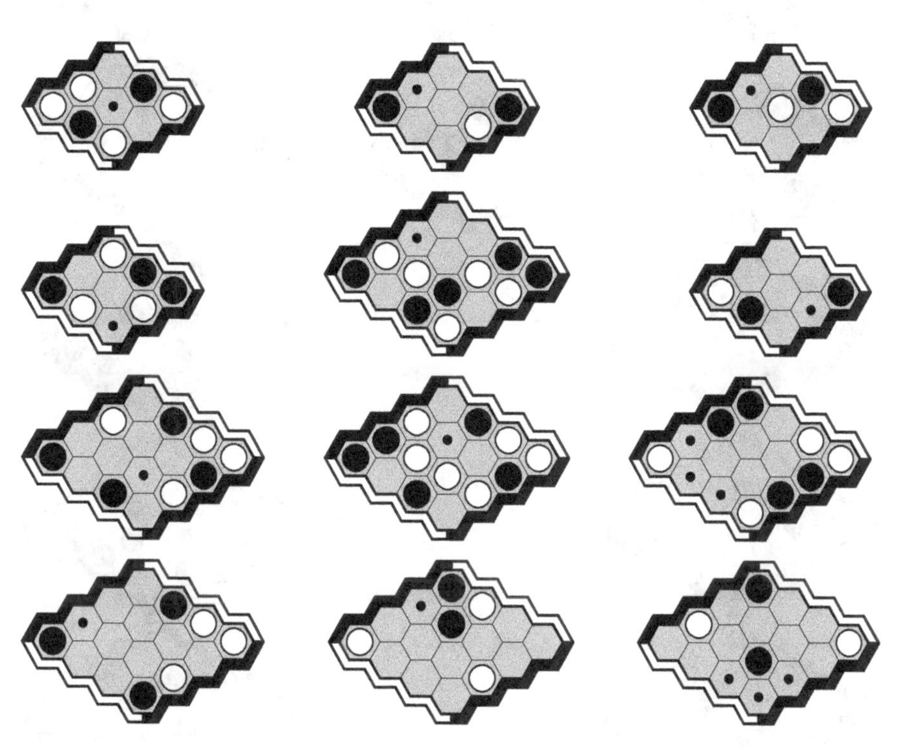

FIGURE D.3: Rex solutions 1-12. All winning Black moves.

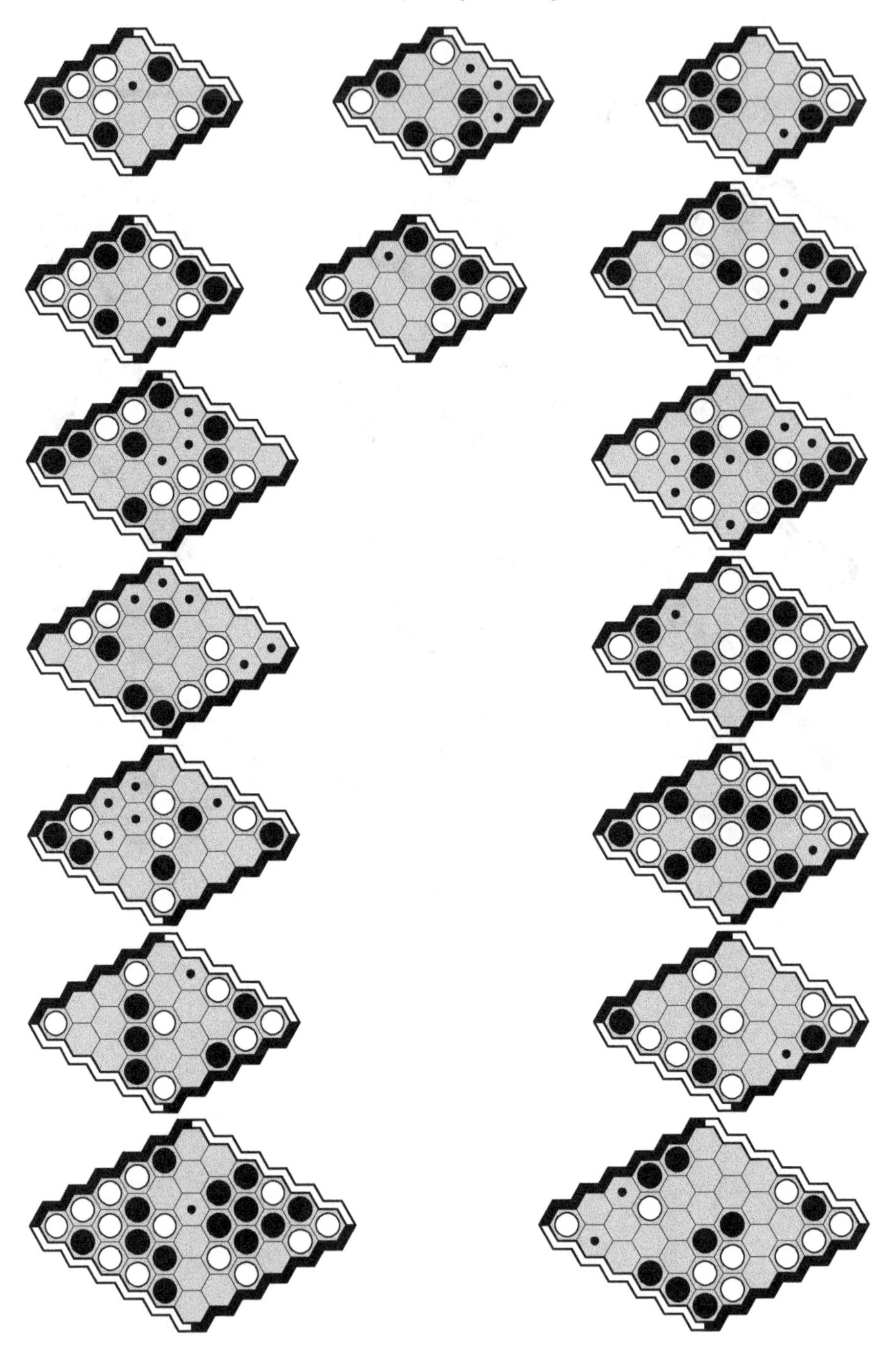

FIGURE D.4: Rex solutions 13-28. All winning Black moves.

Appendix E

Open problems

We shall have to evolve
problem solvers galore –
since each problem we solve
creates ten problems more.

Piet Hein

We exagerate in calling our book the full story of Hex. There are parts of the story that we have not included, there are parts that we do not know, and there are undoubtedly many parts to come.

Here are some open problems that might continue the story.

E.1 Winning cells

For $n \times n$ Hex, the first player has a winning strategy. Hein and his friends knew this from the beginning, as did Nash, for whom it was a defining feature of the game. But little is known about particular strategies. For example, which cells are the first moves of such strategies?

Shannon's electrical circuit and Monte Carlo Tree Search each favor opening near the center or on the short (joining the obtuse corners) diagonal. On boards where the win/loss value of each cell is known — currently up to 9×9 — the two easiest-to-prove winning opening cells are the centermost and the second (and second-last) from the side on the short diagonal.

FIGURE E.1: Problems galore. ©Piet Hein, courtesy Hugo Hein.

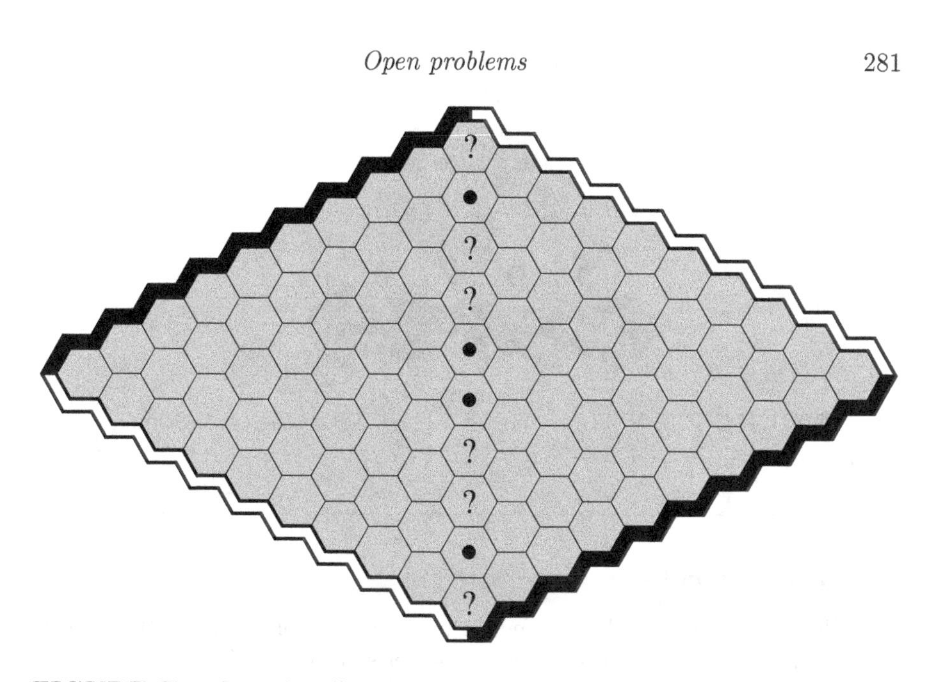

FIGURE E.2: Is each cell on the short diagonal a winning opening move? For this board, for each dot, yes.

Problem E.1 *Winning cells for* $n \times n$ *Hex. Does the centermost cell always win? Does the second cell on the short diagonal always win? Does each cell on the short diagonal always win? See Figure E.2.*

As far as we know — for n up to 10 for the first two problems, for n up to 9 for the third problem — the answer to each question is yes. Can you find a proof by induction for any of these problems?

Starting from the empty board, we know who wins Hex. By contrast, finding the winner of an arbitrary position is P-space complete. What about positions with only one stone?

Problem E.2 *One-stone position hardness. What is the computational complexity of solving a Hex position with only one stone? Are these easier to solve than arbitrary Hex positions?*

FIGURE E.3: White probes of a black 4.3.2 side connection.

E.2 Inferior cells

In Hex, it is useful to know when a move is inferior to (no better than) some other move. For example, when a player's stone is safely connected to the side, some opponent probes of this connection can be inferior to others.

Problem E.3 *4.3.2 probes. Is each White 4.3.2 probe 4-8 in Figure E.3 inferior to at least one of 1-3?*

Henderson showed that White probes 4-8 are inferior when Black preserves the connection [39, 37]. Does the conjecture still hold when the player need not preserve the connection?

E.3 Efficient wins

Problem E.4 *Shortest win. For the $n \times n$ board, what is the minimum number of moves needed to win?*

For n up to 5, counting only the first player's moves, this is 1, 2, 3, 5, 7. Weaver found 6×6 and 7×7 wins needing only 11 and 13 moves respectively: are these the minimum values? Garikai Campbell gave some lower bounds [12].

Problem E.5 *Most compact win. For the $n \times n$ board, what is the minimum number of cells needed to win?*

For example, on the 4×4 board, the centermost-cell opening strategy uses only 11 of the board's 16 cells.

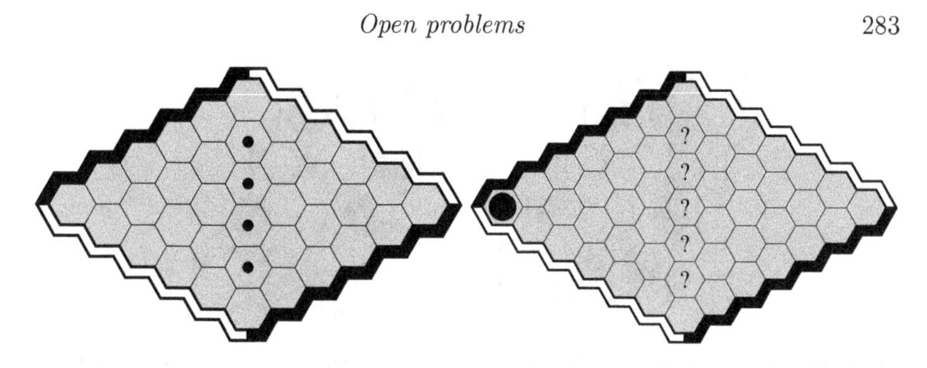

FIGURE E.4: (left) All 6×6 Rex losing opening moves. (right) After Black moves to the acute corner, are these White's only 7×7 Rex losing replies?

E.4 Losing cells in Rex

There is a rough duality between Hex and Rex: easy-to-prove Hex-winning opening moves are often easy-to-prove Rex-losing opening moves. Hayward, Toft and Henderson gave some tips that Young and Hayward implemented in a computer solver [29, 76].

Problem E.6 *Losing cells in $n \times n$ Rex. For n even, does each move to a cell on the short diagonal — except the obtuse corners — lose? And does every other move win?*

For n odd, after a first move to an acute corner, does each move to a cell on the short diagonal — except the obtuse corners — lose? And does every other move win?

For n from 2 to 6, the answer to each question is yes. See Figure E.4.

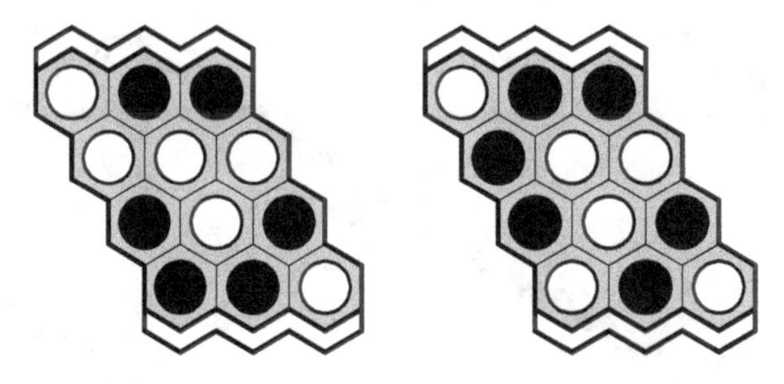

FIGURE E.5: 3×4 Cylindrical Hex. Around is Black, Ends is White. (left) Black wins. (right) White wins.

E.5 Simple heuristics

Solving Hex is hard, but on smaller boards a simple heuristic — say, a cell through which most current flows, or a cell in the most side-to-side winning paths — often finds the best move. Thomas Fischer investigated Shannon's Bridg-it circuit and found positions where it misses the best move [19].

Problem E.7 *Heuristics. Pick a heuristic, and investigate when it finds the best move.*

E.6 Cylindrical Hex

In cylindrical Hex, the board is drawn on the side of a cylinder, like the label on a soup can. We call the two players Ends and Around: the former wants to join the top to the bottom, the latter wants to encircle the cylinder. See Figure E.5. As with Hex, a draw is not possible.

Problem E.8 *Cylindrical Hex. Who wins on a cylinder with circumference m and number of rings n?*

Anatole Beck and Steven Alpern introduced this game and showed that Ends wins when m is even [1]. David Gale conjectured that Ends wins for all m. Samuel C. Huneke, Hayward and Toft confirmed the conjecture for $m = 3$ [42]. For some $m \geq 5$, can you solve it?

Bibliography

[1] S. Alpern and A. Beck. Hex Games and Twist Maps on the Annulus. *American Mathematical Monthly*, 98(9):803–811, 1991.

[2] Vadim V. Anshelevich. The Game of Hex: An Automatic Theorem Proving Approach to Game Programming. In *AAAI Proceedings*, pages 189–194, 2000.

[3] Broderick Arneson, Ryan Hayward, and Philip Henderson. Wolve 2008 wins Hex Tournament. *ICGA*, 32(1):49–53, March 2009.

[4] Broderick Arneson, Ryan B. Hayward, and Philip Henderson. Solving Hex: Beyond Humans. In H.J̃aap van den Herik, Hiroyuki Iida, and Aske Plaat, editors, *Computers and Games 2010 Revised Selected Papers*, volume 6515 of *Lecture Notes in Computer Science*, pages 1–10. Springer, 2011.

[5] Anatole Beck, Michael N. Bleicher, and Donald W. Crowe. *Excursions into Mathematics*. Worth, New York, 1969. republished as [6].

[6] Anatole Beck, Michael N. Bleicher, and Donald W. Crowe. *Excursions into Mathematics, Millennium Edition*. A.K. Peters, Natick, 2000.

[7] Claude Berge. Some remarks about a Hex problem. In David Klarner, editor, *The Mathematical Gardener*, pages 25–28. Wadsworth, Belmont, CA, 1981.

[8] Yngvi Björnsson, Ryan Hayward, Michael Johanson, and Jack van Rijswijck. Dead Cell Analysis in Hex and the Shannon Game. In A. Bondy, J. Fonlupt, J.L. Fouquet, J.C. Fournier, and J.L.R̃amirez Alfonsin, editors, *Graph Theory in Paris*, Trends in Mathematics, pages 45–59. Birkhäuser, Basel, 2006.

[9] Cameron Browne. *Hex Strategy: Making the Right Connections*. AK Peters, Natick, 2000.

[10] Cameron Browne. *Connection Games: Variations on a Theme*. A.K. Peters, Natick, 2005.

[11] Cameron Browne and Frederic Maire. Monte Carlo analysis of a puzzle game. In B. Pfahringer and J. Renz, editors, *AI 2015: Advances in AI*, number 9457 in LNCS. Springer, 2015.

[12] Garikai Campbell. On Optimal Play in the Game of Hex. *Integers: Electronic Journal of Combinatorial Number Theory*, 4:G02, 2004.

[13] Anne Chamberlin. King of Supershape. *Esquire Magazine*, pages 110–111, January 1967.

[14] Vašek Chvátal and Paul Erdös. Biased positional games. *Annals of Discrete Math*, 2:221–229, 1978.

[15] Jack Edmonds. Lehman's Switching Game and a Theorem of Tutte and Nash-Williams. *Journal Research National Bureau Standards B. Math and Physics*, 69B(1-2):73–77, 1965. https://pdfs.semanticscholar.org/8fa1/e25652f341fdec30a68b12ab8dc22f554aed.pdf.

[16] Bert Enderton. Answers to infrequently asked questions about the game of Hex. www.cs.cmu.edu/~hde/hex/hexfaq, 1995-2005.

[17] Ronald J. Evans. A Winning Opening in Reverse Hex. *Recreational Math*, 7(3):189–192, 1974.

[18] Shimon Even and R.Ẽndre Tarjan. A Combinatorial Problem which is Complete in Polynomial Space. *Journal ACM*, 23(4):710–719, 1976.

[19] Thomas Fischer. *Exacte Analyse von Heuristken fuer kombinatorishe Spiele*. PhD thesis, Friedrich-Schiller-Universität Jena, Germany, April 2011. https://althofer.de/dissertation_thomas-fischer.pdf.

[20] David Gale. Game of Hex and the Brouwer Fixed Point Theorem. *American Math Monthly*, 86(10):818–827, 1979.

[21] Martin Gardner. Mathematical Games: Concerning the game of Hex, which may be played on the tiles of the bathroom floor. *Scientific American*, 197(1):145–150, July 1957.

[22] Martin Gardner. Mathematical Games: How to remember numbers by mnemonic devices such as cuff links and red zebras. *Scientific American*, 197(4):130–138, October 1957. [Adds Nash as Hex inventor].

[23] Martin Gardner. Martin Gardner Papers. Stanford University Library, 1957-1997. Game files Hex (box 2 folder 5) and Rex (box 33 folder 5).

[24] Martin Gardner. Mathematical Games: Four mathematical diversions involving concepts of topology. *Scientific American*, 199(4):124–129, October 1958. [Mentions Gale's game].

[25] Martin Gardner. *2nd Scientific American Book of Mathematical Puzzles and Diversions*, chapter Recreational Logic, pages 119–129. Simon and Schuster, New York, 1961. [Addendum discusses Shannon's Birdcage machine].

[26] Martin Gardner. Mathematical Games: Games of strategy for two players: Star Nim, Meander, Dodgem and Rex. *Scientific American*, 232(6):106–111, June 1975.

[27] Olav Harsløf, editor. *Piet Hein Verdensdanskeren.* Gyldendal, Copenhagen, 2005. [With E. Kristiansen, B. Toft, M.L. Lund and C. Steffensen].

[28] Ryan Hayward. Six wins Hex Tournament. *ICGA*, 29(3):163–165, September 2006.

[29] Ryan Hayward, Bjarne Toft, and Philip Henderson. How to Play Reverse Hex. *Discrete Math*, 312:148–156, 2012.

[30] Ryan Hayward and Jack van Rijswijck. Hex and Combinatorics. *Discrete Math*, 306(19-20):2515–2528, 2006.

[31] Ryan B. Hayward. A Note on Domination in Hex. Technical report, University of Alberta, 2003.

[32] Ryan B. Hayward, Yngvi Björnsson, Michael Johanson, Morgan Kan, Nathan Po, and Jack van Rijswijck. Solving 7x7 Hex: Virtual Connections and Game-state Reduction. In H. Jaap van den Herik, Hiroyuki Iida, and Ernst A. Heinz, editors, *Advances in Computer Games*, volume 263, pages 261–278. Kluwer, Boston, 2003.

[33] Piet Hein. Piet Hein Papers. Archive of Anni and Hugo Hein, Middlefart, Denmark.

[34] Piet Hein. *Gruk: 2 Samling.* Politikens Forlag, Copenhagen, Nov 1941.

[35] Piet Hein. *Kumbels Almanak 1942.* Grafisk Forlag, Copenhagen, 1941.

[36] Piet Hein. *Vil De laere Polygon?* Politiken, page 4, December 26 1942.

[37] Philip Henderson. *Playing and Solving Hex.* PhD thesis, University of Alberta, Edmonton, Alberta, Canada, Fall 2010. `https://webdocs.cs.ualberta.ca/~hayward/theses/ph.pdf`.

[38] Philip Henderson, Broderick Arneson, and Ryan B. Hayward. Solving 8x8 Hex. In Craig Boutilier, editor, *IJCAI 2009 Proceedings*, pages 505–510, 2009.

[39] Philip Henderson and Ryan Hayward. Probing the 4-3-2 Edge Template in Hex. In H. van den Herik, X. Xu, Z. Ma, and M.H.M. Winands, editors, *Computers and Games 2008*, number 5131 in LNCS, pages 229–240. Springer, 2008.

[40] Jim Hicks. A Poet with a Slide Rule: Piet Hein Bestrides Art and Science. *Life Magazine*, 61(16):55–66, October 14 1966.

[41] R. Hochberg, C. McDiarmid, and M. Saks. On the bandwidth of triangulated triangles. *Discrete Math*, pages 261–265, 1995.

[42] Samuel C. Huneke, Ryan Hayward, and Bjarne Toft. A Winning Strategy for 3xn Cylindrical Hex. *Discrete Math*, 331:93–97, 2014.

[43] Kate Jones. Kadon Enterprises Inc. `http://www.gamepuzzles.com`, 1998-2017.

[44] David King. Hall of Hexagons, The Game of Hex: Templates. `http://www.drking.org.uk/hexagons/hex/templates.html`, 2007.

[45] Kumbel Kumbell. *77 Gruk*. Politikens Forlag, Copenhagen, Dec 1940.

[46] Jeffrey Lagarias and Daniel Sleator. Who Wins Misère Hex? In Elwyn Berlekamp and Tom Rodgers, editors, *The Mathemagician and Pied Puzzler: A Collection in Tribute to Martin Gardner, editors Elwyn Berlekamp and Tom Rodgers*, chapter 3, pages 237–240. A.K. Peters, 1999.

[47] Alfred Lehman. A solution of the Shannon Switching Game. *J. Soc. Indust. Appl. Math*, 12(4):687–725, 1964.

[48] Alan Levinovitz. The mystery of Go, the ancient game that computers still can't win. `www.wired.com/2014/05/the-world-of-computer-go`, May 2014.

[49] Jens Lindhard. The Jens Lindhard Papers. History of Science Archives, Center for Science Studies, Aarhus University, Denmark.

[50] Anthony Liversidge. Claude Shannon Interview: Father of the Electronic Information Age. *Omni*, page 61ff, August 1987.

[51] Thomas Maarup. Everything You Always Wanted To Know About Hex But Were Afraid To Ask. Master's thesis, University of Southern Denmark, 2005. `http://maarup.net/thomas/hex`.

[52] Gábor Melis and Ryan Hayward. Six wins Hex Tournament. *ICGA*, 26(4):277–280, December 2003.

[53] Steven Meyers. Boxoff: a new solitaire board game. *GAMES*, 37(6):12–13, August 2013.

[54] Sylvia Nasar. *A Beautiful Mind*. Touchstone, New York, 1998.

[55] John Nash. *Non-cooperative games*. PhD thesis, Princeton University, May 1950. `https://rbsc.princeton.edu/sites/default/files/Non-Cooperative_Games_Nash.pdf`.

[56] John Nash. Some Games and Machines for Playing Them. Technical Report D-1164, RAND Corp., February 1952.

[57] Kohei Noshita. Union-Connections and A Simple Readable Winning Way in 7x7 Hex. *Proceedings of 9th Game Programming Workshop in Japan*, pages 72–79, 2004. [Japanese name of journal is Joho Shori Gakkai Shinpojiumu Ronbunshu].

[58] Kohei Noshita. Union-Connections and Straightforward Winning Strategies in Hex. *ICGA Journal*, 28(1):3–12, 2005.

[59] Jakub Pawlewicz and Ryan B. Hayward. Scalable Parallel DFPN Search. In *Computers and Games 2013 Revised Selected Papers*, volume 8427 of *Lecture Notes in Computer Science*, pages 138–150. Springer, 2014.

[60] Jakub Pawlewicz, Ryan B. Hayward, Philip Henderson, and Broderick Arneson. Stronger Virtual Connections in Hex. *IEEE Transactions on Computation Intelligence and AI in Games*, 7(2):156–166, June 2015.

[61] John R. Pierce. *Symbols, Signals and Noise*, pages 10–13. Harper and Brothers, 1961.

[62] Stefan Reisch. Hex ist PSPACE-vollständig. *Acta Informatica*, 15:167–191, 1981.

[63] Jonathan Schaeffer. *One Jump Ahead: Challenging Human Supremacy in Checkers*. Springer-Verlag, New York, 1997.

[64] Craige Schensted and Charles Titus. *Mudcrack Y and Poly-Y*. Neo Press, Peaks Island, Maine, 1975.

[65] Paul Seymour. Hadwiger's conjecture. `web.math.princeton.edu/~pds/papers/hadwiger/paper.pdf`, 2015.

[66] Claude Shannon. Programming a computer for playing chess. *Philosophical Magazine*, 41(314):256–275, March 1950.

[67] Claude E. Shannon. Computers and Automata. *Proc. Inst. Radio Engineers*, 41:1234–1241, 1953.

[68] D Silver, A Huang, CJ Maddison, A Guez, L Sifre, G van den Driessche, J Schrittwieser, I Antonoglou, V Panneershelvam, M Lanctot, S Dieleman, D Grewe, J Nham, N Kalchbrenner, I Sutskever, T Lillicrap, M Leach, K Kavukcuoglu, T Graepel, and D Hassabis. Mastering the Game of Go with Deep Neural Networks and Tree Search. *Nature*, 529(7587):484–489, January 2016.

[69] Bjarne Toft. A Survey of Hadwiger's Conjecture. In G. Chartrand and M. Jacobson, editors, *Surveys in Graph Theory*, pages 249–283. 1996.

[70] C.B. Tompkins. Sperner's Lemma and some extensions. In E.F. Beckenbach, editor, *Applied Combinatorial Mathematics*, chapter 15, pages 416–455. Wiley, 1964.

[71] Jack van Rijswijck. Computer Hex: Are Bees better than Fruitflies? Master's thesis, University of Alberta, Edmonton, 2000.

[72] Robin Wilson. *Four Colours Suffice: How the Map Problem was Solved.* Penguin UK and Princeton University Press, London, 2002. Revised color edition, Princeton Science Library, 2013.

[73] Philip Yam. Profile: Martin Gardner, the mathematical Gamester. *Scientific American*, pages 38–41, December 1995. [Republished May 22 2010].

[74] Jing Yang, Simon Liao, and Mirek Pawlak. On a decomposition method for finding winning strategy in Hex game. In *Proc. 1st Intl. Conf. Appl. and Devt. 21st Century Computer Games ADCOG 2001*, pages 96–111, 2001.

[75] Jing Yang, Simon Liao, and Mirek Pawlak. New Winning and Losing Positions for 7x7 Hex. In J. Schaeffer, M. Müller, and Yngvi Björnsson, editors, *Computers and Games 2002*, number 2883 in LNCS, pages 230–248. Springer, 2003.

[76] Kenny Young and Ryan Hayward. A Reverse Hex Solver. In A. Plaat, W. Klosters, and J. van den Herik, editors, *Computers and Games 2016 Revised Selected Papers*, pages 137–148. Springer, 2016.

[77] Kenny Young and Ryan B. Hayward. A Reverse Hex Solver. In Aske Plaat, Walter A. Kosters, and H. Jaap van der Herik, editors, *Computers and Games 2016 Revised Selected Papers*, volume 10068 of *LNCS*, pages 137–148. Springer, 2016.

Index